Introduction to Photorefractive Nonlinear Optics

WILEY SERIES IN PURE AND APPLIED OPTICS

Founded by Stanley S. Ballard, University of Florida

BEISER · *Holographic Scanning*
BOYD · *Radiometry and The Detection of Optical Radiation*
CATHEY · *Optical Information Processing and Holography*
DELONE AND KRAINOV · *Fundamentals of Nonlinear Optics of Atomic Gases*
DERENIAK AND CROWE · *Optical Radiation Detectors*
DE VANY · *Master Optical Techniques*
GASKILL · *Linear Systems, Fourier Transform, and Optics*
GOODMAN · *Statistical Optics*
HUDSON · *Infrared System Engineering*
JUDD · *Color in Business, Science, and Industry*, Third Edition
KAFRI AND GLATT · *The Physics of Moire Metrology*
KLEIN AND FURTAK · *Optics*, Second Edition
MALACARA · *Optical Shop Testing*, Second Edition
MILONNI AND EBERLY · *Lasers*
NASSAU · *The Physics and Chemistry of Color*
NIETO-VESPERINAS · *Scattering and Diffraction in Physical Optics*
O'SHEA · *Elements of Modern Optical Design*
SALEH AND TEICH · *Fundamentals of Photonics*
SCHUBERT AND WILHELMI · *Nonlinear Optics and Quantum Electronics*
SHEN · *The Principles of Nonlinear Optics*
UDD · *Fiber Optic Sensors: An Introduction for Engineers and Scientists*
VANDERLUGT · *Optical Signal Processing*
VEST · *Holographic Interferometry*
VINCENT · *Fundamentals of Infrared Detector Operation and Testing*
WILLIAMS AND BECKLUND · *Introduction to the Optical Transfer Function*
WYSZECKI AND STILES · *Color Science: Concepts and Methods, Quantitative Data and Formulae*, Second Edition
YAMAMOTO · *Coherence, Amplification, and Quantum Effects in Semiconductor Lasers*
YARIV AND YEH · *Optical Waves in Crystals*
YEH · *Optical Waves in Layered Media*
YEH · *Introduction to Photorefractive Nonlinear Optics*

Introduction to Photorefractive Nonlinear Optics

POCHI YEH
Professor of Electrical and Computer Engineering
University of California, Santa Barbara

A Wiley-Interscience Publication

John Wiley & Sons, Inc.

New York / Chichester / Brisbane / Toronto / Singapore

Wiley Series in Pure and Applied Optics

The Wiley Series in Pure and Applied Optics publishes outstanding books in the field of optics. The nature of these books may be basic ("pure" optics) or practical ("applied" optics). The books are directed towards one or more of the following audiences: researchers in universities, government, or industrial laboratories; practitioners of optics in industry; or graduate-level courses in universities. The emphasis is on the quality of the book and its importance to the discipline of optics.

This text is printed on acid-free paper.

Library of Congress Cataloging in Publication Data:

Yeh, Pochi, 1948–
Introduction to photorefractive nonlinear optics / Pochi Yeh.
p. cm. -- (Wiley series in pure and applied optics)
Includes index.
ISBN 0-471-58692-7
1. Nonlinear optics. 2. Photorefractive materials. I. Title.
II. Series.
QC446.2.Y44 1993
535.2--dc20 92-21978

Printed in the United States of America

10 9 8 7 6 5 4 3 2 1

Foreword

Nonlinear optics, a field of optics that has become important since the invention of the laser, is providing us with both a fundamental understanding of the nature of light and a variety of potential applications. Dr. Pochi Yeh and many other researchers at Rockwell International Science Center have made significant contributions across this broad front and especially in the area of photorefractive effect, which is the subject of this text. As photorefractive materials and the related optical techniques mature, the feasibility of applications will increase and create a demand for a source of information on theory and experiment. This text has been written to fulfill this need of current and future workers in the field.

Rockwell International is honored to have played a key role in the developments leading to this book through the work of Pochi and his colleagues at the Rockwell International Science Center, where he continues as Principal Technical Advisor in Optics. I hope that this text will stimulate interest in this important area and lead to additional advances in science, technology, and industrial applications.

JOSEPH T. LONGO

Corporate Vice President Research
& Vice President and General Manager
Rockwell International Science Center
Thousand Oaks, California

Preface

This book treats the theory and applications of nonlinear optics in photorefractive media. It is intended as a text for an advanced course in the area of modern optics for electrical engineering and applied physics students. Students are assumed to have a general knowledge of electromagnetism, elementary solid state physics, and differential equations. Some additional background in Fourier analysis and linear algebra would be helpful. The primary objectives of this book are to present a clear picture of some of the important phenomena in photorefractive media and to teach the readers how to employ these phenomena for practical applications. Although a number of edited volumes on similar subjects exist, most are monographs which are not suitable as textbooks.

Photorefractive crystals are by far the most efficient nonlinear optical media for wave mixing and phase conjugation with relatively low intensity requirements. In addition, holograms can be recorded and erased in these media. To fully utilize these media for the transmission and manipulation of optical information, we must understand the propagation of electromagnetic waves in these media. Although applications such as optical computing and neural networks are included, the emphasis of the book is on the theory of the propagation of optical waves and the mixing of electromagnetic radiation in these media. An effort is made to bridge the gap between theory and practice through the use of numerical examples based on real situations. Classical electrodynamics is used in dealing with the mixing of waves in photorefractive media. Some concepts of elementary solid state physics are employed to treat the transport of charges in photorefractive crystals.

Starting from the basics of electromagnetic waves and periodic structures, the book treats the physics of the photorefractive effect and the mixing of waves in these media. Optical phase conjugators and photorefractive resonators are then discussed. Some of the fundamental properties of gratings and holograms are covered in Chapters 7 and 8. The next three chapters are devoted to applications in optical information processing, optical interconnection and neural networks. The last chapter takes up the subject of higher order photorefractive effects in optical fibers. Since this is primarily a textbook, the materials covered in the book were chosen for the purpose of teaching the basics and introducing some of the important applications, rather than the historical development or technical review of the subject. A large number of papers have been published during the last 10 years on the subject of photorefractive nonlinear optics. Many of the papers represent such small differences of opinion that discussion of all of them is unnecessary and even undesirable. Thus many of the topics and fragmentary pieces of information are omitted for the purpose of the pedagogic flow of the subject. I apologize to any of my colleagues whose work has not been acknowledged or adequately represented in this book.

I am deeply indebted to Professor Amnon Yariv for introducing the subject of nonlinear optics to me during the years when I was a graduate student at Caltech, and for his continued encouragement and support. The main portion of this book was written at the Rockwell International Science Center. Many of the subjects covered in this book are the result of several years of research and development carried out by my colleagues and myself. I wish to acknowledge the support of the Rockwell International Science Center. I am also grateful to my friends at the Science Center for their technical support and helpful discussions in the preparation of the manuscript. I am particularly indebted to Dr. Claire Gu for her contributions, critical reading, and useful comments on the content of the manuscript. In the Spring of 1991, a preliminary version of the manuscript was used as the text of a graduate course in modern optics which I taught at the University of California at Santa Barbara. The same manuscript was used as the text of a similar course which Professors Ken Hsu and Sien Chi taught at the National Chiao Tung University in Taiwan. I am grateful to my colleagues and students for their comments and useful discussions.

I wish to express my sincere thanks to Mrs. B. J. Ellis for her patient and competent typing of the manuscript. My sincere thanks are also extended to Drs. Joseph Longo, Monte Khoshnevisan and Derek Cheung for their constant support and encouragement. Finally, I am deeply grateful to my wife, Linda, for her patience and support.

POCHI YEH

Thousand Oaks, California
January 1993

Contents

Introduction to Photorefractive Nonlinear Optics

General Introduction

It is known that a beam of light propagates along a straight line in a homogeneous medium. The velocity of propagation is given by c/n where n is the index of refraction of the medium, and c is the velocity of light in a vacuum. At an interface between two media, reflection and refraction occur as a result of the dielectric discontinuity. In linear optics, the index of refraction n is a constant depending on the frequency (or wavelength) of the light. The reflection and refraction are thus independent of the intensity of the light beam.

NONLINEAR OPTICS

The index of refraction of a medium, in general, depends on the arrangement of atoms and the distribution of electrons. Under the illumination of light, the charges (mostly electrons) are driven up and down by the electric field of the light beam. The moving electrons will then radiate and generate a field which is proportional to the incident field. This results in a phase shift in the transmitted field. The phase shift is equivalent to a slower propagation velocity. If the intensity of the incident beam is small so that the electric field of the light is much smaller compared with the intra-atomic field, it can be shown that the phase shift is independent of the intensity. In other words, the index of refraction is independent of the intensity of the light beam. This is the regime of linear optics.

When the intensity of the light beam is comparable to the intra-atomic electric field, the distribution of electrons in the medium can be modified by the radiation. This leads to a different index of refraction. The reflection and refraction at an interface and the propagation through a medium will now depend on the intensity of the incident beam. This is the regime of nonlinear optics. On the fundamental level, nonlinear optics is the study of the interaction of light beams in matter. It is called nonlinear because the material responds nonlinearly to the amplitude of the electric field. We know that the propagation of light beams in material media is determined by the index of refraction. If the index of refraction of the medium can be controlled by the illumination of a beam of light, then we can manipulate the propagation of a beam of light by another beam of light. This can lead to great variety of technological innovation based on control of light by light itself.

In the electromagnetic theory of light, the material response to the illumination of light is described by the following equation:

$$\mathbf{P} = \epsilon_0 \chi^{(1)} \mathbf{E} + \epsilon_0 \chi^{(2)} \mathbf{EE} + \epsilon_0 \chi^{(3)} \mathbf{EEE} + \cdots \tag{1}$$

where **P** is the induced polarization of the medium, ϵ_0 is the dielectric constant of a vacuum. **E** is the electric field of the light and $\chi^{(1)}$, $\chi^{(2)}$, $\chi^{(3)}$ are constants. Linear optics is described by $\chi^{(1)}$ which is related to the index of refraction by

$$n^2 = 1 + \chi^{(1)} \tag{2}$$

The nonlinear optical response of the medium characterized by the parameters $\chi^{(2)}$ and $\chi^{(3)}$ gives rise to numerous interesting phenomena and applications. The second-order term $\chi^{(2)}\mathbf{EE}$ is responsible for the second harmonic generation (frequency doubling), for sum- and difference-frequency generation, and for parametric amplification and oscillation. The third order term $\chi^{(3)}\mathbf{EEE}$ is responsible for such diverse phenomena as third harmonic generation, Raman and Brillouin scattering, self-focusing, and optical phase conjugation.

For low light intensities, the higher order terms are very small and the optics is adequately described by the first term $\epsilon_0 \chi^{(1)} \mathbf{E}$. The second order term can be viewed as a modification to the first term by

$$\Delta\chi^{(1)} = \chi^{(2)} \mathbf{E} \tag{3}$$

For typical optical media, $\chi^{(1)}$ is of the order of 1. For ordinary light intensity similar to the sunlight at sea level, the electric field of the light is of the order of 10 V/cm and $\Delta\chi^{(1)}$ is of the order of 10^{-8}–10^{-10}. For laser beams with moderate power, the associated electric field is comparable to an intra-atomic electric field of about 10×10^6 V/cm. Illumination of appropriate materials by such a laser beam induces a significant change in the index of refraction which may affect the propagation of the light beam.

Prior to 1961, many of the nonlinear optical properties in matter were not perceived. Scientific research indicated that the transmission, reflection, and refraction of light in transparent materials were affected neither by the intensity of the light nor by the presence of another beam. This is the regime of linear optics and the principle of superposition applies. Transparent materials are characterized by an index of refraction which was thought to be independent of the intensity of the light. Thus in the regime of linear optics, scientists concluded that the output intensity in light would always be directly proportional to the input intensity and the frequency of the output light would always equal that of the input light in the rest frame of the material. Nonlinear optical phenomena have disproven these earlier conclusions, giving us the tools to manipulate and control laser beam frequency and intensity, and thereby opening a broad new realm of optical applications.

One reason for these early, linear interpretations of light was that scientists were not yet able to create a coherent (i.e., highly monochromatic) beam of light that would give us a reasonable nonlinear effect; only with the first laser were we able to detect significant evidence of nonlinear phenomena. In 1961, an early second harmonic generation experiment gave the first experimental confirmation for nonlinear theories. A ruby laser with a wavelength of 0.6943 μm was focused on the front surface of a crystalline quartz plate. The emergent radiation was examined with a spectrometer and was found to contain radiation at twice the input frequency (i.e., at wavelength = 0.34715 μm).

Since the early days of harmonic generation experiments, the field of nonlinear optics has grown rapidly. Many different nonlinear optical phenomena have been discovered. These include optical parametric mixing, stimulated Brillouin scattering, phase conjugation, wave mixing, etc. As indicated, these nonlinear optical phenomena require intense laser beams. The intensity requirements become even more stringent for high order processes (e.g., third order processes).

OPTICAL PHASE CONJUGATION

One of the most important applications of nonlinear optics is the generation of phase conjugate waves. If an electromagnetic radiation is described by

$$E = E_0 \cos[\omega t - kz - \phi(z)] \tag{4}$$

where E_0 is a constant, ω is the angular frequency, k is the wave number and ϕ is a phase. The propagation of such an electromagnetic wave can be viewed as the spatial motion of the *wave fronts* defined as

$$kz + \phi(z) = \text{constants} \tag{5}$$

The phase conjugate wave of the electromagnetic wave given by Eq. (4) can be written

$$E_c = E_0 \cos[\omega t + kz + \phi(z)] \tag{6}$$

We note that the wave fronts of E_c are identical to those of E. These two waves form a conjugate pair. They have exactly the same wave fronts, but move in opposite direction. In principle, the phase conjugate wave can be generated by using a mirror whose surface is identical to that of the wave front of the incident wave. This technique, known as *adaptive optics*, requires the prior knowledge of the wave front of the incoming radiation. Using deformable mirrors driven by mechanical actuators, the surface of the mirror can be made to be identical to that of the incoming wave front. An interferometry technique can be employed to measure the wave front of the incoming radiation. The results are then used to prepare the deformable mirrors.

In a 1972 experiment it was observed that the wave front of back scattered radiation in stimulated Brillouin scattering (SBS) is a time-reversed replica of that of the incident pump beam. Radiation from a single mode pulsed ruby laser was directed into an SBS cell. The back scattered radiation was examined and the wave fronts were found to be identical to those of the incident radiation. It was discovered that the phase conjugate wave in SBS has an intensity gain which is twice of that of the nonconjugate waves. Such a difference in the intensity exponential gain coefficients translates into an enormous difference in the intensity of these waves. A few years later, it was proposed and demonstrated that optical four-wave mixing (FWM) can be employed to generate phase conjugate waves. In optical FWM, a nonlinear medium is pumped by a pair of counterpropagating waves. When a third beam is incident into the medium, a fourth wave is generated which is exactly the phase conjugate of the incident one. The process can be interpreted as real-time holography. In the medium, the incident beam and one of the pump beams form an interference intensity pattern which depends on the product of the field amplitudes of the two waves. In some nonlinear media the index of refraction can be written

$$n = n_0 + n_2 I \tag{7}$$

where n_0, n_2 are constants and I is the intensity. The constant n_2 is known as the Kerr coefficient. For optical glass n_2 is 3.2×10^{-16} cm^2/W. As a result of the nonlinear response of the medium, an index grating is formed. This index grating is read out by the other pump beam, and the diffracted beam is exactly the phase conjugate of the incident beam.

Both SBS and FWM are third order processes and require intense radiation of the order of many MW/cm^2 for efficient operations. Nonlinear optical techniques for phase conjugation do not require the prior knowledge of the wave fronts of the incoming radiation. Since optical four-wave mixing in nonlinear media is equivalent to real-time holography, any holographic media can be employed for the generation of phase conjugate waves. In this book, we deal with a new class of materials which can be employed for efficient generation of phase conjugate waves or wave mixing using optical intensities of the order of W/cm^2.

PHOTOREFRACTIVE NONLINEAR OPTICS

The photorefractive effect is a phenomenon in which the local index of refraction is changed by the spatial variation of light intensity. In the early days of laser modulation works, it was discovered that $LiNbO_3$ crystals became "optically damaged" after irradiation by a laser beam. This turned out to be the spatial index variation caused by the photorefractive effect which caused a distortion of the wave front. The photorefractive effect has since been observed in many electro-optic crystals, including $LiNbO_3$, $BaTiO_3$, SBN, KTN, KNSBN, BSO, BGO, GaAs, InP, and CdTe. It is believed that there are impurities in the crystals. The energy levels of electrons in these impurities are somewhere in the middle of the bandgap of the crystals. When the crystal is illuminated with electromagnetic radiation, some of the electrons can be excited into the conduction band provided the photon energy is large enough. The impurities that can be ionized and provide electrons are called donors. Once these donors are ionized, they can serve as trap sites which capture electrons. The electrons can migrate and get trapped at these sites. The trapped electrons can be re-excited. They will not be re-excited provided they are in the dark region. Thus a space-charge separation is created when the crystal is illuminated with a spatially carrying intensity pattern. The charge separation leads to a space-charge field. Such a field induces a change in index of refraction via the Pockels effect (linear electro-optic effect). Thus a hologram can be recorded by two-beam interference inside a photorefractive medium. The hologram or index grating formed by the two-beam interference inside the photorefractive medium can be erased by illuminating the hologram with light. Thus dynamic holography is possible using photorefractive materials. Photorefractive materials such as $LiNbO_3$, $BaTiO_3$, SBN, BSO, and are by far the most efficient media for the generation of optical phase conjugation and real-time holography using relatively low intensity levels (e.g., 1 W/cm^2).

In addition to the efficient holographic response, beam coupling known as two-wave mixing occurs naturally in photorefractive media. When two beams of coherent radiation intersect inside a photorefractive medium, a stationary index grating is formed. This index grating is spatially shifted by $\pi/2$ relative to the intensity pattern. Such a spatial phase shift leads to nonreciprocal energy transfer when these two beams propagate through the medium. The unique property of nonreciprocal energy transfer can be employed for many applications, including laser beam clean-up, photorefractive resonators, nonreciprocal transmission window, biased elements for laser gyros, self-pumped phase conjugators, mutually pumped phase conjugators, optical interconnection, neural networks, phase conjugate interferometry, etc. In this book, we start from the basic principles of photorefractive effect, we then discuss generic processes such as two-wave mixing and four-wave mixing. Phase conjugators and resonators are then treated. The latter part of the book is devoted to applications.

1

Electromagnetic Waves in Crystals

This book deals with the propagation and coupling of electromagnetic waves in photorefractive crystals and their applications in optics, including wave mixing, phase conjugation, dynamic holography, optical computing, interconnection, and neural networks. The photorefractive effect is an electro-optical phenomenon in which a change in the index of refraction of the medium is produced by the presence of an internal space-charge field. The space-charge field inside the crystal is due to the nonuniform illumination of light and is a result of photoexcitation and the subsequent redistribution of charge carriers in the medium. In the first chapter, we briefly review some of the most important properties of the electromagnetic waves and their propagation in electro-optic crystals. These background materials are used frequently in the book and included for completeness and as a ready source of reference.

We begin by reviewing Maxwell's equations and their plane wave solutions. We then discuss the propagation of plane waves in anisotropic crystals. These are followed by a discussion of the linear electro-optic effect and optical activity. MKS units will be used throughout the book.

1.1 MAXWELL'S EQUATIONS AND ELECTROMAGNETIC ENERGY

All phenomena in electromagnetism, including the propagation of electromagnetic waves, are described by the following set of equations (known as Maxwell's equations):

$$\nabla \times \mathbf{E} + \frac{\partial \mathbf{B}}{\partial t} = 0 \tag{1.1-1}$$

$$\nabla \times \mathbf{H} - \frac{\partial \mathbf{D}}{\partial t} = \mathbf{J} \tag{1.1-2}$$

$$\nabla \cdot \mathbf{D} = \rho \tag{1.1-3}$$

$$\nabla \cdot \mathbf{B} = 0 \tag{1.1-4}$$

where **E** is the electric field vector, **H** is the magnetic field vector, **B** is the magnetic induction vector, **D** is the electric displacement vector, **J** is the current density vector, and ρ is the charge density. These four field vectors are linked by the following material equations:

$$\mathbf{D} = \epsilon_0 \mathbf{E} + \mathbf{P} \tag{1.1-5}$$

$$\mathbf{B} = \mu_0 \mathbf{H} + \mathbf{M} \tag{1.1-6}$$

where ϵ_0, μ_0 are constants, $\mathbf{P}$ is the electric polarization and $\mathbf{M}$ is the magnetic polarization. These two constants ϵ_0, μ_0 are known as the dielectric permittivity and permeability of the vacuum. All the materials described in this book are assumed to be nonmagnetic. In other words, they are pure dielectric with $\mathbf{M} = 0$. The electric polarization vector $\mathbf{P}$ represents the response of the material under the influence of an electric field. In the linear regime when the electric field is small relative to the intra-atomic field, the electric polarization vector is proportional to the electric field vector and is often written as

$$\mathbf{P} = \epsilon_0 \chi \mathbf{E} \tag{1.1-7}$$

where χ is the electric susceptibility of the medium. In anisotropic media like some of the electro-optic crystals, χ is a tensor of rank 2 (or a 3×3 matrix).

If we substitute Eq. (1.1-7) for $\mathbf{P}$ in Eq. (1.1-5), we obtain

$$\mathbf{D} = \epsilon_0 (1 + \chi) \mathbf{E} \tag{1.1-8}$$

Equation (1.1-8) is often written as

$$\mathbf{D} = \epsilon \mathbf{E} \tag{1.1-9}$$

with ϵ defined as

$$\epsilon = \epsilon_0 (1 + \chi) \tag{1.1-10}$$

which is called the dielectric constant of the medium. Again, in anisotropic media, ϵ is a tensor of rank 2 (or a 3×3 matrix).

The charge density and the current density may be regarded as the source for the generation of the electromagnetic field. In optics, one often deals with electromagnetic fields in regions where both the charge density and the current density are zero. In fact, if we set $\mathbf{J} = 0$ and $\rho = 0$ in Maxwell's equations, we find that nonzero solutions for the field vectors (e.g., $\mathbf{E}$ and $\mathbf{H}$) exist. This means that electromagnetic field can exist in the absence of any charges and currents. Electromagnetic fields occurring in media without any charges and currents are called electromagnetic waves (or electromagnetic radiations).

The energy density U and power flow $\mathbf{S}$ associated with the electromagnetic field are given by [1, 2]

$$U = \tfrac{1}{2}(\mathbf{E} \cdot \mathbf{D} + \mathbf{B} \cdot \mathbf{H}) \tag{1.1-11}$$

$$\mathbf{S} = \mathbf{E} \times \mathbf{H} \tag{1.1-12}$$

1.2 WAVE EQUATIONS AND MONOCHROMATIC PLANE WAVES

We now consider electromagnetic field in media where both charge density and current density are zero. We will derive the wave equations and the monochromatic plane wave solutions. Using Eqs. (1.1-6, 1.1-9) with $\mathbf{M} = 0$ (nonmagnetic media), we can rewrite Maxwell's equations in terms of the field vectors **E** and **H**

$$\nabla \times \mathbf{E} + \mu_0 \frac{\partial \mathbf{H}}{\partial t} = 0 \tag{1.2-1}$$

$$\nabla \times \mathbf{H} - \epsilon \frac{\partial \mathbf{E}}{\partial t} = 0 \tag{1.2-2}$$

$$\nabla \cdot \epsilon \mathbf{E} = 0 \tag{1.2-3}$$

$$\nabla \cdot \mathbf{H} = 0 \tag{1.2-4}$$

By eliminating the field vector **H** from these equations, we obtain

$$\nabla \times (\nabla \times \mathbf{E}) + \mu_0 \epsilon \frac{\partial^2 \mathbf{E}}{\partial t^2} = 0 \tag{1.2-5}$$

If we further assume that the medium is homogeneous and isotropic, then Eq. (1.2-3) reduced to $\nabla \cdot \mathbf{E} = 0$. Using the following identity

$$\nabla \times (\nabla \times \mathbf{E}) = \nabla(\nabla \cdot \mathbf{E}) - \nabla^2 \mathbf{E} \tag{1.2-6}$$

Eq. (1.2-5) can be written

$$\nabla^2 \mathbf{E} - \mu_0 \epsilon \frac{\partial^2 \mathbf{E}}{\partial t^2} = 0 \tag{1.2-7}$$

which is known as the wave equation. A similar wave equation for **H** given as follows can also be derived:

$$\nabla^2 \mathbf{H} - \mu_0 \epsilon \frac{\partial^2 \mathbf{H}}{\partial t^2} = 0 \tag{1.2-8}$$

It is important to note that these two standard wave equations are derived under the assumption that the medium is homogeneous and isotropic. The wave equations are satisfied by the well-known monochromatic plane wave solution

$$\mathbf{E} = \mathbf{E}_0 e^{i(\omega t - \mathbf{k}\cdot\mathbf{r})} \tag{1.2-9}$$

$$\mathbf{H} = \mathbf{H}_0 e^{i(\omega t - \mathbf{k}\cdot\mathbf{r})} \tag{1.2-10}$$

where ω is the angular frequency and $\mathbf{k}$ is the wave vector. They are related by

$$k = |\mathbf{k}| = \omega\sqrt{\mu_0 \epsilon} \tag{1.2-11}$$

In Eqs. (1.2-9) and (1.2-10), $\mathbf{E}_0$, $\mathbf{H}_0$ are two constants which are related according to Maxwell's equations (1.2-1)–(1.2-4) by the following relationship:

$$\mathbf{k}\cdot\mathbf{E}_0 = \mathbf{k}\cdot\mathbf{H}_0 = 0 \tag{1.2-12}$$

$$\mathbf{H}_0 = \frac{1}{\omega\mu_0}\mathbf{k}\times\mathbf{E}_0 \tag{1.2-13}$$

According to the above relationships, the triad $(\mathbf{k}, \mathbf{E}_0, \mathbf{H}_0)$ forms a set of mutually orthogonal vectors.

The phase velocity of the plane waves described by Eqs. (1.2-9) and (1.2-10) is given by

$$v = \frac{\omega}{k} = \frac{1}{\sqrt{\mu_0 \epsilon}} \tag{1.2-14}$$

which depends on the dielectric constant of the medium. In a vacuum, the phase velocity is

$$c = \frac{1}{\sqrt{\mu_0 \epsilon_0}} = 2.997930 \times 10^8 \text{ m/s} \tag{1.2-15}$$

The index of refraction of an isotropic medium is defined as

$$n = \frac{c}{v} = \left(\frac{\epsilon}{\epsilon_0}\right)^{1/2} \tag{1.2-16}$$

We must keep in mind that the index of refraction of a medium depends on the frequency of the wave. In other words, each frequency component of the electromagnetic waves propagates at its own phase velocity in material media. This is known as the dispersion in optics.

In Eqs. (1.2-9) and (1.2-10), the field vectors are expressed in terms of complex exponential functions. We must keep in mind that both field vectors are physical quantities and thus they must be real. Although both the real and the imaginary parts of Eqs. (1.2-9) and (1.2-10) are solutions to the wave equations, only the real part has physical meaning. Thus, when the field vectors are written in complex forms, we must remember that only the real parts are physically measurable. In other words, Eqs. (1.2-9) and (1.2-10) actually mean

$$\mathbf{E} = \mathbf{E}_0 \cos(\omega t - \mathbf{k} \cdot \mathbf{r}) \tag{1.2-17}$$

$$\mathbf{H} = \mathbf{H}_0 \cos(\omega t - \mathbf{k} \cdot \mathbf{r}) \tag{1.2-18}$$

The use of complex forms greatly simplifies the mathematical manipulations as far as linear operations are concerned. However, we must be careful when the products of field vectors are involved. For example, the time-averaged field energy density and Poynting vector are given by [1, 2]

$$U = \tfrac{1}{4}\,\mathrm{Re}(\mathbf{E}^* \cdot \mathbf{D} + \mathbf{B}^* \cdot \mathbf{H}) \tag{1.2-19}$$

$$\mathbf{S} = \tfrac{1}{2}\,\mathrm{Re}(\mathbf{E}^* \times \mathbf{H}) \tag{1.2-20}$$

where * indicates complex conjugation. For the plane waves given by Eq. (1.2-9), the time-averaged energy density and Poynting vector are reduced to

$$U = \tfrac{1}{2}\epsilon(\mathbf{E}^* \cdot \mathbf{E}) \tag{1.2-21}$$

$$\mathbf{S} = \frac{\mathbf{k}}{2\omega\mu}\,(\mathbf{E}^* \cdot \mathbf{E}) \tag{1.2-22}$$

1.2.1 Parabolic Wave Equation

The monochromatic plane wave solutions given by Eqs. (1.2-9) and (1.2-10) are idealized electromagnetic waves. They are simple and useful in terms of deriving fundamental physical phenomena. They can be used as the basis for the expansion of any electromagnetic waves. In practice, we often deal with highly directional monochromatic waves (e.g., laser beams) which can be viewed as a linear superposition of many monochromatic plane waves with almost identical wave vectors. For these monochromatic waves, the wave equations reduce to Helmholtz equation

$$\nabla^2 E + k^2 E = 0 \tag{1.2-23}$$

with

$$k^2 = \omega^2 \mu_0 \epsilon \tag{1.2-24}$$

and E can be any component of the field vectors.

Although, we can always expand these highly directional monochromatic waves in terms of the plane wave components, it is often more convenient to write

$$E = A(x, y, z)e^{-ikz} \tag{1.2-25}$$

where $A(x, y, z)$ is the complex amplitude that depends on position and z is measured along the central direction of propagation. For highly directional beams, the amplitude $A(x, y, z)$ is a relatively slow-varying function on the scale of $\lambda = 2\pi/k$, as compared with the rapidly varying factor $\exp(-ikz)$. If we substitute Eq. (1.2-25) for E in the Helmholtz equation, we obtain

$$\nabla^2 A - 2ik \frac{\partial A}{\partial z} = 0 \tag{1.2-26}$$

or equivalently,

$$\frac{\partial^2 A}{\partial x^2} + \frac{\partial^2 A}{\partial y^2} + \frac{\partial^2 A}{\partial z^2} - 2ik \frac{\partial A}{\partial z} = 0 \tag{1.2-27}$$

Let $a \gg \lambda$ be the transverse dimension of the beam at any point in z, then $(\partial^2 A/\partial x^2 + \partial^2 A/\partial y^2)$ is of the order of A/a^2. For a Gaussian beam, a is the spot size of the beam. If we neglect the term $\partial^2 A/\partial z^2$, then it follows directly from Eq. (1.2-27) that $\partial A/\partial z$ is of the order of A/ka^2. In other words, the field amplitude A changes significantly only over a distance of $dz = ka^2$. Thus, the second order term $\partial^2 A/\partial z^2$ is of the order of $A/(ka^2)^2$, which is a factor of $(ka)^2$ smaller than the remaining terms. For highly directional monochromatic waves with $a \gg \lambda$, we thus obtain the following parabolic wave equation:

$$\nabla_\perp^2 A - 2ik \frac{\partial A}{\partial z} = 0 \tag{1.2-28}$$

where $\nabla_\perp^2 A = (\partial^2/\partial x^2 + \partial^2/\partial y^2)$. It can be shown that plane waves and spherical waves are solutions of the parabolic wave equation (see Problem 1.1). The $\nabla_\perp^2 A$ term represents the diffraction of electromagnetic waves in homogeneous medium. The parabolic wave equation is sometimes written as

$$\frac{\partial A}{\partial z} + \frac{i}{2k} \nabla_\perp^2 A = 0 \tag{1.2-29}$$

We note that this is a first order differential in z. A boundary condition of $A(x, y, z_o)$ at any plane $z = z_o$ is enough to uniquely determine $A(x, y, z)$ at all points in space.

1.3 WAVE PROPAGATION IN ANISOTROPIC MEDIA

Most of the photorefractive media that are discussed in this book are anisotropic crystals (e.g., $BaTiO_3$, $LiNbO_3$, $Sr_xBa_{1-x}Nb_2O_6$). In these media, the dielectric permittivity ϵ is a symmetric tensor of rank 2. In the principal coordinate, such a dielectric tensor has the following diagonalized form:

$$\epsilon = \begin{bmatrix} \epsilon_x & 0 & 0 \\ 0 & \epsilon_y & 0 \\ 0 & 0 & \epsilon_z \end{bmatrix} = \epsilon_0 \begin{bmatrix} n_x^2 & 0 & 0 \\ 0 & n_y^2 & 0 \\ 0 & 0 & n_z^2 \end{bmatrix} \tag{1.3-1}$$

where ϵ_x, ϵ_y, ϵ_z are the principal dielectric constants, and n_x, n_y, n_z are the principal indices of refraction.

Wave propagation in these media is still governed by Maxwell's equations and plane wave solutions still exist. It is important to note that $\nabla \cdot \mathbf{E} = 0$ is no longer true even if the medium is homogeneous. This leads to a unique relationship between the angular frequency ω and the wave vector $\mathbf{k}$ of a plane wave. If we substitute Eq. (1.2-9) for the electric field vector $\mathbf{E}$ in Eq. (1.2-5), we obtain

$$\mathbf{k} \times (\mathbf{k} \times \mathbf{E}_0) + \mu_0 \epsilon \omega^2 \mathbf{E}_0 = 0 \tag{1.3-2}$$

This equation not only determines the relationship between the angular frequency ω and the wave vector $\mathbf{k}$, but also determines the polarization state of the plane wave. It is a vector equation which is equivalent to a set of three homogeneous linear equations with three unknowns E_{0x}, E_{0y}, E_{0z}. Using Eq. (1.3-1) and carrying out the vector multiplication in Eq. (1.3-2), we obtain the following matrix equation:

$$\begin{bmatrix} \omega^2\mu_0\epsilon_x - k_y^2 - k_z^2 & k_x k_y & k_x k_z \\ k_y k_x & \omega^2\mu_0\epsilon_y - k_x^2 - k_z^2 & k_y k_z \\ k_z k_x & k_z k_y & \omega^2\mu_0\epsilon_z - k_x^2 - k_y^2 \end{bmatrix} \begin{bmatrix} E_{0x} \\ E_{0y} \\ E_{0z} \end{bmatrix} = 0 \tag{1.3-3}$$

where k_x, k_y, k_z are components of the wave vector along the principal axes. Nontrivial solutions exist only when the determinant of the matrix is zero. This leads to the following relation between the angular frequency ω and the wave vector $\mathbf{k}$:

$$\begin{vmatrix} \omega^2\mu_0\epsilon_x - k_y^2 - k_z^2 & k_x k_y & k_x k_z \\ k_y k_x & \omega^2\mu_0\epsilon_y - k_x^2 - k_z^2 & k_y k_z \\ k_z k_x & k_z k_y & \omega^2\mu_0\epsilon_z - k_x^2 - k_y^2 \end{vmatrix} = 0 \tag{1.3-4}$$

For a given set of principal dielectric constants ϵ_x, ϵ_y, ϵ_z and an angular frequency ω, Eq. (1.3-4) represents a surface in three-dimensional space (momentum space), wave vector space, or $\mathbf{k}$-space. This is known as the normal surface which consists of an inner shell and an outer shell. These two shells are enclosed and have, in general, four intersection points. The two lines that go through the origin and these points are known as the optical axes. Figure 1.1 shows one octant of a general normal surface. For most

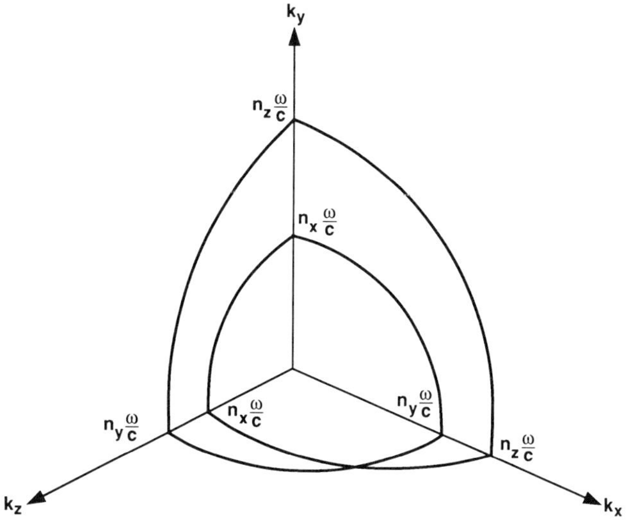

Figure 1.1. One octant of a general normal surface.

transparent media, with the difference among the principal dielectric constants being very small (i.e., $\epsilon_x \approx \epsilon_y \approx \epsilon_z$ or $n_x \approx n_z$), these two shells are almost spherical except near the intersection points.

We now draw a straight line from the origin and along the direction of propagation. This line intersects with the normal surface, in general, at two points. These two k values correspond to two different phase velocities ω/k of propagation along the given direction. If we write

$$\mathbf{k} = n\mathbf{s}(\omega/c) \tag{1.3-5}$$

where $\mathbf{s}$ is a unit vector along the direction of propagation and n is to be determined, then each of these points corresponds to a phase velocity of c/n. This parameter n may be viewed as the index of refraction associated with each of the modes of propagation and is called the eigen-index of refraction. If we substitute Eq. (1.3-5) for $\mathbf{k}$ in Eq. (1.3-4) and carry out the multiplication, we obtain

$$\frac{s_x^2}{n^2 - n_x^2} + \frac{s_y^2}{n^2 - n_y^2} + \frac{s_z^2}{n^2 - n_z^2} = \frac{1}{n^2}, \tag{1.3-6}$$

where s_x, s_y, s_z are components of the unit vector $\mathbf{s}$ along the principal axes. This is known as the Fresnel equation of wave normals and can be solved for the eigenvalues of the index of refraction (or eigenindices of refraction). Eq. (1.3-6) is a quadratic equation in n^2. Therefore, for each direction of propagation (a set of s_x, s_y, s_z), two solutions for n^2 can be obtained from Eq. (1.3-6). Let n_1^2 and n_2^2 be the solutions to the Fresnel equation.

To obtain the solution for the corresponding electric field vector, we substitute Eq. (1.3-5) for **k** and use the values of n, one at a time, in Eq. (1.3-3). This leads to [1, 2]:

$$\mathbf{E}_1 = \begin{bmatrix} \dfrac{s_x}{n_1^2 - n_x^2} \\ \dfrac{s_y}{n_1^2 - n_y^2} \\ \dfrac{s_z}{n_1^2 - n_z^2} \end{bmatrix}, \quad \mathbf{E}_2 = \begin{bmatrix} \dfrac{s_x}{n_2^2 - n_x^2} \\ \dfrac{s_y}{n_2^2 - n_y^2} \\ \dfrac{s_z}{n_2^2 - n_z^2} \end{bmatrix} \tag{1.3-7}$$

where n_1, n_2 are the eigen-indices of refraction and $\mathbf{E}_1$, $\mathbf{E}_2$ are the corresponding electric field vectors. Strictly speaking, these two vectors are, in general, not orthogonal to the wave vector **k** and are not mutually orthogonal. However, for most transparent media with $n_x \approx n_y \approx n_z$, these three vectors are almost mutually orthogonal.

Although, we have general solutions for the normal modes of propagation, it is useful to examine the propagation for the special case when the propagation is in one of the coordinate planes, especially for the cases when all the principal indices of refraction n_x, n_y, n_z are different. For example, let us consider the propagation in the yz-plane, if we set $k_x = 0$, Eq. (1.3-3) becomes

$$\begin{bmatrix} \omega^2\mu_0\epsilon_x - k_y^2 - k_z^2 & 0 & 0 \\ 0 & \omega^2\mu_0\epsilon_y - k_z^2 & k_y k_z \\ 0 & k_z k_y & \omega^2\mu_0\epsilon_z - k_y^2 \end{bmatrix} \begin{bmatrix} E_{0x} \\ E_{0y} \\ E_{0z} \end{bmatrix} = 0 \tag{1.3-8}$$

and the secular equation (1.3-4) can be written

$$\left(\frac{k_y^2}{n_x^2} + \frac{k_z^2}{n_x^2} - \frac{\omega^2}{c^2}\right)\left(\frac{k_y^2}{n_z^2} + \frac{k_z^2}{n_y^2} - \frac{\omega^2}{c^2}\right) = 0 \tag{1.3-9}$$

Equation (1.3-9) consists of a circle and an ellipse in yz-plane of the momentum space ($k_y k_z$-plane). This is the intersection of the normal surface with yz-coordinate plane. The circle corresponds to an eigen-index of refraction of n_x and an electric field vector of

$$\mathbf{E} = \begin{bmatrix} 1 \\ 0 \\ 0 \end{bmatrix} \tag{1.3-10}$$

This electric field is perpendicular to the plane of propagation (yz-plane).

The ellipse corresponds to an eigen-index of refraction of

$$\frac{1}{n^2} = \frac{s_y^2}{n_z^2} + \frac{s_z^2}{n_y^2} \tag{1.3-11}$$

where s_y, s_z are the directional cosines of propagation in the yz-plane. The electric field vector for this mode can be obtained directly from the secular equation and is given by

$$\mathbf{E} = \begin{bmatrix} 0 \\ \dfrac{s_z}{n_y^2} \\ -\dfrac{s_y}{n_z^2} \end{bmatrix} \tag{1.3-12}$$

This electric field is parallel to the plane of propagation (yz-plane). It can be shown that this electric field vector is tangent to the ellipse. In fact, it is generally true that the electric field vectors of the normal modes of propagation are tangent to the normal surface.

In summary, there are two modes of propagation in a coordinate plane which is perpendicular to one of the principal axes. The electric field of one of the modes is parallel to the principal axis with an eigen-index of refraction equal to the principal index of refraction associated with the axis. The electric field of the other mode is perpendicular to the axis and is given by Eq. (1.3-12) with an eigen-index of refraction given by eq. (1.3-11). This electric field is almost transverse to the direction of propagation for most transparent media with $n_x \approx n_y \approx n_z$.

Propagations in other coordinate planes are similar. The intersection of the normal surface with the coordinate planes always consists of a circle and an ellipse. The circle and the ellipse intersect only in one coordinate plane. This plane is determined by the relative magnitude of the principal indices of refraction n_x, n_y, n_z. For $n_x < n_y < n_z$, the circle and the ellipse intersect in the xz-plane. Figure 1.2 shows the intersection of the normal surface with each of the coordinate planes.

1.3.1 Uniaxial Crystals

Anisotropic media are classified into three categories. In the most general case when the principal indices of refraction n_x, n_y, n_z are all different, there are two optic axes. The crystals are said to be biaxial. In many of the photorefractive crystals discussed in this book, it happens that two of the principal indices of refraction are equal. This is a result of crystal symmetry. In this case, the normal surface [Eq. (1.3-4)] reduces to

$$\left(\frac{k_x^2 + k_y^2}{n_e^2} + \frac{k_z^2}{n_o^2} - \frac{\omega^2}{c^2}\right)\left(\frac{k^2}{n_o^2} - \frac{\omega^2}{c^2}\right) = 0 \tag{1.3-13}$$

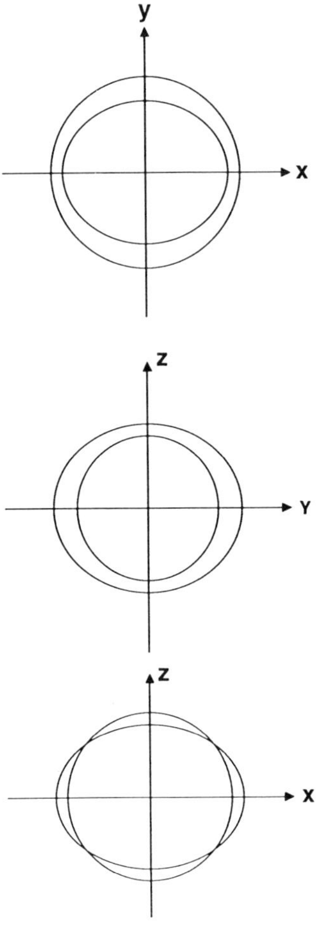

Figure 1.2. Intersection of normal surface with coordinate planes.

where $n_e = n_z$ and $n_o = n_x = n_y$. Crystals in this category exhibit tetragonal, hexagonal or trigonal symmetry. The z-axis is chosen as the axis of symmetry.

The normal surface in this case consists of a sphere and an ellipsoid of revolution. These two sheets of the normal surfaces touch at two points on the z-axis. The z-axis is therefore the only optic axis, and the crystals are said to be uniaxial. Figure 1.3 shows the intersection of the normal surface with the yz-plane. In uniaxial crystals, n_o is called the ordinary index of refraction, and n_e is called the extraordinary index of refraction. The sphere corresponds to a propagation with an eigen-index of refraction of n_o, regardless of the direction of propagation. Such a mode of propagation is called an ordinary wave. The ellipsoid of revolution corresponds to an extraordinary wave whose eigen-index of refraction depends on the direction of propagation. In the event when all three principal indices of

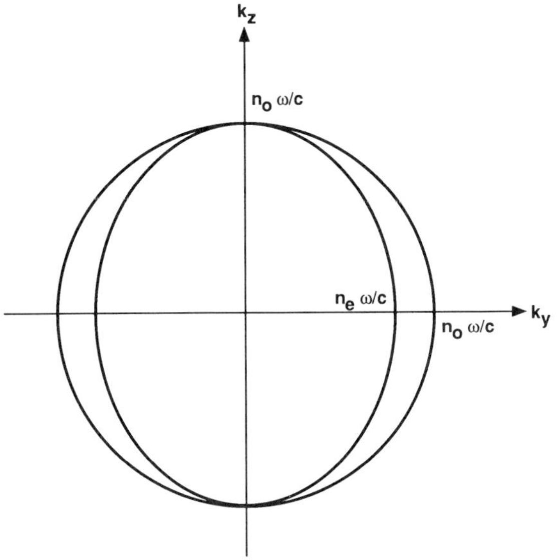

Figure 1.3. Normal surface of a uniaxial crystal in the *yz*-plane.

refraction are the same, the two sheets of the normal surface degenerate to a single sphere, and the crystals are said to be optically isotropic.

Because of the uniaxial symmetry, there is no loss of generality to consider propagation in the *yz*-plane. Let θ be the angle between the wave vector and the *c*-axis (the *z*-axis). The eigen-index of refraction of the ordinary wave is always n_o and the corresponding electric field vector is given, according to Eq. (1.3-10), by

$$\mathbf{E}_o = \begin{bmatrix} 1 \\ 0 \\ 0 \end{bmatrix}, \quad O\text{-wave} \tag{1.3-14}$$

which is perpendicular to the *c*-axis. The eigen-index of refraction of the extraordinary wave and the corresponding electric field vector are given, according to Eqs. (1.3-11) and (1.3-12), by

$$\frac{1}{n_e^2(\theta)} = \frac{\sin^2\theta}{n_e^2} + \frac{\cos^2\theta}{n_o^2} \tag{1.3-15}$$

$$\mathbf{E}_e = \begin{bmatrix} 0 \\ \dfrac{\cos\theta}{n_o^2} \\ -\dfrac{\sin\theta}{n_e^2} \end{bmatrix}, \quad E\text{-wave} \tag{1.3-16}$$

respectively. We note that the electric field vector of the ordinary wave (O-wave) is always perpendicular to the c-axis and the wave vector $\mathbf{k}$. The electric field of the extraordinary wave (E-wave), however, always lies in the plane formed by the c-axis and the wave vector $\mathbf{k}$. For most transparent media with small birefringence $(n_e - n_o) \ll n_{o,e}$, $\mathbf{E}_e$ is almost perpendicular to the direction of propagation. In this case, both electric field vectors are parallel to their corresponding displacement field vector $\mathbf{D}$.

1.3.2 The Index Ellipsoid

We have shown that the eigen-indices of refraction and the electric field vector of the normal modes of propagation can be obtained directly from Maxwell's equations. There is an equivalent approach known as the method of index ellipsoid. Index ellipsoid is a surface in three-dimensional space defined as

$$\sum \eta_{ij} \xi_i \xi_j = 1 \tag{1.3-17}$$

where the η_{ij} are components of the impermeability tensor, defined as the inverse of the dielectric tensor,

$$\eta \epsilon = \epsilon_0 \tag{1.3-18}$$

In tensor notation, the above equation can be written $\eta_{ij}\epsilon_{jk} = \epsilon_0 \delta_{ik}$. In the principal coordinate, the index ellipsoid is written

$$\frac{x^2}{n_x^2} + \frac{y^2}{n_y^2} + \frac{z^2}{n_z^2} = 1 \tag{1.3-19}$$

where n_x^2, n_y^2, n_z^2 are the principal indices of refraction. The index ellipsoid can be used to find the eigen-indices of refraction n_1, n_2 and the corresponding displacement vectors $\mathbf{D}_1$, $\mathbf{D}_2$ of a plane wave propagating along any direction $\mathbf{s}$. This is done by the following prescription (see Figure 1.4): Find the intersection ellipse between the index ellipsoid and a plane which passes through the center of the ellipsoid and is perpendicular to the direction of propagation $\mathbf{s}$. The length of the major and minor axes of the intersection ellipse are given by $2n_1$, $2n_2$ where n_1, n_2 are the eigen-indices of refraction. These axes are parallel, respectively, to the direction of the vectors $\mathbf{D}_1$, $\mathbf{D}_2$ of the normal modes.

The method of index ellipsoid can be shown to be equivalent to the method discussed earlier in this section by examining Maxwell's equation. Using Eq. (1.3-5) and the following relationship between $\mathbf{E}_0$ and $\mathbf{D}_0$,

$$\mathbf{E}_0 = \frac{1}{\epsilon_0} \eta \mathbf{D}_0 \tag{1.3-20}$$

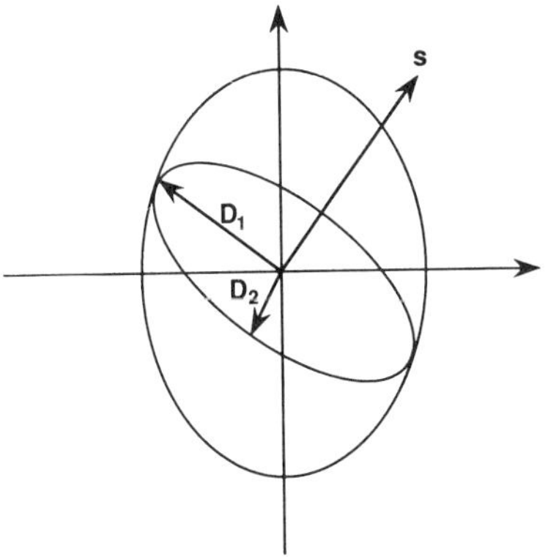

Figure 1.4. The method of index ellipsoid. The inner ellipse is the intersection of the index ellipsoid with a plane perpendicular to the direction of propagation **s**. The principal axes of the ellipse, $\mathbf{D}_1$ and $\mathbf{D}_2$, determine the polarization state of the normal modes. The lengths of the principal axes are exactly two times the eigen-indices of refraction of the normal modes.

we obtain, according to Eq. (1.3-2),

$$\mathbf{s} \times (\mathbf{s} \times \eta \mathbf{D}_0) + \frac{1}{n^2} \mathbf{D}_0 = 0 \tag{1.3-21}$$

where we recall that **s** is a unit vector along the direction of propagation.

Since the displacement vector **D** is always perpendicular to the direction of propagation (i.e., $\mathbf{s} \cdot \mathbf{D} = 0$), it is convenient to use a coordinate system in which the unit vector **s** is given by

$$\mathbf{s} = \begin{bmatrix} 0 \\ 0 \\ 1 \end{bmatrix} \tag{1.3-22}$$

In this coordinate system, Eq. (1.3-21) can be written

$$\begin{bmatrix} \eta_{11} & \eta_{12} & \eta_{13} \\ \eta_{21} & \eta_{22} & \eta_{23} \\ 0 & 0 & 0 \end{bmatrix} \mathbf{D}_0 = \frac{1}{n^2} \mathbf{D}_0 \tag{1.3-23}$$

The transverse nature of the displacement field vector ($\mathbf{s} \cdot \mathbf{D} = 0$) dictates that the third component of $\mathbf{D}_0$ is always zero. Thus, Eq. (1.3-23) is further simplified and becomes

$$\eta_t \mathbf{D}_0 = \frac{1}{n^2} \mathbf{D}_0 \tag{1.3-24}$$

where η_t is the transverse impermeability tensor

$$\eta_t = \begin{bmatrix} \eta_{11} & \eta_{12} \\ \eta_{21} & \eta_{22} \end{bmatrix} \tag{1.3-25}$$

We note that η_t is a symmetric tensor with $\eta_{21} = \eta_{12}$. According to this equation, the displacement field vectors of the normal modes are eigenvectors of the transverse impermeability tensor with eigenvalues $1/n^2$. Since η_t is a symmetric tensor, these eigenvectors are orthogonal.

The mathematical equation describing the intersection ellipse between the index ellipsoid and the plane normal to the unit vector **s** can be obtained by setting $\xi_3 = 0$ in Eq. (1.3-17). This leads to the following equation for the intersection ellipse:

$$\eta_{11}\xi_1\xi_1 + 2\eta_{12}\xi_1\xi_2 + \eta_{22}\xi_2\xi_2 = 1 \tag{1.3-26}$$

By comparing with Eq. (1.3-24), we note that the coefficients of this ellipse are exactly identical to the component of the transverse impermeability tensor. A coordinate transformation which diagonalizes the quadratic form (1.3-26) will also diagonalize the transverse impermeability matrix η_t. The eigenvectors of this tensor are therefore along the principal axes of the ellipse. The lengths of the principal axes determine the values of n. This proves the equivalence of the method of the index ellipsoid and the method discussed earlier in this section.

For uniaxial media, the index ellipsoid is an ellipsoid of revolution around the z-axis. Using the method of index ellipsoid, it is relatively easy to show that the ordinary wave has an eigen-index of refraction of n_o, with a displacement field vector given by

$$\mathbf{d}_o = \frac{\mathbf{k} \times \mathbf{c}}{|\mathbf{k} \times \mathbf{c}|} \tag{1.3-27}$$

where **c** is a unit vector parallel to the c-axis (z-axis). The same method also yields the eigen-index of refraction $n_e(\theta)$, and the following displacement field vector for the extraordinary wave

$$\mathbf{d}_e = \frac{\mathbf{d}_o \times \mathbf{k}}{|\mathbf{d}_o \times \mathbf{k}|} \tag{1.3-28}$$

We note that the triad $(\mathbf{k}, \mathbf{d}_o, \mathbf{d}_e)$ forms a set of mutually orthogonal vectors.

The eigen-index of refraction n can be obtained by using the following formula provided the eigenvector is known. According to Eqs. (1.3-23) and (1.3-24), the expectation value of the impermeability tensor is given by

$$\langle \eta_t \rangle = \langle \mathbf{d} | \eta_t | \mathbf{d} \rangle = \frac{1}{n^2} \tag{1.3-29}$$

where **d** is a normalized displacement field vector. If **d** is an eigenvector, then the above equation yields exactly the eigen-index of refraction. In addition, the above equation is equivalent to the following because of the transverse nature of the displacement field vector

$$\langle \eta \rangle = \langle \mathbf{d} | \eta | \mathbf{d} \rangle = \frac{1}{n^2} \tag{1.3-30}$$

The last equation is very useful in evaluating the eigen-index of refraction when only the direction of polarization is known.

1.4 ELECTRO-OPTIC EFFECTS

We consider the propagation of electromagnetic radiation in crystals in the presence of a dc (or low frequency) applied electric field. It was shown earlier that normal modes of propagation in anisotropic media can be determined by the dielectric permittivity tensor ϵ or equivalently, the method of the index ellipsoid. It is known that, in certain types of crystals, the application of an applied electric field results in a change in the dielectric permittivity tensor ϵ, or equivalently a change in both the dimension and the orientation of the index ellipsoid. This is referred to as the electro-optic effect. In photorefractive crystals, the dc electric field is produced by the charge carriers which are generated by photoexcitation due to the presence of light beams.

The electro-optic effect is defined, traditionally, in terms of the change in impermeability tensor

$$\Delta\eta_{ij} = \Delta\left(\frac{1}{n^2}\right)_{ij} = r_{ijk}E_k + s_{ijkm}E_kE_m \tag{1.4-1}$$

where E_k, E_m are components of the applied electric field. In Eq. (1.4-1), summation over repeated indices is assumed and terms higher than the quadratic are neglected. The constants r_{ijk}, s_{ijkm} are called linear electro-optic coefficients and quadratic electro-optic coefficients, respectively.

According to the definition Eq. (1.3-18), the change in dielectric tensor $\Delta\epsilon$ is related to the change in impermeability tensor $\Delta\eta$ by (see Problem 1.4)

$$\Delta\epsilon = -\frac{\epsilon\,\Delta\eta\,\epsilon}{\epsilon_0} \tag{1.4-2}$$

In the principal coordinate in which the dielectric tensor is diagonal, the change can be written

$$\Delta\epsilon_{ij} = -\epsilon_0 n_i^2 n_j^2\,\Delta\eta_{ij} \tag{1.4-3}$$

where n_i, n_j are the principal indices of refraction.

We shall limit our discussion to the linear electro-optic effect (the Pockels effect) because that is responsible for most of the phenomena in photorefractive crystals. The linear electro-optic coefficients r_{ijk} are components of a tensor of rank 3. Although there are 27 components, the symmetric property of ϵ and η reduces the independent numbers to 18. The symmetry leads to the interchangeability of the first two indices i and j. In other words,

$$r_{ijk} = r_{jik} \tag{1.4-4}$$

As a result of this symmetry, it is convenient to introduce the contracted indices to abbreviate the notations. They are defined by

$$\begin{aligned}
1 &= (11) = (xx) \\
2 &= (22) = (yy) \\
3 &= (33) = (zz) \\
4 &= (23) = (32) = (yz) = (zy) \\
5 &= (31) = (13) = (zx) = (xz) \\
6 &= (12) = (21) = (xy) = (yx)
\end{aligned} \tag{1.4-5}$$

Using the contracted indices, we can write $r_{Ik} = r_{ijk}$, or specifically

$$\begin{aligned}
r_{1k} &= r_{11k} \\
r_{2k} &= r_{22k} \\
r_{3k} &= r_{33k} \\
r_{4k} &= r_{23k} = r_{32k} \\
r_{5k} &= r_{31k} = r_{13k} \\
r_{6k} &= r_{12k} = r_{21k}
\end{aligned} \tag{1.4-6}$$

with $k = 1, 2, 3$ (or x, y, z).

Using the contracted notation, the electro-optic coefficients are often written in terms of a 6×3 matrix. It is important to note that such a contracted notation is only for the purpose of convenience. These matrix elements do not have the usual transformation property of a regular tensor.

Although there are, in general, 18 independent components of the linear electro-optic tensor, some of these components are zero or identical in many of the photorefractive crystals. This is a result of the point group symmetry in crystals. The symmetry consideration dictates that physical constants must remain invariant under the point group transformation. Each of these transformations provides a linear relationship between the electro-optic coefficients. Thus the number of independent components decreases as the symmetry increases. Table 1.1 lists electro-optic coefficients in contracted notation for all crystal symmetry classes. Table 1.2 lists the measured electro-optic coefficients for some selected crystals.

We want to point out that all components of the electro-optic coefficients r_{ijk} are zero for the crystals which possess inversion symmetry. The operation of inversion in crystals is to move every atom at position $\mathbf{r}$ to a new position $-\mathbf{r}$ with respect to the inversion center. Carrying out the mathematical transformation (i.e., $x \rightarrow -x$, $y \rightarrow -y$, $z \rightarrow -z$), we obtain

Table 1.1 Electro-Optic Coefficients in Contracted Notation for All Crystal Symmetry Classes[a]

Centrosymmetric ($\bar{1}$, $2/m$, mmm, $4/m$, $4/mmm$, $\bar{3}$, $\bar{3}m$, $6/m$, $6/mmm$, $m3$, $m3m$)

$$\begin{bmatrix} 0 & 0 & 0 \\ 0 & 0 & 0 \\ 0 & 0 & 0 \\ 0 & 0 & 0 \\ 0 & 0 & 0 \\ 0 & 0 & 0 \end{bmatrix}$$

Triclinic

1

$$\begin{bmatrix} r_{11} & r_{12} & r_{13} \\ r_{21} & r_{22} & r_{23} \\ r_{31} & r_{32} & r_{33} \\ r_{41} & r_{42} & r_{43} \\ r_{51} & r_{52} & r_{53} \\ r_{61} & r_{62} & r_{63} \end{bmatrix}$$

Monoclinic

$2 \quad (2 \parallel x_2)$ and $2 \quad (2 \parallel x_3)$

$$\begin{bmatrix} 0 & r_{12} & 0 \\ 0 & r_{22} & 0 \\ 0 & r_{32} & 0 \\ r_{41} & 0 & r_{43} \\ 0 & r_{52} & 0 \\ r_{61} & 0 & r_{63} \end{bmatrix} \begin{bmatrix} 0 & 0 & r_{13} \\ 0 & 0 & r_{23} \\ 0 & 0 & r_{33} \\ r_{41} & r_{42} & 0 \\ r_{51} & r_{52} & 0 \\ 0 & 0 & r_{63} \end{bmatrix}$$

$m \quad (m \perp x_2)$ and $m \quad (m \perp x_3)$

$$\begin{bmatrix} r_{11} & 0 & r_{13} \\ r_{21} & 0 & r_{23} \\ r_{31} & 0 & r_{33} \\ 0 & r_{42} & 0 \\ r_{51} & 0 & r_{53} \\ 0 & r_{62} & 0 \end{bmatrix} \begin{bmatrix} r_{11} & r_{12} & 0 \\ r_{21} & r_{22} & 0 \\ r_{31} & r_{32} & 0 \\ 0 & 0 & r_{43} \\ 0 & 0 & r_{53} \\ r_{61} & r_{62} & 0 \end{bmatrix}$$

Orthorhombic

222 and $2mm$

$$\begin{bmatrix} 0 & 0 & 0 \\ 0 & 0 & 0 \\ 0 & 0 & 0 \\ r_{41} & 0 & 0 \\ 0 & r_{52} & 0 \\ 0 & 0 & r_{63} \end{bmatrix} \begin{bmatrix} 0 & 0 & r_{13} \\ 0 & 0 & r_{23} \\ 0 & 0 & r_{33} \\ 0 & r_{42} & 0 \\ r_{51} & 0 & 0 \\ 0 & 0 & 0 \end{bmatrix}$$

Table 1.1 (*Continued*)

Tetragonal

4	$\bar{4}$	422
$\begin{bmatrix} 0 & 0 & r_{13} \\ 0 & 0 & r_{13} \\ 0 & 0 & r_{33} \\ r_{41} & r_{51} & 0 \\ r_{51} & -r_{41} & 0 \\ 0 & 0 & 0 \end{bmatrix}$	$\begin{bmatrix} 0 & 0 & r_{13} \\ 0 & 0 & -r_{13} \\ 0 & 0 & 0 \\ r_{41} & -r_{51} & 0 \\ r_{51} & r_{41} & 0 \\ 0 & 0 & r_{63} \end{bmatrix}$	$\begin{bmatrix} 0 & 0 & 0 \\ 0 & 0 & 0 \\ 0 & 0 & 0 \\ r_{41} & 0 & 0 \\ 0 & -r_{41} & 0 \\ 0 & 0 & 0 \end{bmatrix}$

$4mm$	$\bar{4}2m(2 \parallel x_1)$
$\begin{bmatrix} 0 & 0 & r_{13} \\ 0 & 0 & r_{13} \\ 0 & 0 & r_{33} \\ 0 & r_{42} & 0 \\ r_{42} & 0 & 0 \\ 0 & 0 & 0 \end{bmatrix}$	$\begin{bmatrix} 0 & 0 & 0 \\ 0 & 0 & 0 \\ 0 & 0 & 0 \\ r_{41} & 0 & 0 \\ 0 & r_{41} & 0 \\ 0 & 0 & r_{63} \end{bmatrix}$

Trigonal

3	32
$\begin{bmatrix} r_{11} & -r_{22} & r_{13} \\ -r_{11} & r_{22} & r_{13} \\ 0 & 0 & r_{33} \\ r_{41} & r_{51} & 0 \\ r_{51} & -r_{41} & 0 \\ -r_{22} & -r_{11} & 0 \end{bmatrix}$	$\begin{bmatrix} r_{11} & 0 & 0 \\ -r_{11} & 0 & 0 \\ 0 & 0 & 0 \\ r_{41} & 0 & 0 \\ 0 & r_{41} & 0 \\ 0 & -r_{11} & 0 \end{bmatrix}$

$3m(m \perp x_1)$	$3m(m \perp x_2)$
$\begin{bmatrix} 0 & -r_{22} & r_{13} \\ 0 & r_{22} & r_{13} \\ 0 & 0 & r_{33} \\ 0 & r_{51} & 0 \\ r_{51} & 0 & 0 \\ -r_{22} & 0 & 0 \end{bmatrix}$	$\begin{bmatrix} r_{11} & 0 & r_{13} \\ -r_{11} & 0 & r_{13} \\ 0 & 0 & r_{33} \\ 0 & r_{51} & 0 \\ r_{51} & 0 & 0 \\ 0 & -r_{11} & 0 \end{bmatrix}$

Hexagonal

6	$6mm$	622
$\begin{bmatrix} 0 & 0 & r_{13} \\ 0 & 0 & r_{13} \\ 0 & 0 & r_{33} \\ r_{41} & r_{51} & 0 \\ r_{51} & -r_{41} & 0 \\ 0 & 0 & 0 \end{bmatrix}$	$\begin{bmatrix} 0 & 0 & r_{13} \\ 0 & 0 & r_{13} \\ 0 & 0 & r_{33} \\ 0 & r_{51} & 0 \\ r_{51} & 0 & 0 \\ 0 & 0 & 0 \end{bmatrix}$	$\begin{bmatrix} 0 & 0 & 0 \\ 0 & 0 & 0 \\ 0 & 0 & 0 \\ r_{41} & 0 & 0 \\ 0 & -r_{41} & 0 \\ 0 & 0 & 0 \end{bmatrix}$

Table 1.1 (*Continued*)

$\bar{6}$	$\bar{6}m2(m \perp x_1)$	$\bar{6}m2(m \perp x_2)$
$\begin{bmatrix} r_{11} & -r_{22} & 0 \\ -r_{11} & r_{22} & 0 \\ 0 & 0 & 0 \\ 0 & 0 & 0 \\ 0 & 0 & 0 \\ -r_{22} & -r_{11} & 0 \end{bmatrix}$	$\begin{bmatrix} 0 & -r_{22} & 0 \\ 0 & r_{22} & 0 \\ 0 & 0 & 0 \\ 0 & 0 & 0 \\ 0 & 0 & 0 \\ -r_{22} & 0 & 0 \end{bmatrix}$	$\begin{bmatrix} r_{11} & 0 & 0 \\ -r_{11} & 0 & 0 \\ 0 & 0 & 0 \\ 0 & 0 & 0 \\ 0 & 0 & 0 \\ 0 & -r_{11} & 0 \end{bmatrix}$

Cubic

$\bar{4}3m, 23$	432
$\begin{bmatrix} 0 & 0 & 0 \\ 0 & 0 & 0 \\ 0 & 0 & 0 \\ r_{41} & 0 & 0 \\ 0 & r_{41} & 0 \\ 0 & 0 & r_{41} \end{bmatrix}$	$\begin{bmatrix} 0 & 0 & 0 \\ 0 & 0 & 0 \\ 0 & 0 & 0 \\ 0 & 0 & 0 \\ 0 & 0 & 0 \\ 0 & 0 & 0 \end{bmatrix}$

[a]The symbol over each matrix is the conventional symmetry-group designation.

$$Ir_{ijk} = -r_{ijk} \tag{1.4-7}$$

where I is the inversion operation. By virtue of the symmetry, the crystals must remain the same. For the same reason the electro-optic coefficients also remain the same. In other words,

$$Ir_{ijk} = r_{ijk} \tag{1.4-8}$$

Equations (1.4-7) and (1.4-8) can be satisfied simultaneously only if $r_{ijk} = 0$. This proves that linear electro-optic coefficients must vanish in centrosymmetric crystals. In fact, this property applies to all third rank tensors.

Using the contracted notation, the equation for the index ellipsoid in the presence of an applied electric field can be written

$$\left(\frac{1}{n_x^2} + r_{1k}E_k\right)x^2 + \left(\frac{1}{n_y^2} + r_{2k}E_k\right)y^2 + \left(\frac{1}{n_z^2} + r_{3k}E_k\right)z^2 + 2yzr_{4k}E_k + 2zxr_{5k}E_k + 2xyr_{6k}E_k = 1 \tag{1.4-9}$$

where the E_k ($k = 1, 2, 3$ or x, y, z) are components of the applied electric field and summation over repeated index k is assumed. If we set $E = 0$ in Eq. (1.4-9), the new index ellipsoid reduces to the unperturbed index ellipsoid given by Eq. (1.3-19).

Table 1.2 Properties of Selected Electro-Optic Crystals[a]

Substance	Symmetry	Wavelength (μm)	E–O Coefficients (10^{-12} m/V)	Index of Refraction	Dielectric Constant
CdTe	$\bar{4}3m$	1.0	(T) $r_{41} = 4.5$	$n = 2.84$	(S) $\epsilon = 9.4$
		1.3		$n = 2.82$	
		3.39	(T) $r_{41} = 6.8$		
		10.6	(T) $r_{41} = 6.8$	$n = 2.60$	
		23.35	(T) $r_{41} = 5.5$	$n = 2.58$	
		27.95	(T) $r_{41} = 5.0$	$n = 2.53$	
GaAs	$\bar{4}3m$	0.9	$r_{41} = 1.1$	$n = 3.60$	(S) $\epsilon = 13.2$
					(T) $\epsilon = 12.3$
		1.15	(T) $r_{41} = 1.43$	$n = 3.43$	
		3.39	(T) $r_{41} = 1.24$	$n = 3.3$	
		10.6	(T) $r_{41} = 1.51$	$n = 3.3$	
InP	$\bar{4}3m$	1.06	(S) $r_{41} = -1.34$	$n = 3.29$	$\epsilon = 12.55$
		1.50	(S) $r_{41} = -1.68$	$n = 3.17$	
BSO ($Bi_{12}SiO_{20}$)	23	0.63	(T) $r_{41} = 5$	$n = 2.54$	(T) $\epsilon = 56$
BGO ($Bi_{12}GeO_{20}$)	23	0.63	(T) $r_{41} = 3.4$	$n = 2.54$	(T) $\epsilon = 40$
		0.666	(T) $r_{41} = 3.22$		
BTO ($Bi_{12}TiO_{20}$)	23		(T) $r_{41} = 5.17$	$n = 2.25$	(T) $\epsilon = 47$
CdS	$6mm$	0.589	(T) $r_{51} = 3.7$	$n_o = 2.501$	(T) $\epsilon_1 = 9.35$
			(T) $r_c = 4$	$n_e = 2.519$	(T) $\epsilon_3 = 10.33$

CdS	$6mm$	0.633	(T) $r_{51} = 1.6$	$n_o = 2.460$	(S) $\epsilon_1 = 9.01$
			(T) $r_c = 4.8$	$n_e = 2.477$	(S) $\epsilon_3 = 9.53$
		1.15	(T) $r_{13} = 3.1$	$n_o = 2.320$	
			(T) $r_{33} = 3.2$	$n_e = 2.336$	
			(T) $r_{51} = 2.0$		
			(T) $r_c = 6.2$		
		3.39	(T) $r_{13} = 3.5$	$n_o = 2.276$	
			(T) $r_{33} = 2.9$	$n_e = 2.292$	
			(T) $r_{51} = 2.0$		
			(T) $r_c = 6.4$		
		10.6	(T) $r_{13} = 2.45$	$n_o = 2.226$	
			(T) $r_{33} = 2.75$	$n_e = 2.239$	
			(T) $r_{51} = 1.7$		
			(T) $r_c = 5.2$		
$LiNbO_3$	$3m$	0.633	(T) $r_{13} = 9.6$	$n_o = 2.286$	(T) $\epsilon_1 = \epsilon_2 = 78$
			(T) $r_{22} = 6.8$	$n_e = 2.200$	(T) $\epsilon_3 = 32$
			(T) $r_{33} = 30.9$		(S) $\epsilon_1 = \epsilon_2 = 43$
			(T) $r_{51} = 32.6$		(S) $\epsilon_3 = 28$
			(T) $r_c = 21.1$		
			(S) $r_{13} = 8.6$		
			(S) $r_{22} = 3.4$		
			(S) $r_{33} = 30.8$		
			(S) $r_{51} = 28$		
		1.15	(T) $r_{22} = 5.4$	$n_o = 2.229$	
			(T) $r_c = 19$	$n_e = 2.150$	
		3.39	(T) $r_{22} = 3.1$	$n_o = 2.136$	
			(T) $r_c = 18$	$n_e = 2.073$	
			(S) $r_{33} = 28$		
			(S) $r_{22} = 3.1$		

Table 1.2 *(Continued)*

Substance	Symmetry	Wavelength (μm)	E–O Coefficients (10^{-12} m/V)	Index of Refraction	Dielectric Constant
$LiNbO_3$	$3m$	3.39	(S) $r_{13} = 6.5$		
			(S) $r_{51} = 23$		
$LiTaO_3$	$3m$	0.633	(T) $r_{13} = 8.4$	$n_o = 2.176$	(T) $\epsilon_1 = \epsilon_2 = 51$
			(T) $r_{33} = 30.5$	$n_e = 2.180$	(T) $\epsilon_3 = 45$
			(T) $r_{22} = -0.2$		(S) $\epsilon_1 = \epsilon_2 = 41$
			(T) $r_c = 22$		(S) $\epsilon_3 = 43$
			(S) $r_{13} = 7.5$		
			(S) $r_{33} = 33$		
			(S) $r_{51} = 20$		
			(S) $r_{22} = 1$		
		3.39	(S) $r_{33} = 27$	$n_o = 2.060$	
			(S) $r_{13} = 4.5$	$n_e = 2.065$	
			(S) $r_{51} = 15$		
			(S) $r_{22} = 0.3$		
KDP	$\bar{4}2m$	0.546	(T) $r_{41} = 8.77$	$n_o = 1.5115$	(T) $\epsilon_1 = \epsilon_2 = 42$
			(T) $r_{63} = 10.3$	$n_e = 1.4698$	(T) $\epsilon_3 = 21$
		0.633	(T) $r_{41} = 8$	$n_o = 1.5074$	(S) $\epsilon_1 = \epsilon_2 = 44$
			(T) $r_{63} = 11$	$n_e = 1.4669$	(S) $\epsilon_3 = 21$
		3.39	(T) $r_{63} = 9.7$		
			(T) $n_o^3 r_{63} = 33$		
$BaTiO_3$ ($5°C < T < 130°C$)	$4mm$	0.546	(T) $r_{51} = 1640$	$n_o = 2.437$	(T) $\epsilon_1 = \epsilon_2 = 3600$
			(T) $r_c = 108$	$n_e = 2.365$	(T) $\epsilon_3 = 135$
			(S) $r_{51} = 820$		(S) $\epsilon_1 = \epsilon_2 = 2300$
			(S) $r_c = 23$		(S) $\epsilon_3 = 60$
		0.633	(S) $r_c = 19$	$n_o = 2.41$	
			(S) $r_{33} = 28$	$n_e = 2.36$	
			(S) $r_{13} = 8$		

KTN $KTa_{1-x}Nb_xO_3$ $x = 0.35$	$4mm$	0.633	(T) $r_{51} = 8000(T_c - 28)$ (T) $r_c = 500(T_c - 28)$ (T) $r_{51} = 3000(T_c - 16)$ (T) $r_c = 700(T_c - 16)$	$n_o = 2.318$ $n_e = 2.277$ $n_o = 2.318$ $n_e = 2.281$	$\epsilon_1 = \epsilon_2 = 450$ $\epsilon_3 = 900$
SBN:60	$4mm$		$r_{33} = 420$		$\epsilon_3 = 800$
SBN:75 $Sr_{0.75}Ba_{0.25}Nb_2O_6$ (~330 K)	$4mm$	0.633	(T) $r_c = 1410$ (T) $r_{33} = 1340$ (T) $r_{13} = 67$ (T) $r_{51} = 42$ (S) $r_c = 1090$	$n_o = 2.3117$ $n_e = 2.2987$	$\epsilon_3 = 3400$ (15 MHz) $\epsilon_1 = 700$
$KNbO_3$	$2mm$	0.633	(T) $r_{13} = 28$ (T) $r_{23} = 1.3$ (T) $r_{33} = 64$ (T) $r_{42} = 380$ (T) $r_{51} = 105$ (S) $r_{33} = 25$ (S) $r_{13} = 10$ (S) $r_{23} = 2$ (S) $r_{42} = 270$ (S) $r_{51} = 23$	$n_1 = 2.280$ $n_2 = 2.329$ $n_3 = 2.169$	(S) $\epsilon_1 = 37$ (S) $\epsilon_2 = 780$ (S) $\epsilon_3 = 24$ (T) $\epsilon_1 = 160$ (T) $\epsilon_2 = 1000$ (T) $\epsilon_3 = 55$
KNSBN (BSKNN-2) $(Ba_{2-x}Sr_xK_{1-y}Na_yNb_5O_{15})$ $x = 1.5$, $y = 0.5$ $T_c = 170$–$210°C$			$r_{51} = 500$		$\epsilon_1 = 700$

[a](T) = low frequency; (S) = high frequency; $r_c = r_{33} - n_o^3 r_{13}/n_e^3$; T_c = Curie temperature.

By using a coordinate transformation, a new set of principal axes can be found in which the quadratic form Eq. (1.4-9) becomes diagonal. This is known as the principal-axes transformation of the quadratic form. As a result of the diagonalization, the principal indices of refraction will be obtained. This information is useful for the design of electro-optic modulators and devices. We illustrate this by examining the electro-optic effect in some of the most useful crystals in the following examples.

Example: Electro-Optic Effect in GaAs Consider the specific example of a gallium arsenide crystal, GaAs. This is a cubic crystal of the zincblende structure (point group $\bar{4}3m$). According to Table 1.1, the electro-optic coefficients are given in the form

$$r_{lk} = \begin{bmatrix} 0 & 0 & 0 \\ 0 & 0 & 0 \\ 0 & 0 & 0 \\ r_{41} & 0 & 0 \\ 0 & r_{41} & 0 \\ 0 & 0 & r_{41} \end{bmatrix} \tag{1.4-10}$$

so that the only nonvanishing components are $r_{52} = r_{63} = r_{41}$. Let the applied electric field be (E_x, E_y, E_z), the change in the impermeability tensor can be written

$$\Delta\eta = \begin{bmatrix} 0 & r_{41}E_z & r_{41}E_y \\ r_{41}E_z & 0 & r_{41}E_x \\ r_{41}E_y & r_{41}E_x & 0 \end{bmatrix} \tag{1.4-11}$$

and the corresponding change in the dielectric tensor is given by

$$\Delta\epsilon = -\epsilon_0 n^4 \begin{bmatrix} 0 & r_{41}E_z & r_{41}E_y \\ r_{41}E_z & 0 & r_{41}E_x \\ r_{41}E_y & r_{41}E_x & 0 \end{bmatrix} \tag{1.4-12}$$

where n is the index of refraction of the medium.

Using Eqs. (1.4-9) and (1.4-10), the index ellipsoid can be written

$$\frac{x^2}{n^2} + \frac{y^2}{n^2} + \frac{z^2}{n^2} + 2yzr_{41}E_x + 2zxr_{41}E_y + 2xyr_{41}E_z = 1 \tag{1.4-13}$$

where we notice that the presence of the electric field causes the appearance of cross terms. These are the terms with yz, zx, xy. This means that the major axes of the ellipsoid are no longer along the x, y, and z crystal axes. To illustrate this, let us choose the case when the electric field is along the z-axis, and the equation for the index ellipsoid becomes

$$\frac{x^2}{n^2} + \frac{y^2}{n^2} + \frac{z^2}{n^2} + 2xyr_{41}E_z = 1 \tag{1.4-14}$$

By examining the equation, we find that the z-axis remains a principal axis with the corresponding principal index unchanged. To see the effect of the cross term xy on the index ellipsoid, let us focus our attention on the xy-plane. Figure 1.5 shows the intersection of the ellipsoid with the xy-plane. We note that the presence of the term $2xyr_{41}E_z$ causes the circle to deform into an ellipse. The new principal axes of the ellipse are 45° relative to the crystal axes. In other words, Eq. (1.4-14) can be diagonalized by using the following coordinate transformation

$$\begin{aligned} x &= x' \cos 45° - y' \sin 45° \\ y &= x' \sin 45° + y' \cos 45° \\ z &= z' \end{aligned} \tag{1.4-15}$$

Substituting Eq. (1.4-15) into Eq. (1.4-14), we obtain

$$\frac{x'^2}{n'^2_x} + \frac{y'^2}{n'^2_y} + \frac{z'^2}{n'^2_z} = 1 \tag{1.4-16}$$

with $n'_z = n$ and

$$\frac{1}{n'^2_x} = \frac{1}{n^2} + r_{41}E_z \tag{1.4-17}$$

$$\frac{1}{n'^2_y} = \frac{1}{n^2} - r_{41}E_z \tag{1.4-18}$$

Equation (1.4-16) shows that indeed x', y', and z are the principal axes when

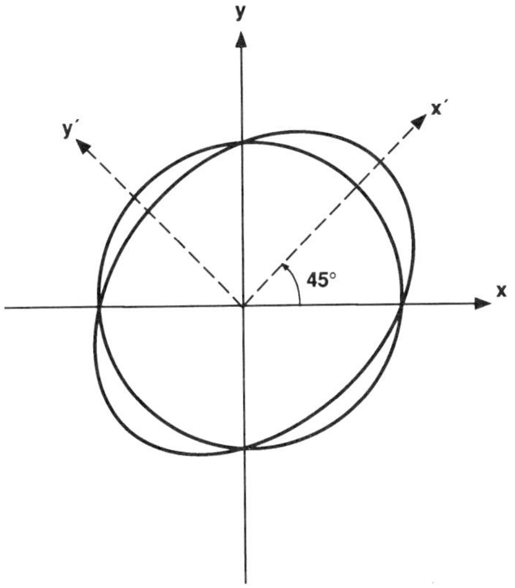

Figure 1.5. Index ellipsoid of zincblende cubic crystals (e.g., GaAs, InP) in the presence of an applied field. The field causes the deformation of the ellipsoid and creates a new set of principal axes (x', y').

an electric field is applied along the z-axis. The new principal indices of refraction are given, approximately, by

$$n_{x'} = n - \tfrac{1}{2} n^3 r_{41} E_z \tag{1.4-19}$$

$$n_{y'} = n + \tfrac{1}{2} n^3 r_{41} E_z \tag{1.4-20}$$

In arriving at the above results, we assumed that $n^2 r_{41} E_z \ll 1$ and used the following differential relationship:

$$\Delta n = -\frac{1}{2} n^3 \Delta\left(\frac{1}{n^2}\right) \tag{1.4-21}$$

Similar results are obtained if the electric field is applied along any other crystal axes.

Example: Electro-Optic Effect in BaTiO$_3$ Consider next the electro-optic effect in barium titanate crystal which has a point group symmetry of 4 mm. The nonvanishing electro-optic coefficients are given, according to Table 1.1, by

$$\begin{bmatrix} 0 & 0 & r_{13} \\ 0 & 0 & r_{13} \\ 0 & 0 & r_{33} \\ 0 & r_{42} & 0 \\ r_{42} & 0 & 0 \\ 0 & 0 & 0 \end{bmatrix} \tag{1.4-22}$$

where $r_{51} = r_{42}$. Let the applied electric field be written $\mathbf{E} = (E_x, E_y, E_z)$ the change in the impermeability tensor can be written

$$\Delta\eta = \begin{bmatrix} r_{13}E_z & 0 & r_{42}E_x \\ 0 & r_{13}E_z & r_{42}E_y \\ r_{42}E_x & r_{42}E_y & r_{33}E_z \end{bmatrix} \tag{1.4-23}$$

and the corresponding change in the dielectric tensor is given by

$$\Delta\epsilon = -\epsilon_0 \begin{bmatrix} n_o^4 r_{13} E_z & 0 & n_o^2 n_e^2 r_{42} E_x \\ 0 & n_o^4 r_{13} E_z & n_o^2 n_e^2 r_{42} E_y \\ n_o^2 n_e^2 r_{42} E_x & n_o^2 n_e^2 r_{42} E_y & n_e^4 r_{33} E_z \end{bmatrix} \tag{1.4-24}$$

where n_o, n_e are the indices of refraction of the medium. A special case of interest in photorefractive two-beam coupling involves a space-charge field in the xz-plane. In this case, the electric field can be written

$$E_x = E \sin\theta\,, \quad E_y = 0\,, \quad E_z = E\cos\theta \tag{1.4-25}$$

where θ is the angle of the direction of the electric field relative to the c-axis. Using Eq. (1.4-25), the change in the impermeability tensor becomes

$$\Delta\eta = \begin{bmatrix} r_{13}E\cos\theta & 0 & r_{42}E\sin\theta \\ 0 & r_{13}E\cos\theta & 0 \\ r_{42}E\sin\theta & 0 & r_{33}E\cos\theta \end{bmatrix} \tag{1.4-26}$$

According to Eq. (1.4-26), the y-axis remains a principal axis. As a result of the off-diagonal terms $\Delta\eta_{xz}$ and $\Delta\eta_{zx}$, the crystal axes x and z are no longer principal axes. A coordinate transformation can be used to obtain the principal axes and the corresponding principal indices of refraction. If we consider the propagation of a plane wave in the xz-plane, we can obtain the eigen-indices of refraction for the normal modes of propagation by directly finding the expectation value of the impermeability tensor. Let α be the angle of propagation of the plane wave relative to the c-axis, the displacement field vectors of the normal modes are

$$\mathbf{d}_o = \begin{bmatrix} 0 \\ 1 \\ 0 \end{bmatrix}, \qquad \mathbf{d}_e = \begin{bmatrix} \cos\alpha \\ 0 \\ -\sin\alpha \end{bmatrix} \tag{1.4-27}$$

with the corresponding eigen-indices of refraction given by

$$n = n_o$$

and

$$\frac{1}{n_e^2(\alpha)} = \frac{\sin^2\alpha}{n_e^2} + \frac{\cos^2\alpha}{n_o^2} \tag{1.4-28}$$

In the presence of the applied electric field, the displacement field vectors given by Eq. (1.4-27) are still eigenvectors because the y-axis remains a principal axis. The eigen-indices of refraction, however, are changed. These changes can be easily obtained by using Eq. (1.3-29), or equivalently

$$\langle \Delta\eta \rangle = \langle \mathbf{d}|\Delta\eta|\mathbf{d} \rangle = \Delta\left(\frac{1}{n^2}\right) \tag{1.4-29}$$

Substituting Eq. (1.4-27) for $\mathbf{d}$ in Eq. (1.4-29), we obtain the following changes in the eigen-indices of refraction

$$\begin{aligned} \Delta\left(\frac{1}{n^2}\right) &= r_{13}E\cos\theta \\ \Delta\left(\frac{1}{n_e^2(\alpha)}\right) &= E(r_{13}\cos\theta\cos^2\alpha - 2r_{42}\sin\theta\sin\alpha\cos\alpha + r_{33}\cos\theta\sin^2\alpha) \end{aligned} \tag{1.4-30}$$

1.5 OPTICAL ACTIVITY

Some of the photorefractive media (e.g., BSO, BGO) are found to cause a rotation of the plane of polarization of linearly polarized light passing through them. This is known as *optical activity* and was first observed in quartz. In these media, the normal modes of propagation are no longer linearly polarized. Figure 1.6 illustrates the rotation of the plane of polarization as the light propagates in an optically active medium. The amount of rotation is proportional to the distance travelled in the medium.

The phenomenon arises from the double refraction of circularly polarized waves and must be carefully distinguished from that of ordinary birefringence, the double refraction of linearly polarized waves in crystals. Optical activity is empirically characterized by the specific rotation ρ, the degree of angular rotation per unit path length, which is related to the difference in refractive indices of the left- and right-handed circularly polarized light waves.

Gases, liquids and solids can exhibit optical activity, but its application to optical devices is most common for crystals, particularly α-quartz, in which the effect was first discovered. Some compounds such as cane sugar exhibit optical activity in the crystalline state and in solution, whereas other compounds such as sodium chlorate are optically active only in the crystalline state. Camphor readily demonstrates optical activity in its vapor, solution and crystalline forms.

Optical activity in isotropic crystals is a type of double refraction where the normal modes of light propagation are two circularly polarized waves of opposite handedness; that is, these two polarization states are transmitted through the crystal unchanged, whereas a plane-polarized wave may not be so transmitted. Plane-polarized light entering an optically active crystal is decomposed into two waves of right- and left-handed circular polarizations which travel through the medium at different velocities. This velocity difference brings about a phase difference δ between the two waves which,

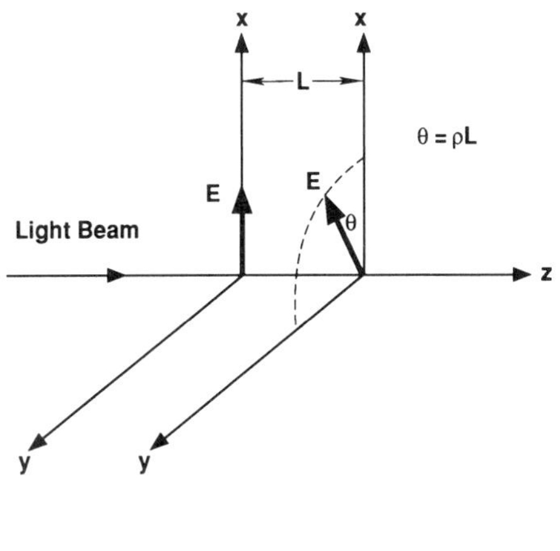

Figure 1.6. Rotation of the plane of polarization of an incident light beam in optically active media. The angle of rotation θ is proportional to the length L.

upon exiting the crystal, is superposed to produce plane-polarized light having its polarization plane rotated by an angle of $\delta/2$ with respect to that of the entering light. The phase difference between the two waves after propagating a distance L is the difference between the refractive indices for left-handed (n_-) and right-handed (n_+) circular polarizations multiplied by the factor $2\pi L/\lambda$, where λ is the vacuum wavelength. Thus the specific rotation, which equals $\delta/2L$, is

$$\rho = \pi(n_- - n_+)/\lambda \tag{1.5-1}$$

The plane of polarization turns in the same sense as the circularly polarized wave that travels with the greater phase velocity. A substance is called dextro-rotatory, or right-handed, if the sense of rotation of the plane of polarization is counterclockwise as viewed by an observer facing the approaching light beam. If the sense of rotation is clockwise, the substance is called levo-rotatory, or left-handed.

Optical activity may occur in some crystals in combination with ordinary double refraction. When both effects are present, the optical activity, which is a weak effect ($|n_- - n_+| = 7 \times 10^{-5}$ for α-quartz), is easily overwhelmed by the finite linear birefringence (for quartz, $n_e - n_o = 9 \times 10^{-3}$ where n_e and n_o are the extraordinary and ordinary refractive index of the crystal, respectively). Thus, the normal modes are nearly plane-polarized for α-quartz, except around the optic axis. For optically active cubic crystals, however, such as sodium chlorate and bismuth germanium oxide, which do not have ordinary double refraction, the normal modes are circularly polarized in all crystalline directions.

For plane-wave propagation in a homogeneous medium, the material equation for such an optically active material is often written as

$$\mathbf{D} = \epsilon \mathbf{E} + i\epsilon_0 \mathbf{G} \times \mathbf{E} \tag{1.5-2}$$

where ϵ is the dielectric tensor in the absence of optical activity, and $\mathbf{G}$ is a vector parallel to the direction of propagation and is called the gyration vector. The typical magnitude of $\mathbf{G}$ is $\sim 10^{-4}$ and is dimensionless (see Table 1.3). The vector product $\mathbf{G} \times \mathbf{E}$ can always be represented by the product of an antisymmetric tensor $[G]$ with $\mathbf{E}$. The matrix elements of this antisymmetric tensor $[G]$ are given by

$$\begin{aligned} [G]_{23} &= -[G]_{32} = -G_x \\ [G]_{31} &= -[G]_{13} = -G_y \\ [G]_{12} &= -[G]_{21} = -G_z \end{aligned} \tag{1.5-3}$$

Thus, Eq. (1.5-2) becomes

$$\mathbf{D} = (\epsilon + i\epsilon_0[G])\mathbf{E} \tag{1.5-4}$$

Table 1.3 Rotatory Power of Selected Photorefractive Crystals

Crystal	Symmetry	Wavelength (nm)	ρ (deg/mm)	G (rad)
BGO	23	488	46	3.2×10^{-4}
($Bi_{12}GeO_{20}$)		515	39	2.8×10^{-4}
		633	21	1.9×10^{-4}
BSO	23	488	45	3.1×10^{-4}
($Bi_{12}SiO_{20}$)		515	38	2.8×10^{-4}
		633	21	1.9×10^{-4}
BTO	23	488	14	8.5×10^{-5}
($Bi_{12}TiO_{20}$)		515	11	7.1×10^{-5}
		633	6.3	5.0×10^{-5}

It is often convenient to define a new dielectric tensor as

$$\epsilon' = \epsilon + i\epsilon_0[G] \tag{1.5-5}$$

This new dielectric tensor is Hermitian, that is, $\epsilon'_{ij} = \epsilon'^{*}_{ji}$. We can substitute this dielectric tensor ϵ' into Eq. (1.3-2) and solve for the normal modes of propagation.

For isotropic or cubic media, we can arbitrarily choose the z-axis as the direction of propagation. The dielectric tensor including the effect of optical activity can be written

$$\epsilon' = \epsilon_0 \begin{bmatrix} n^2 & -iG & 0 \\ iG & n^2 & 0 \\ 0 & 0 & n^2 \end{bmatrix} \tag{1.5-6}$$

where n is the index of refraction, G is the magnitude of gyration constant.

Let the electric field of the plane wave be written

$$\mathbf{E} = \mathbf{E}_0 e^{i(\omega t - kz)} \tag{1.5-7}$$

where $\mathbf{E}_0$ and k are to be determined. Substituting the elements of the dielectric tensor Eq. (1.5-6) into Eq. (1.3-2), we obtain

$$\begin{bmatrix} \left(\frac{\omega}{c} n\right)^2 - k^2 & -i\frac{\omega^2}{c^2} G & 0 \\ i\frac{\omega^2}{c^2} G & \left(\frac{\omega}{c} n\right)^2 - k^2 & 0 \\ 0 & 0 & \left(\frac{\omega}{c} n\right)^2 \end{bmatrix} \begin{bmatrix} E_{0x} \\ E_{0y} \\ E_{0z} \end{bmatrix} = 0 \tag{1.5-8}$$

The eigenvalues for the wave number k are given by

$$k = \frac{\omega}{c}(n^2 \pm G)^{1/2} \simeq \frac{\omega}{c}\left(n \pm \frac{G}{2n}\right) \tag{1.5-9}$$

where we assume $G \ll n$. The corresponding eigenvectors of these normal modes are

$$\mathbf{E}_L = \frac{1}{\sqrt{2}}\begin{bmatrix} 1 \\ i \\ 0 \end{bmatrix}, \qquad \mathbf{E}_R = \frac{1}{\sqrt{2}}\begin{bmatrix} 1 \\ -i \\ 0 \end{bmatrix} \tag{1.5-10}$$

We note that the normal modes are circularly polarized waves. These two circularly polarized modes are orthogonal.

$\mathbf{E}_R$ is the polarization state of the right-handed circularly polarized mode with a wave number of

$$k_R = \frac{\omega}{c}\left(n - \frac{G}{2n}\right) \tag{1.5-11}$$

$\mathbf{E}_L$ is that of the left-handed circularly polarized mode with a wave number of

$$k_L = \frac{\omega}{c}\left(n + \frac{G}{2n}\right) \tag{1.5-12}$$

The rotation angle is related to the difference in these wave numbers and is given by

$$\theta = \rho L = \tfrac{1}{2}(k_L - k_R)L \tag{1.5-13}$$

Using these equations, the specific rotatory power can be written

$$\rho = \frac{\pi G}{\lambda n} \tag{1.5-14}$$

The rotation of the plane of polarization can also be understood in terms of coupled-mode theory. If we treat $i\epsilon_0 \mathbf{G} \times \mathbf{E}$ as a small perturbation (see Problem 1.9) and assume a solution of the form $\mathbf{E}\exp[i(\omega t - kz)]$, we obtain, from Eqs. (1.2-23), (1.2-24) and (1.5-2)

$$\frac{d}{dz}\mathbf{E} = \frac{\pi}{\lambda n}\mathbf{G} \times \mathbf{E} \tag{1.5-15}$$

where n is the index of refraction of the medium (or n_o if the medium is uniaxially birefringent) in the absence of optical activity. According to this equation, the linearly polarized electric field vector $\mathbf{E}$ gyrates (or rotates) as the light travels in the medium. The angle of rotation per unit length of propagation, also called specific rotatory power ρ, is given by

$$\rho = \frac{\pi}{\lambda n} G , \qquad (1.5\text{-}16)$$

where G is the magnitude of $\mathbf{G}$ (i.e., $G = |\mathbf{G}|$). Table 1.3 lists the rotatory power and the gyration constant for some materials.

PROBLEMS

1.1 Solutions of Parabolic Wave Equation

a. Consider a plane wave solution of the parabolic wave Eq. (1.2-28). Let $A(x, y, z) = C(z) \exp(-ipx - iqy)$, show that

$$C(z) = C \exp[i(p^2 + q^2)z/(2k)]$$

b. Show that

$$A(x, y, z) = (1/z) \exp[-ikz - ik(x^2 + y^2)/(2z)]$$

is a solution of the parabolic wave equation.

c. Show that the solution in (**b**) is an approximation of the spherical wave when $z \gg x, y$

$$A(x, y, z) = (1/r) \exp(-ikr) , \quad \text{with } r = (x^2 + y^2 + z^2)^{1/2}$$

when $z \gg x, y$.

d. Show that the solution in (**b**) can be obtained by the following:

$$A(x, y, z) = \int\int C(z)\, dp\, dq$$

Physically this means that the spherical wave consists of plane wave components with equal amplitudes. The following integral may be useful:

$$\int \exp(-\alpha x^2 - \beta x)\, dx = \left(\frac{\pi}{4\alpha}\right)^{1/2} \exp\left(\frac{\beta^2}{4\alpha}\right)$$

where α and β are constants with $\mathrm{Re}[\alpha] > 0$.

1.2 Wave Propagation in a Lithium Niobate Crystal. Consider the propagation of a laser beam (HeNe at $\lambda = 0.6328\ \mu$m) in a lithium niobate crystal which is uniaxial and has the following indices of refraction: $n_o = 2.286$, $n_e = 2.200$. Let the plane of propagation be the yz-plane, where z-axis is the c-axis and xyz are the principal axes. Let the electric field of the light be written

$$\mathbf{E} \exp[i(\omega t - \mathbf{k} \cdot \mathbf{r})]$$

a. Let θ be the angle between the c-axis and the wave vector $\mathbf{k}$, show that the electric field vector of the normal modes can be written

$$\mathbf{E}_o = (1, 0, 0)$$
$$\mathbf{E}_e = (0, \cos\theta/n_o^2, -\sin\theta/n_e^2)$$

b. Show that both electric field vectors are tangent to the normal surface and are mutually orthogonal.

c. Evaluate the difference in phase velocity at $\theta = 45°$.

d. Show that $\mathbf{E}_e$ is not exactly orthogonal to the wavevector $\mathbf{k}$. Derive an expression for the angle between these two vectors.

e. Evaluate the deviation from 90° at $\theta = 45°$.

f. Using $\mathbf{D} = \epsilon\mathbf{E}$, show that

$$\mathbf{D}_o = (1, 0, 0)$$
$$\mathbf{D}_e = (0, \cos\theta, -\sin\theta)$$

within a factor of proportionality.

g. Derive $\mathbf{D}_e \eta \mathbf{D}_e = 1/n^2$, where n^2 is given by Eq. (1.3-15).

1.3 Solutions of the Fresnel Equation. The Fresnel equation is given by Eq. (1.3-6). Let

$$n_x^2 = \epsilon_x/\epsilon_0, \quad n_y^2 = \epsilon_y/\epsilon_0, \quad n_z^2 = \epsilon_z/\epsilon_0$$

and

$$0 < n_x < n_y < n_z$$

a. Show that the Fresnel equation is quadratic in n^2.

b. Let n_1^2 and n_2^2 be the roots of the Fresnel equation with $n_1^2 < n_2^2$. show that

$$n_x^2 < n_1^2 < n_y^2 < n_2^2 < n_z^2$$

Hint: Evaluate the sign of the Fresnel equation near the four singularities, $n^2 = 0$, n_x^2, n_y^2, n_z^2, and then use the continuity of the function between the neighboring singularities.

$$f(n^2) = s_x^2/(n^2 - n_x^2) + s_y^2/(n^2 - n_y^2) + s_z^2/(n^2 - n_z^2) - 1/n^2$$

1.4 Relationship Between $\Delta\epsilon$ and $\Delta\eta$

a. Using the definition $\eta = \epsilon_0/\epsilon$, show that $\Delta\epsilon = -\epsilon\,\Delta\eta\,\epsilon/\epsilon_0$. Here we recall that η, ϵ, $\Delta\eta$, $\Delta\epsilon$ are 3×3 matrices (or tensors of rank 2). (Hint: Neglect second order term $\Delta\eta\,\Delta\epsilon$).

b. Show that in the principal coordinate

$$\Delta\epsilon_{ij} = -\epsilon_0 n_i^2 n_j^2 \sum r_{ijk} E_k$$

where the summation is over the index k, and the n_i are principal indices.

1.5 Electro-Optic Effect in KDP Crystals. Consider the case of electro-optic effect in KDP crystals.

a. Let the dc electric field be applied along the z-axis. Derive the change in dielectric impermeability tensor $\Delta\eta$.

b. Using the relationship in Problem 1.4, derive the corresponding change in dielectric tensor $\Delta\epsilon$.

c. Repeat (**a**) and (**b**) for the case when the dc applied field is along the x-axis.

1.6 Electro-Optic Effect in Zincblende Crystals (e.g., GaAs, InP)

a. Show that $\Delta\epsilon_{ij} = -\epsilon_0 n^4 \Sigma\, r_{ijk} E_k$, where summation is over the index k.

b. Let the dc applied field be along the z-axis, write down the change in dielectric tensor $\Delta\epsilon_{ij}$.

c. Diagonalize the matrix derived in (**b**) and derive expressions for the principal components of the dielectric tensor $\epsilon_{x'}$, $\epsilon_{y'}$, $\epsilon_{z'}$.

d. Let the field strength be 10 KV/cm, $n = 3.6$, $r_{41} = 1.1$ pm/V, find the change in the index of refraction.

1.7 Symmetry and Electro-Optic Effect. Using symmetry argument, show that germanium and silicon crystals do not exhibit linear electro-optic effect.

1.8 Electro-Optic Effect in Tetragonal $BaTiO_3$ Crystal (4 mm). The electro-optic coefficients in contracted notations are given by

$$\begin{bmatrix} 0 & 0 & r_{13} \\ 0 & 0 & r_{13} \\ 0 & 0 & r_{33} \\ 0 & r_{42} & 0 \\ r_{42} & 0 & 0 \\ 0 & 0 & 0 \end{bmatrix}$$

where we note that $r_{42} = r_{51}$. The indices of refraction are n_o and n_e.

a. Write down $\Delta\eta$, $\Delta\epsilon$ in terms of the applied electric field which is given by $\mathbf{E}^{dc} = (E_x, E_y, E_z)$.

b. Find the phase velocities and polarization states (**D** vectors) of normal modes in the absence of applied field ($\mathbf{E}^{dc} = 0$) for propagation along the direction $\mathbf{s} = (0, \sin\theta, \cos\theta)$.

c. Repeat (**b**) for $\mathbf{E}^{dc} = (0, 0, E^{dc})$.

d. Repeat (**b**) for $\mathbf{E}^{dc} = (0, E^{dc}, 0)$.

e. Repeat (**b**) for $\mathbf{E}^{dc} = (E^{dc}, 0, 0)$.

1.9 Equation of Gyration. The gyration of the electric field vector in isotropic media or along the c-axis of uniaxial crystals can be described by using Eq. (1.2-23) and Eq. (1.5-2).

a. Show that for monochromatic waves, the wave equation can be written as

$$\frac{d^2}{dz^2}\mathbf{E} + \left(\frac{\omega}{c} n\right)^2 \mathbf{E} + i\left(\frac{\omega}{c}\right)^2 \mathbf{G} \times \mathbf{E} = 0 ,$$

where $n^2 = \epsilon/\epsilon_0$.

b. Show that substitution of $\mathbf{E}(z)\exp(-ikz)$ for $\mathbf{E}$ in the wave equation in (**a**), with $k = n\omega/c$, yields

$$\mathbf{E}''(z) - 2ik\mathbf{E}'(z) + i\left(\frac{\omega}{c}\right)^2 \mathbf{G} \times \mathbf{E} = 0 ,$$

where a prime indicates differentiation with respect to z.

c. Derive the equation of gyration (1.5-15) by neglecting the second-order derivative in (**b**).

REFERENCES

1. See, for example, P. Yeh, Optical Waves in Layered Media (Wiley, New York, 1988).
2. See, for example, A. Yariv and P. Yeh, Optical Waves in Crystals (Wiley, New York, 1984).
3. J. D. Jackson, Classical Electrodynamics (Wiley, New York, 1975).

2

Wave Propagation in Periodic Media

Many of the important and useful phenomena associated with photorefractive effects involve the scattering of light from gratings or holograms formed in photorefractive crystals. These include wave mixing, phase conjugation, dynamic holography, etc. The gratings or holograms in these media are manifested by a periodic variation of the index of refraction. In this chapter, we discuss some general properties of electromagnetic propagation in periodic media. We start from some basic concepts of Bragg scattering, including conservation of momentum and energy. We then describe a coupled mode theory for both transmission gratings and reflection gratings. This is followed by a discussion on Raman–Nath diffraction and finite gratings. Near the end of the chapter, we introduce the method of grating integral.

2.1 PERIODIC MEDIA

Generally speaking, periodic media are any structures whose dielectric constants, reflecting the translational symmetry, are periodic functions of position:

$$\epsilon(\mathbf{x}) = \epsilon(\mathbf{x} + \mathbf{a}) \tag{2.1-1}$$

where **a** is a constant vector. In the case of one-dimensional periodic media, Eq. (2.1-1) becomes

$$\epsilon(z) = \epsilon(z + \Lambda) \tag{2.1-2}$$

or

$$n^2(z) = n^2(z + \Lambda) \tag{2.1-3}$$

where z is a measure of position and Λ is the periodicity. According to this general definition, periodic media include the following: a phase grating which consists of a thin glass plate ruled with an array of parallel equidistant grooves, a reflecting grating which consists of a flat metal surface ruled with an array of parallel equidistant grooves, a volume hologram which consists of a homogeneous medium with an index of refraction modulated periodically, a three-dimensional periodic array of atoms such as crystals.

A photorefractive material can become a periodic medium provided the index of refraction is modulated periodically. This can be achieved by the illumination of light with a spatially periodic intensity pattern. The index of refraction of periodic media is often written

$$n(z) = n_0 + n_1 p(z) \tag{2.1-4}$$

where n_0 and n_1 are constants and $p(z)$ is a periodic function of z

$$p(z) = p(z + \Lambda) \tag{2.1-5}$$

with

$$\max\{|p(z)|\} = 1 \tag{2.1-6}$$

and

$$\int_0^{\Lambda} p(z)\, \mathrm{d}z = 0 \tag{2.1-7}$$

In Eq. (2.1-4), n_0 is the index of refraction of the material without spatial modulation and n_1 is the amplitude of the index modulation. n_1 is often referred to as the depth of index modulation. In most of the photorefractive materials covered in this book, n_1 is of the order of 10^{-3}–10^{-5}.

We now consider the scattering of a monochromatic plane wave from a periodic medium (see Figure 2.1). The study of such scattering benefits from early works on X-ray diffraction in solids. Consider an extreme situation when the index modulation is lumped to an array of equidistant planes as depicted in Figure 2.1. In addition, we assume that these planes are infinite so that reflection from these planes are specular (i.e., mirrorlike with angle of reflection equal to angle of incidence). Each of the planes reflects only a very small fraction of the incident plane wave. The scattered light consists of linear superposition of all these reflected plane waves. The diffracted beams are found when all these reflected plane waves add up constructively.

Let Λ be the spacing between these planes. This is also the period of the index variation. The path difference for rays reflected from two adjacent planes is $2\Lambda \sin\theta$, where θ is the angle between the ray and the planes. Constructive interference occurs when the path difference is an integral number of wavelengths λ/n in the medium, so that

$$2\Lambda \sin\theta = N(\lambda/n) \tag{2.1-8}$$

where $n = n_0$ is the spatially averaged index of refraction of the medium and

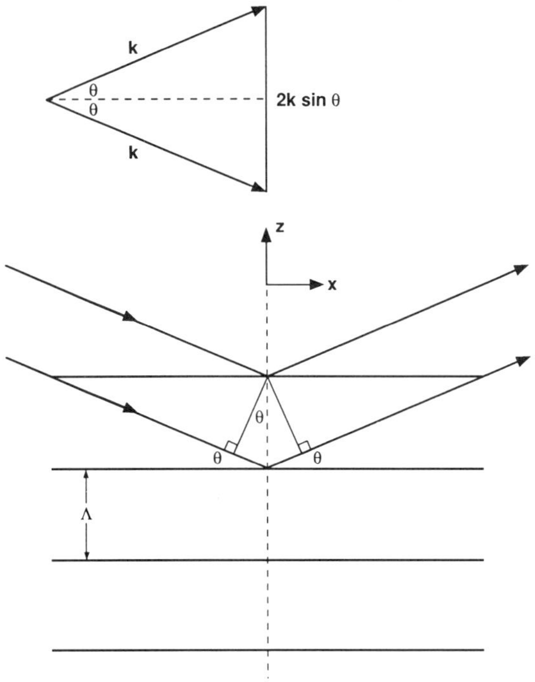

Figure 2.1. Scattering of monochromatic plane wave from a periodic medium.

N is an integer. This is known as the Bragg law. Although reflection from each plane is specular, beam diffraction occurs only for certain values of θ which obey the Bragg law so that reflections from all planes add up in phase.

Equation (2.1-8) can also be written

$$2k \sin \theta = N \frac{2\pi}{\Lambda} \tag{2.1-9}$$

where k is the wave number of the light beam in the medium ($k = n2\pi/\lambda$). The term $2\pi/\Lambda$ is known as grating wave number and is often written as

$$K = \frac{2\pi}{\Lambda} \tag{2.1-10}$$

The left side of Eq. (2.1-9), $2k \sin \theta$, is the change of the wave vector upon diffraction from the periodic medium. Thus the Bragg law can be interpreted as simply the conservation of momentum. Upon diffraction from a periodic medium (grating), the change of wave vector is exactly an integral number of the grating wave vector.

In fact, if we decompose the periodic function $p(x)$ in Eq. (2.1-4) into its Fourier components, we obtain

$$n(z) = n_0 + n_1 \sum_m a_m \exp\left(-i \frac{2\pi}{\Lambda} mz\right) \tag{2.1-11}$$

where a_m is the mth Fourier component of the periodic index variation. We note that the mth Fourier component has a wave number of $m(2\pi/\Lambda)$. Each of the Fourier component contributes to a Bragg diffraction order according to Eq. (2.1-9). Although the Bragg law is derived by assuming an infinite medium, the results Eqs. (2.1-8) and (2.1-9) are valid provided the dimensions of these reflection planes are much larger than the beam size. Under these conditions, the periodic medium is called a thick grating (or volume grating).

Referring to Figure 2.2, we now consider the case of a thin grating. In a thin grating, the transverse dimension of the periodic index variation is relatively small compared with the beam size and/or wavelength of light. When a plane wave is incident into the periodic medium, diffraction from each of the planes occurs in addition to the specular reflection. The diffraction from each of these planes is a result of the finite size of the planes. This allows the scattered light to be directed along an angle θ' which can be different from the incident angle θ. Using a similar argument leading to Eq. (2.1-8), we obtain the following condition for constructive interference:

$$\Lambda \sin\theta + \Lambda \sin\theta' = N(\lambda/n) \tag{2.1-12}$$

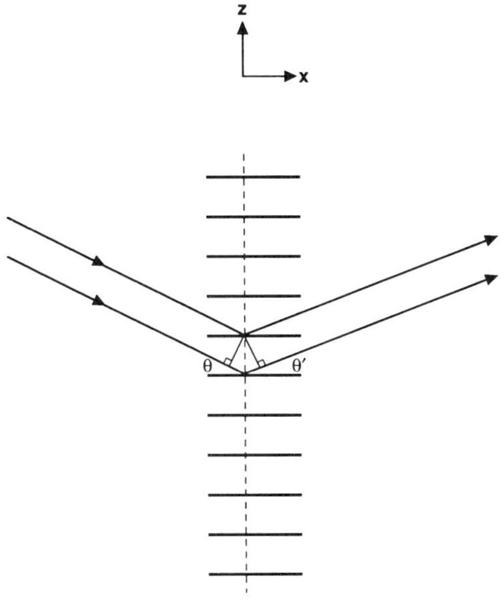

Figure 2.2. Diffraction from a thin grating.

where N is an integer. According to this equation, there are a series of diffraction orders for any given angle of incidence. This is the main difference between a thin grating and a thick grating. In a thick grating, there is only one diffraction order. In addition, diffraction occurs only when the incidence angle satisfies the Bragg law.

2.2 COUPLED MODE THEORY

In the previous section, we derived the Bragg law of diffraction in periodic media based on conditions that lead to constructive interference. The results are useful for determining the angle of the incident and diffracted beams. In order to answer questions such as the intensities of the diffracted beam, we need to solve for the amplitude of the diffracted waves. This is done by carrying out the so-called coupled-mode analysis.

Referring to Figure 2.3, we consider a periodic medium with an index of refraction given by

$$n = n_0 + n_1 \cos Kz \tag{2.2-1}$$

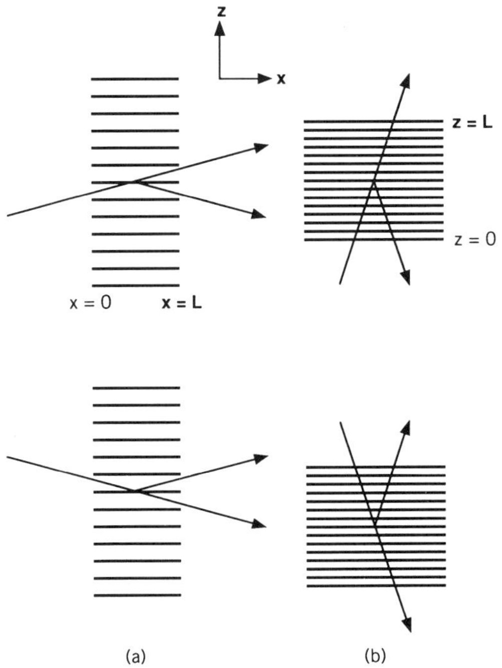

Figure 2.3. Configurations for Bragg diffraction: (a) transmission grating; (b) reflection grating.

where n_0, n_1 are constants and K is the grating wave number. We assume that the index variation is purely sinusoidal.

Let the electric field of the incident and diffracted waves be written

$$E = A_1 \exp[i(\omega t - \mathbf{k}_1 \cdot \mathbf{r})] + A_2 \exp[i(\omega t - \mathbf{k}_2 \cdot \mathbf{r})] \qquad (2.2\text{-}2)$$

where $\mathbf{k}_1$ and $\mathbf{k}_2$ are wave vectors, ω is the frequency of the light waves, A_1 and A_2 are amplitudes of the waves. Here we assume that the electric fields are polarized along the y-axis. They are called s-waves (or TE waves) because the electric field vectors are perpendicular to the plane of incident (xz-plane). The electric field Eq. (2.2-2) is a solution to the wave equation in the absence of the periodic index variation. The wave vectors are related to the angular frequency by

$$|\mathbf{k}_1| = |\mathbf{k}_2| = n_0 \omega / c \qquad (2.2\text{-}3)$$

In the presence of the periodic variation in the index of refraction, the two waves are coupled and the amplitudes A_1 and A_2 are functions of position. Let the xz-plane be the plane of incidence (i.e., the plane formed by $\mathbf{k}_1$ and $\mathbf{K}$). Conservation of momentum requires that $\mathbf{k}_2$ must also lie in this plane. The electric field can thus be written

$$E = A_1 \exp[i(\omega t - \alpha_1 x - \beta_1 z)] + A_2 \exp[i(\omega t - \alpha_2 x - \beta_2 z)] \qquad (2.2\text{-}4)$$

where β_1 and β_2 are the z-components of the wave vectors $\mathbf{k}_1$ and $\mathbf{k}_2$, respectively, and $\alpha_{1,2}$ are the components of these wave vectors parallel to the wave front of the grating. The amplitudes A_1 and A_2 are in general functions of both x and z. However, there are a number of cases in which the interaction configuration requires that the amplitudes A_1 and A_2 are functions of either x or z only. The coupled-mode analysis is thus divided into two interaction configurations which are illustrated in Figure 2.3.

2.2.1 Transmission Gratings

For the case of transmission gratings as shown in Figure 2.3(a), the grating spacing (i.e., period Λ) is often larger than the wavelength λ. The Bragg condition necessary for strong coupling requires that the angle between the incident beam and the wave front of the grating be small (i.e., $\theta \ll 1$). For the sake of simplicity in analyzing the coupling, we assume that the grating is infinite in the z-direction so that the amplitudes A_1 and A_2 are functions of x as required by the boundary conditions.

We now substitute the electric field into the following wave equation:

$$\left(\nabla^2 + \frac{\omega^2}{c^2} n^2\right) E = 0 \qquad (2.2\text{-}5)$$

where n is the index of refraction of the periodic medium as given by Eq. (2.2-1). This leads, according to Eqs. (2.2-1) and (2.2-3), to

$$\sum_{j=1,2}\left[\left(\frac{d^2}{dx^2}-2i\alpha_j\frac{d}{dx}\right)A_j\right]\exp[i(\omega t-\alpha_j x-\beta_j z)]$$
$$=-\frac{\omega^2}{c^2}2n_0n_1\cos Kz\sum_{j=1,2}A_j\exp[i(\omega t-\alpha_j x-\beta_j z)] \tag{2.2-6}$$

In arriving at Eq. (2.2-6), we have neglected the second order term n_1^2 which is of the order of 10^{-4} as compared with n_0n_1. The second order derivatives d^2A_j/dx^2 can also be neglected for the same reason. Equation (2.2-6) is thus written

$$-2i\alpha_1\frac{dA_1}{dx}\exp(-i\alpha_1x-i\beta_1z)-2i\alpha_2\frac{dA_2}{dx}\exp(-i\alpha_2x-i\beta_2z)$$
$$=-\frac{\omega^2}{c^2}n_0n_1[\exp(-iKz)+\exp(iKz)][A_1\exp(-i\alpha_1x-i\beta_1z)$$
$$+A_2\exp(-i\alpha_2x-i\beta_2z)] \tag{2.2-7}$$

where we have removed the common factor $\exp(i\omega t)$.

By scalar-multiplying Eq. (2.2-7) with $\exp(-i\alpha_j x-i\beta_j z)$, $j=1,2$, one at a time and integrating over z, we obtain the following set of coupled equations:

$$\frac{dA_1}{dx}=-i\kappa_{12}A_2e^{-i\,\Delta\alpha\,x}$$
$$\frac{dA_2}{dx}=-i\kappa_{21}A_1e^{i\,\Delta\alpha\,x} \tag{2.2-8}$$

where

$$\kappa_{12}=\frac{\pi n_1}{\lambda\cos\theta_1},\qquad \kappa_{21}=\frac{\pi n_1}{\lambda\cos\theta_2}$$
$$\Delta\alpha=\alpha_2-\alpha_1=\frac{2\pi}{\lambda}n_0(\cos\theta_2-\cos\theta_1) \tag{2.2-9}$$

provided the following condition is satisfied:

$$\beta_2=\beta_1\pm K \tag{2.2-10}$$

In the above derivation, incidence angle θ_1 and diffraction angle θ_2 are introduced so that

$$\alpha_1 = \frac{2\pi}{\lambda} n_0 \cos\theta_1 , \qquad \alpha_2 = \frac{2\pi}{\lambda} n_0 \cos\theta_2$$
$$\beta_1 = \frac{2\pi}{\lambda} n_0 \sin\theta_1 , \qquad \beta_2 = \frac{2\pi}{\lambda} n_0 \sin\theta_2 \tag{2.2-11}$$

According to Eqs. (2.2-10) and (2.2-11), the diffraction angle θ_2 is related to the incidence angle by

$$\sin\theta_2 = \sin\theta_1 \pm \frac{\lambda}{n_0\Lambda} \tag{2.2-12}$$

where we have used $K = 2\pi/\Lambda$.

In the coupled Eqs. (2.2-8), $\Delta\alpha$ represents a phase mismatch. The coupling and the exchange of energy between the beams depend on the phase mismatch. We consider first the case of perfect phase matching.

a. $\Delta\alpha = 0$. Phase matching occurs when $\cos\theta_2 = \cos\theta_1$, according to Eqs. (2.2-9). The solution $\theta_2 = \theta_1$ is a trivial one which corresponds to unscattered light. For scattered light, $\theta_2 \neq \theta_1$. Thus phase matching for Bragg scattering requires that $\theta_2 = -\theta_1$. In addition, these angles must also satisfy Eq. (2.2-12). This leads to

$$\theta_1 = -\theta_2 = \pm\sin^{-1}\left(\frac{\lambda}{2n_0\Lambda}\right) \tag{2.2-13}$$

This angle is known as the Bragg angle and is often written

$$\theta_B = \sin^{-1}\left(\frac{\lambda}{2n_0\Lambda}\right) \tag{2.2-14}$$

At this angle of incidence, the coupled Eqs. (2.2-8) become

$$\frac{dA_1}{dx} = -i\kappa A_2 , \qquad \frac{dA_2}{dx} = -i\kappa A_1 \tag{2.2-15}$$

with

$$\kappa = \frac{\pi n_1}{\lambda \cos\theta_B} \tag{2.2-16}$$

The solutions are

$$A_1(x) = A_1(0)\cos\kappa x - iA_2(0)\sin\kappa x$$
$$A_2(x) = A_2(0)\cos\kappa x - iA_1(0)\sin\kappa x \tag{2.2-17}$$

where $A_1(0)$ and $A_2(0)$ are the amplitudes at $x = 0$.

In the special case of a single wave incident at $x = 0$, i.e., $A_2(0) = 0$, the solutions become

$$A_1(x) = A_1(0)\cos\kappa x\,, \qquad A_2(x) = -iA_1(0)\sin\kappa x \tag{2.2-18}$$

We note that

$$|A_1(x)|^2 + |A_2(x)|^2 = |A_1(0)|^2 \tag{2.2-19}$$

so that the total power carried by both light waves is conserved. If the interaction distances between the two beams L is such that $\kappa L = \pi/2$, the total power of the incident beam is transferred into the diffracted beam.

Since this process is used in a large number of technological and scientific applications, it may be worthwhile to gain some appreciation for the diffraction efficiencies possible using known periodic media. The fraction of the power of the incident beam transferred in a distance L into the diffracted beam is given, using (2.2-18)

$$\eta = \frac{I_{\text{diffracted}}}{I_{\text{incident}}} = \frac{|A_2(L)|^2}{|A_1(0)|^2} = \sin^2\kappa L \tag{2.2-20}$$

Figure 2.4 plots the beam powers $|A_1|^2$, $|A_2|^2$ as functions of κx where x is the interaction distance. We note that the diffraction efficiency can be anywhere between 0 and 1. This means that a complete power transfer is possible between the two waves.

Example Consider a periodic variation of the index of refraction inside a $BaTiO_3$ crystal with $n_0 = 2.5$ and $n_1 = 10^{-4}$. Assuming a period of $\Lambda = 1\ \mu$m and a wavelength of $\lambda = 0.5\ \mu$m for the incident beam, we find that the Bragg angle for phase-matched incidence is $\theta_B = 5.74°$. Note that this angle

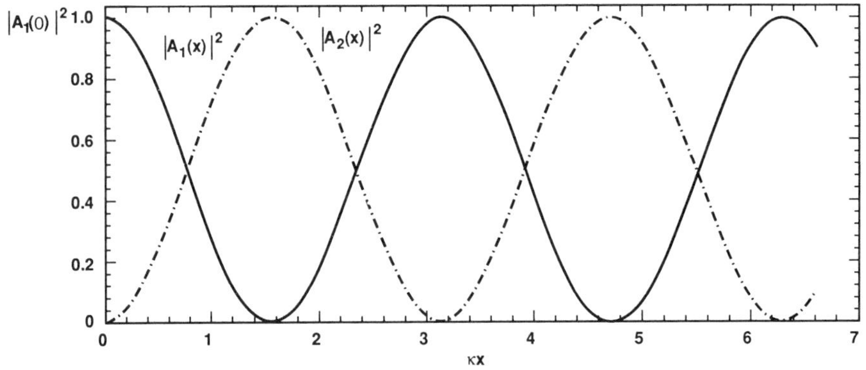

Figure 2.4. Beam powers $|A_1|^2$ and $|A_2|^2$ versus κx in a phase matched Bragg scattering.

is measured in the medium. The coupling constant is $\kappa = 6.31\ \text{cm}^{-1}$. Thus an interaction length of $L = 0.2$ cm would yield a diffraction efficiency of $\eta = \sin^2 \kappa L = 0.91$.

b. $\Delta\alpha \neq 0$. We now consider the case when there is a phase mismatch. This occurs when the incident angle is slightly deviated from the Bragg angle θ_B. Let the incident angle be written

$$\theta_1 = -\theta_B + \Delta\theta \tag{2.2-21}$$

where $\Delta\theta$ is the small deviation and θ_B is the Bragg angle as given by Eq. (2.2-14). The Bragg condition Eq. (2.2-12) requires that

$$\theta_2 = \theta_B + \Delta\theta \tag{2.2-22}$$

Therefore, the momentum mismatch $\Delta\alpha$ becomes, according to Eq. (2.2-9),

$$\Delta\alpha = -2k\,\Delta\theta \sin\theta_B = -K\,\Delta\theta \tag{2.2-23}$$

where $k = 2\pi n_0/\lambda$. Figure 2.5 shows the conservation of momentum in Bragg scatterings.

We need to solve the coupled Eqs. (2.2-8) with $\Delta\alpha \neq 0$. By multiplying the first equation with $\exp(i\,\Delta\alpha\,x)$ on both sides, we obtain

$$e^{i\,\Delta\alpha\,x}\frac{dA_1}{dx} = -i\kappa_{12}A_2 \tag{2.2-24}$$

Carrying out a differentiation with respect to x on both sides of Eq. (2.2-24)

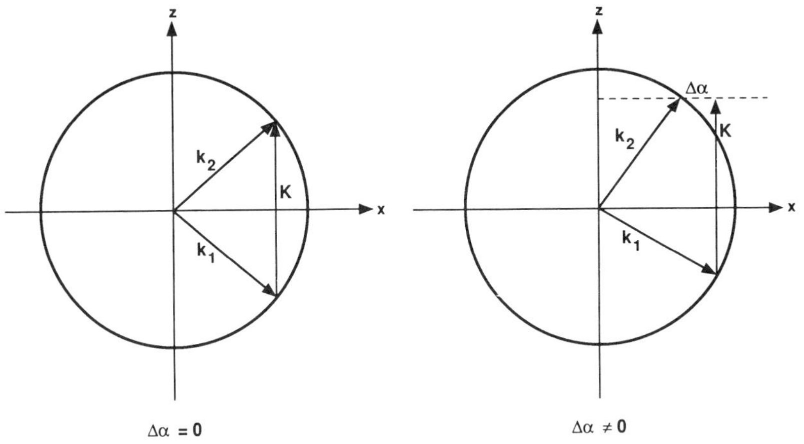

Figure 2.5. (a) Bragg scattering with $\Delta\alpha = 0$; (b) Bragg scattering with momentum mismatch ($\Delta\alpha \neq 0$).

and using Eqs. (2.2-8), we arrive at the following differential equation:

$$A_1'' + i\,\Delta\alpha\, A_1' + \kappa^2 A_1 = 0 \tag{2.2-25}$$

where a prime indicates a differentiation with respect to x and κ^2 is given by

$$\kappa^2 = \kappa_{12}\kappa_{21} = \left(\frac{\pi n_1}{\lambda}\right)^2 \frac{1}{\cos\theta_1 \cos\theta_2} \tag{2.2-26}$$

A general solution to Eq. (2.2-25) is given by

$$A_1(x) = \exp\left(-i\,\frac{\Delta\alpha}{2}\,x\right)[C_1 \sin sx + C_2 \cos sx] \tag{2.2-27}$$

where C_1 and C_2 are constants and

$$s^2 = \kappa^2 + \left(\frac{\Delta\alpha}{2}\right)^2 \tag{2.2-28}$$

Solution for $A_2(x)$ is obtained using Eqs. (2.2-24) and (2.2-27) and is given by

$$\begin{aligned} A_2(x) = \exp\left(i\,\frac{\Delta\alpha}{2}\,x\right)\Bigg[&\left(\frac{\Delta\alpha}{2\kappa_{12}}\,C_1 - \frac{is}{\kappa_{12}}\,C_2\right)\sin sx \\ &+ \left(\frac{is}{\kappa_{12}}\,C_1 + \frac{\Delta\alpha}{2\kappa_{12}}\,C_2\right)\cos sx\Bigg] \end{aligned} \tag{2.2-29}$$

The constants C_1 and C_2 are determined by the boundary conditions. In the case of a single beam incidence at $x = 0$ (i.e., $A_2(0) = 0$), these constants are

$$C_1 = i\,\frac{\Delta\alpha}{2s}\,A_1(0), \qquad C_2 = A_1(0) \tag{2.2-30}$$

Substitution of these constants into Eqs. (2.2-27) and (2.2-29) leads to

$$\begin{aligned} A_1(x) &= A_1(0)\exp\left(-i\,\frac{\Delta\alpha}{2}\,x\right)\left[i\,\frac{\Delta\alpha}{2s}\sin sx + \cos sx\right] \\ A_2(x) &= -iA_1(0)\exp\left(i\,\frac{\Delta\alpha}{2}\,x\right)\frac{\kappa_{21}}{s}\sin sx \end{aligned} \tag{2.2-31}$$

We note that

$$|A_1(x)|^2 \cos\theta_1 + |A_2(x)|^2 \cos\theta_2 = |A_1(0)|^2 \cos\theta_1 \tag{2.2-32}$$

which means that the total energy flux along the x-direction is conserved. The factors $\cos\theta_1$ and $\cos\theta_2$ in Eq. (2.2-32) are important because these two waves propagate along different directions with $\cos\theta_2 \neq \cos\theta_1$. The diffraction efficiency η is defined as

$$\eta = \frac{|A_2(L)|^2 \cos\theta_2}{|A_1(0)|^2 \cos\theta_1} \tag{2.2-33}$$

where L is the interaction length. Using Eqs. (2.2-31), (2.2-26), and (2.2-8), we obtain

$$\eta = \frac{\kappa^2}{\kappa^2 + \left(\frac{\Delta\alpha}{2}\right)^2} \sin^2\left\{\kappa L\left[1 + \left(\frac{\Delta\alpha}{2\kappa}\right)^2\right]^{1/2}\right\} \tag{2.2-34}$$

Inspection of Eq. (2.2-31) or (2.2-34) reveals that power transfer or diffraction efficiency can never be complete (see Figure 2.6). This is a result of the momentum mismatch. The maximum diffraction efficiency can be written, using Eqs. (2.2-34) and (2.2-23),

$$\eta_{\max} = \frac{1}{1 + \left(\frac{K\,\Delta\theta}{2\kappa}\right)^2} \tag{2.2-35}$$

The maximum diffraction efficiency drops to 1/2 when the angular deviation is

$$\Delta\theta = \Delta\theta_{1/2} = \frac{2\kappa}{K} = \frac{\kappa\Lambda}{\pi} \tag{2.2-36}$$

This may be interpreted as the angular aperture of the Bragg diffraction.

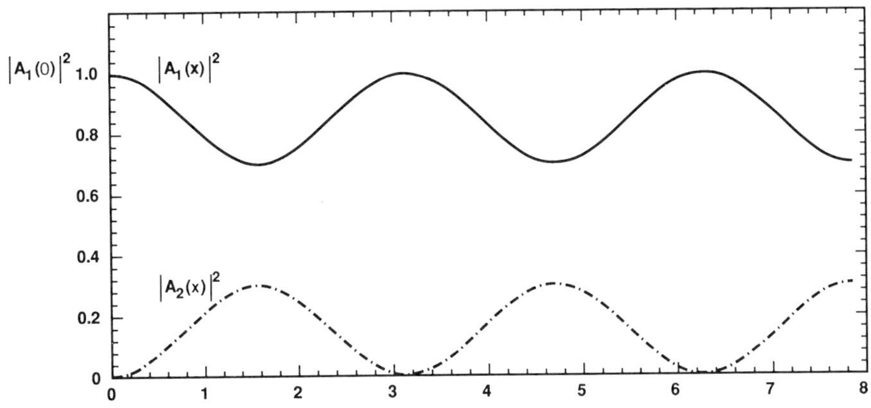

Figure 2.6. Beam powers $|A_1|^2$ and $|A_2|^2$ with $\Delta\alpha \neq 0$.

Figure 2.7 plots the diffraction efficiency as a function of the momentum mismatch. Using the parameter given in the previous example with $\Lambda = 1\ \mu\text{m}$, $\kappa = 6.31\ \text{cm}^{-1}$, this angle is $\Delta\theta_{1/2} = 2 \times 10^{-4}$ rad.

The momentum mismatch can also be a result of wavelength deviation. Let the incident angle be fixed at

$$\theta_1 = -\theta_B = -\sin^{-1}\left(\frac{\lambda_0}{2n_0\Lambda}\right) \tag{2.2-37}$$

where λ_0 is a wavelength. This is Bragg angle for $\lambda = \lambda_0$ and $\Delta\alpha = 0$. However, if the wavelength of the incident wave is deviated slightly from λ_0, then the angle of incidence θ_1 is no longer a Bragg angle. The diffraction angle θ_2 is determined by Eq. (2.2-12) which becomes

$$\sin\theta_2 = -\frac{\lambda_0}{2n_0\Lambda} + \frac{\lambda}{n_0\Lambda} \tag{2.2-38}$$

Let

$$\lambda = \lambda_0 + \Delta\lambda \tag{2.2-39}$$

where $\Delta\lambda$ is the wavelength deviation. The diffraction angle θ_2 becomes

$$\sin\theta_2 = \frac{\lambda_0}{2n_0\Lambda} + \frac{\Delta\lambda}{n_0\Lambda} \tag{2.2-40}$$

If we assume $\Delta\lambda \ll \lambda$, θ_2 can be written

$$\theta_2 = \theta_B + \frac{\Delta\lambda}{n_0\Lambda\cos\theta_B} \tag{2.2-41}$$

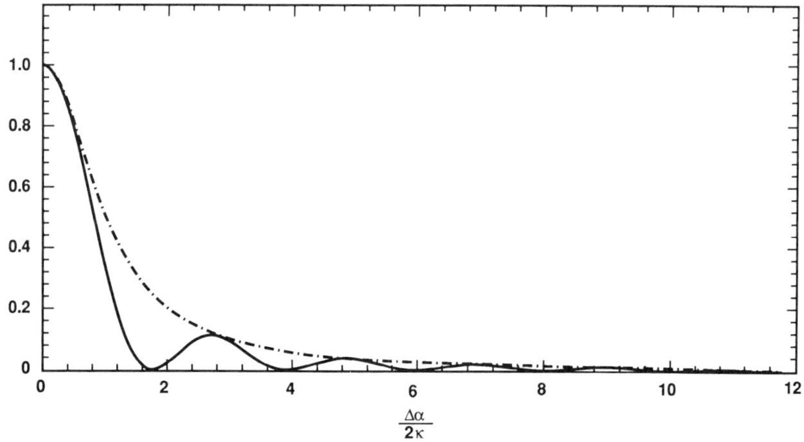

Figure 2.7. Diffraction efficiency (solid curve) as a function of the phase mismatch $\Delta\alpha$.

We note that although the incidence angle is fixed, the diffraction angle shifts away from θ_B as the wavelength deviates from λ_0. This deviation leads to a momentum mismatch of

$$\Delta\alpha = -\frac{\pi}{\Lambda}\frac{\Delta\lambda}{n_0\Lambda\cos\theta_B} \tag{2.2-42}$$

Figure 2.8 shows the momentum mismatch as a result of wavelength detuning. According to Eq. (2.2-34), the diffraction efficiency drops to 1/2 when $|\Delta\alpha| = 2\kappa$ which leads to

$$\Delta\lambda_{1/2} = \frac{2\kappa\Lambda}{\pi} n_0\Lambda\cos\theta_B \tag{2.2-43}$$

where $\Delta\lambda_{1/2}$ may be interpreted as a measure of the spectral bandwidth.

Using the parameters in the previous example with $\kappa = 6.31\ \text{cm}^{-1}$, $\Lambda = 1\ \mu\text{m}$, $n_0 = 2.5$ and $\theta_B = 5.74°$, we obtain a spectral bandwith of $\Delta\lambda_{1/2} = 10^{-3}\ \mu\text{m}$ (or $\Delta\lambda_{1/2} = 10$ Å).

2.2.2 Reflection Gratings

We now consider the case of reflection gratings as shown in Figure 2-3(b). The wave propagation analysis in this case is relatively simple because the

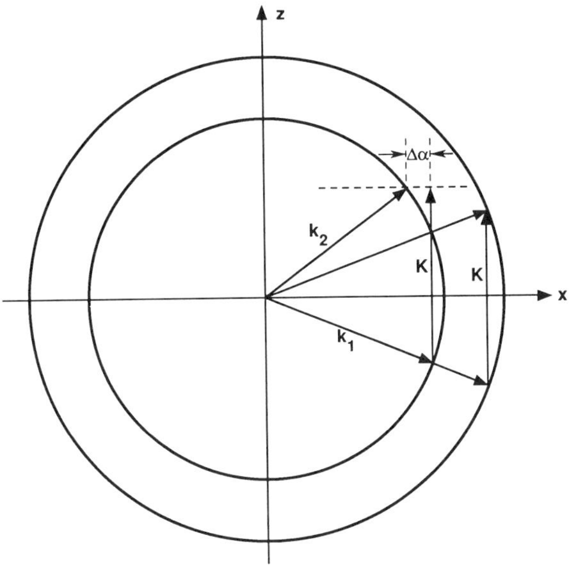

Figure 2.8. Momentum mismatch due to wavelength deviation. The outer normal surface corresponds Bragg matched diffraction as $\lambda = \lambda_0$. When the periodic medium is illuminated with a beam of longer wavelength $\lambda > \lambda_0$, momentum mismatch occurs. This is illustrated in the inner normal surface.

medium is homogeneous in both the x and y directions. The boundary condition requires that $\alpha_2 = \alpha_1$ and the mode amplitudes A_1 and A_2 are functions of z only. Thus, the electric field can be written

$$E = \{A_1(z) \exp[i(\omega t - \beta_1 z)] + A_2(z) \exp[i(\omega t - \beta_2 z)]\} \exp(-i\alpha x) \tag{2.2-44}$$

where β_1 and β_2 are the z-components of the wave vectors and α is the component of the wave vectors parallel to the wave fronts of the grating.

According to Eq. (2.2-11), the condition $\alpha_2 = \alpha_1$ implies that $\theta_2 = -\theta_1$. By following a procedure similar to the one leading to Eqs. (2.2-8), a set of coupled equations is obtained

$$\frac{d}{dz} A_1 = -i\kappa A_2 e^{-i\,\Delta\beta\, z}, \qquad \frac{d}{dz} A_2 = i\kappa A_1 e^{i\,\Delta\beta\, z} \tag{2.2-45}$$

where

$$\Delta\beta = \beta_2 - \beta_1 \pm K \tag{2.2-46}$$

and

$$\kappa = \frac{\pi n_1}{\lambda \sin\theta_1} \tag{2.2-47}$$

An inspection of Eqs. (2.2-45) reveals that a prerequisite for a continuous cumulative transfer of energy between the two waves is that

$$\beta_2 = \beta_1 \pm K \tag{2.2-48}$$

Using $\theta_2 = -\theta_1$ and Eq. (2.2-11), condition (2.2-48) can be written

$$\frac{4\pi}{\lambda} n_0 \sin\theta_1 = \pm K \tag{2.2-49}$$

We note that Eq. (2.2-49) is equivalent to the Bragg law (2.1-8) with $N = 1$. Higher Bragg diffraction orders with $N \neq 1$ do not occur here because the index of refraction as given by Eq. (2.2-1) is purely sinusoidal. In what follows, we solve for the amplitudes.

a. $\Delta\beta = 0$. In the case of perfect phase matching, the coupled Eqs. (2.2-45) become

$$\frac{d}{dz} A_1 = -i\kappa A_2, \qquad \frac{d}{dz} A_2 = i\kappa A_1 \tag{2.2-50}$$

with

$$\kappa = \frac{2\pi n_1 n_0 \Lambda}{\lambda^2} \tag{2.2-51}$$

In arriving at Eqs. (2.2-50) and (2.2-51), we assume $\theta_1 > 0$. Equations (2.2-50) can be easily integrated. The solution with the boundary condition $A_2(L) = 0$ can be written

$$\begin{aligned} A_1(z) &= A_1(0) \frac{\cosh \kappa(z-L)}{\cosh \kappa L} \\ A_2(z) &= iA_1(0) \frac{\sinh \kappa(z-L)}{\cosh \kappa L} \end{aligned} \tag{2.2-52}$$

We note that

$$|A_1(z)|^2 - |A_2(z)|^2 = \frac{|A_1(0)|^2}{\cosh^2 \kappa L} \tag{2.2-53}$$

which indicates that the net power flux along the z-direction is conserved. Figure 2.9 shows the power of these waves as functions of z. The diffraction efficiency η defined as

$$\eta = \frac{|A_2(0)|^2}{|A_1(0)|^2} \tag{2.2-54}$$

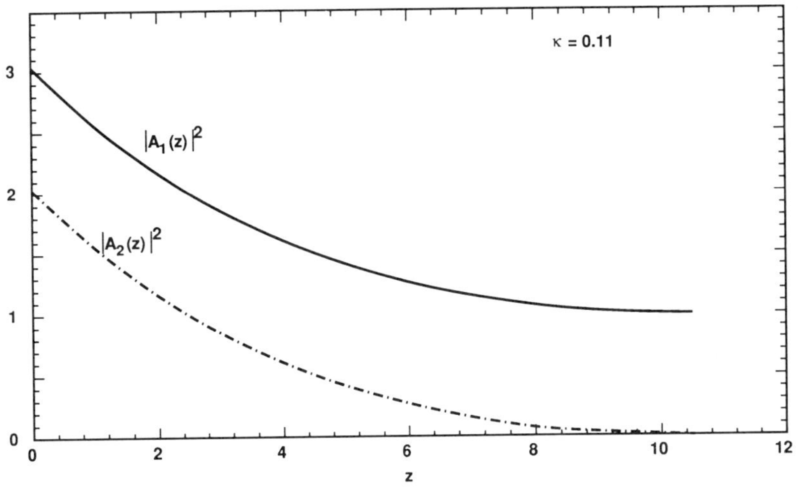

Figure 2.9. Beam powers $|A_1|^2$ and $|A_2|^2$ versus z in a phase matched Bragg scattering from a reflection grating. The incident beam power $|A_1|^2$ decays exponentially donating power to the reflected beam which grows exponentially along the $-z$-direction.

is given by

$$\eta = \tanh^2 \kappa L \tag{2.2-55}$$

We find that the diffraction efficiency is an increasing function of κL and approaches unity as κL becomes very large. At $\kappa L = 1$, the diffraction efficiency is $\eta = \tanh^2 1 = 0.58$. Figure 2.10 plots the diffraction efficiency as a function of κL.

Example Consider a periodic variation of the index of refraction inside a $BaTiO_3$ crystal with $n_0 = 2.5$ and $n_1 = 5.0 \times 10^{-5}$. If the medium is to be used as a reflection grating at $\theta = \pi/2$, the period and the wavelength are related by

$$2n_0\Lambda = \lambda$$

For $\lambda = 0.5$ μm, the period Λ should be $\Lambda = 0.1$ μm. Using Eq. (2.2-51), we obtain the coupling constant $\kappa = 3.15\ \text{cm}^{-1}$. For an interaction length of $L = 0.5$ cm, the diffraction efficiency is $\eta = 0.84$, according to Eq. (2.2-55).

b. $\Delta\beta \neq 0$. We now consider the case of Bragg scattering in a reflection grating with a phase mismatch. The solutions of the coupled Eqs. (2.2-45) subject to the boundary condition $A_2(L) = 0$, are given by

$$\begin{aligned} A_1(z) &= A_1(0) \exp\left(-i \frac{\Delta\beta}{2} z\right) \left[\frac{s \cosh s(L-z) - i \dfrac{\Delta\beta}{2} \sinh s(L-z)}{s \cosh sL - i \dfrac{\Delta\beta}{2} \sinh sL} \right] \\ A_2(z) &= A_1(0) \exp\left(i \frac{\Delta\beta}{2} z\right) \left[\frac{-i\kappa \sinh s(L-z)}{s \cosh sL - i \dfrac{\Delta\beta}{2} \sinh sL} \right] \end{aligned} \tag{2.2-56}$$

where

$$s^2 = |\kappa|^2 - \left(\frac{\Delta\beta}{2}\right)^2 \tag{2.2-57}$$

Again, the solution (2.2-56) reveals that maximum coupling of power occurs at $\Delta\beta = 0$. The fraction of power transferred (diffraction efficiency η) in an interaction distance L is given by

$$\eta = \left| \frac{A_2(0)}{A_1(0)} \right|^2 = \frac{|\kappa|^2 \sinh^2 sL}{s^2 \cosh^2 sL + \left(\dfrac{\Delta\beta}{2}\right)^2 \sinh^2 sL} \tag{2.2-58}$$

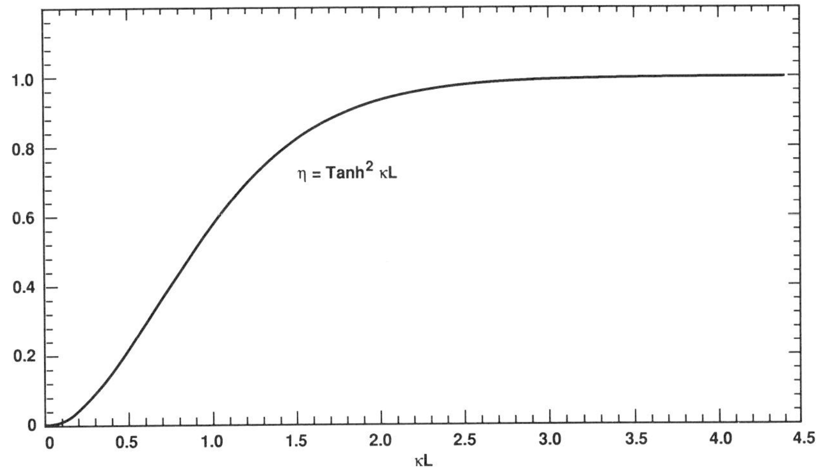

Figure 2.10. Diffraction efficiency of a reflection grating as a function of κL.

The diffraction efficiency η is plotted as a function of $\Delta\beta\, L$ in Figure 2.11. We note that η drops significantly when $\Delta\beta > 2\kappa$. Using Eq. (2.2-48) and $\theta_2 = -\theta_1$, the momentum mismatch can be written

$$\Delta\beta = -2\beta_1 \pm K \qquad (2.2\text{-}59)$$

For $\theta_1 > 0$, phase matching occurs at $2\beta_1 = K$ and the momentum mismatch is written

$$\Delta\beta = \frac{2\pi}{\Lambda} - \frac{4\pi}{\lambda} n_0 \sin\theta_1 \qquad (2.2\text{-}60)$$

Momentum mismatch can be due to either the angle of incidence or the wavelength detuning.

The bandwidth of a reflection grating is determined by $\Delta\beta = \pm 2\kappa$. The region $|\Delta\beta| < 2\kappa$ is known as the stop band because the incident wave decays exponentially in the grating, according to Eqs. (2.2-56) and (2.2-57). Let the angle of incidence θ_1 be fixed at the Bragg angle such that

$$\Delta\beta = \frac{2\pi}{\Lambda} - \frac{4\pi}{\lambda_0} n_0 \sin\theta_B = 0 \qquad (2.2\text{-}61)$$

where λ_0 is the center wavelength. If the wavelength of light is detuned from λ_0 such that $\lambda = \lambda_0 + \Delta\lambda$, then the stop band $\Delta\beta = \pm 2\kappa$ will lead to a spectral band of

$$\frac{\Delta\lambda}{\lambda_0} = \frac{n_1}{2n_0 \sin^2\theta_B} = \pm \frac{2n_1 n_0 \Lambda^2}{\lambda_0^2} = \pm \frac{\kappa\Lambda}{\pi} \qquad (2.2\text{-}62)$$

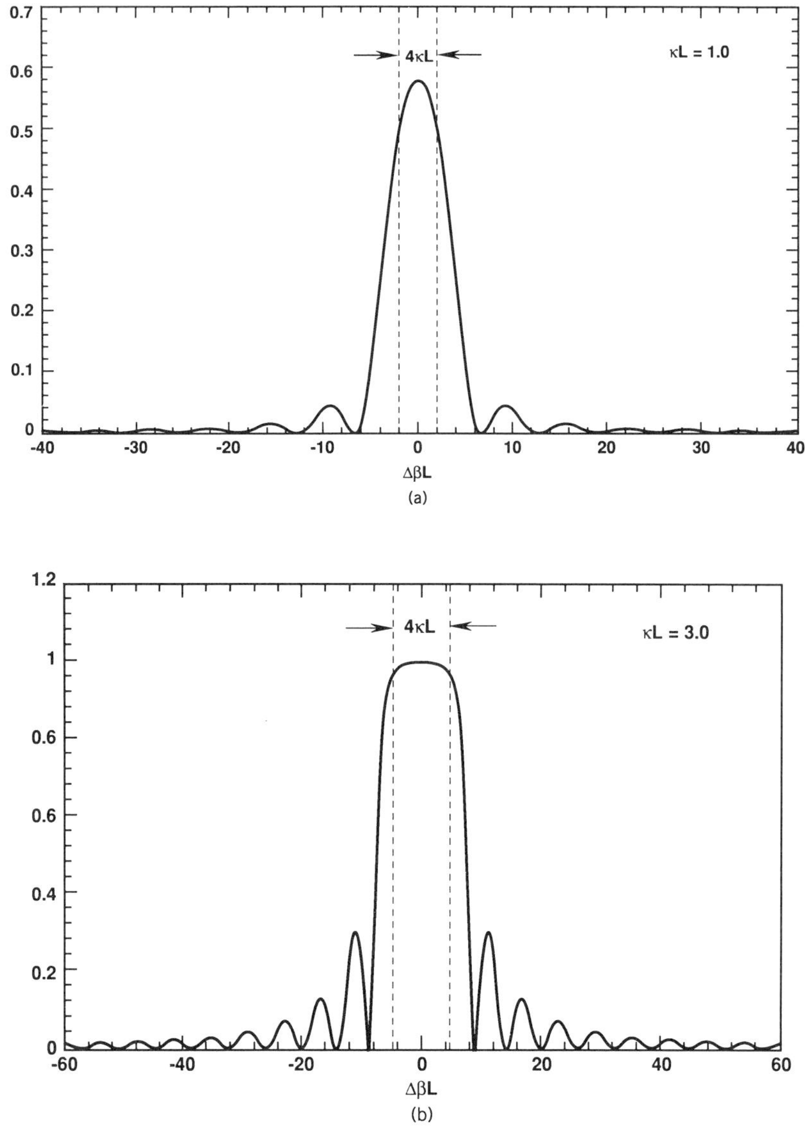

Figure 2.11. Diffraction efficiency η of a reflection grating as a function of the momentum mismatch $\Delta\beta L$. The stop band is indicated by the vertical dashed lines. (a) $\kappa L = 1.0$, (b) $\kappa L = 3.0$.

A similar derivation will lead to an angular width of

$$\Delta\theta_1 = \pm \frac{n_1}{2n_0 \sin\theta_B \cos\theta_B} \tag{2.2-63}$$

provided $\theta_B \neq 0$ or $\pi/2$. In the case when the Bragg angle is at normal incidence (i.e., $\theta_B = \pi/2$), the angular width is given by

$$\Delta\theta_1 = \pm\left(\frac{n_1}{n_0}\right)^{1/2} \tag{2.2-64}$$

2.3 RAMAN–NATH DIFFRACTION

We have thus far treated the diffraction of light by a grating of infinite extent. In the particle picture, a monochromatic plane wave of infinite extent is represented by a quantum with a well-defined momentum and energy. The Bragg diffraction is viewed as a single reflection which involves a change of momentum by $2\pi/\Lambda$. This fundamental process can occur only when both the energy and momentum are conserved. Since the grating is stationary, the conservation of energy and momentum requires that the wave vectors of the photons and the grating form an isosceles (see Figure 2.1). In this Bragg diffraction, the incident wave at the Bragg angle, $\theta_B = \sin^{-1}(\lambda/2n\Lambda)$, is diffracted by the grating with a change of momentum. Can the diffracted wave get scattered at a larger angle? The answer is no for the case of a grating of infinite extent, because the conservation of energy and momentum cannot be satisfied simultaneously. This is illustrated in Figure 2.12(b). The wave vector 0 represents the incident wave at the Bragg angle θ_B. The wave vector 1 represents the diffracted wave that results from a single Bragg scattering. To be scattered again with the same change of wave vector **K**, the momentum conservation cannot be satisfied [Figure 2.12(b)]. Also shown in Figure 2.12(a) are multiple or successive diffraction processes which involve gratings with slightly different wave vectors. In the latter case, both the energy and momentum are conserved. Thus, we conclude that multiple scattering processes cannot occur when the grating wave vector is uniquely defined, which corresponds to a grating of infinite

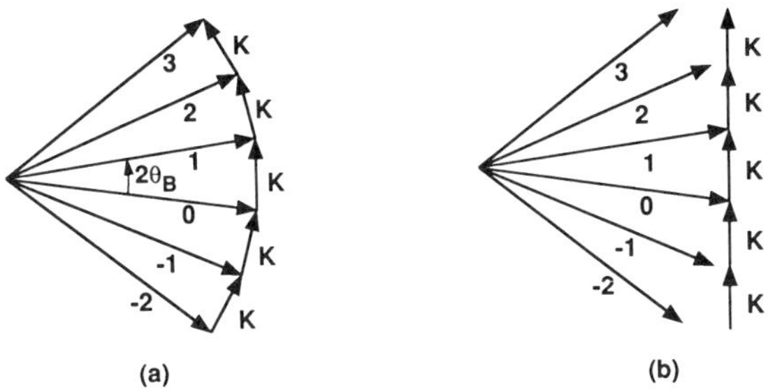

Figure 2.12. Wave vector diagrams for multiple scattering. (a) Multiple scatterings are allowed when the grating wave vector has an angular distribution; (b) multiple scatterings are forbidden when the grating wave vector is well defined.

extent. Multiple scattering processes can occur only when there is an angular distribution of the grating wave vectors, **K**. The latter corresponds to the case when the grating is of finite extent.

Let us now investigate the criterion for multiple scattering. Inspection of Figure 2.12(a) reveals that multiple scattering will occur if the angular distribution of the grating wave vector, **K**, is large compared to the Bragg angle, θ_B. Let L be the width of the grating; the angle of spreading of the grating wave vector is determined by $\Delta K\, L = \pm\pi$ which leads to

$$\psi \simeq \frac{\Lambda}{L} \tag{2.3-1}$$

where Λ is the period of the grating. The Bragg angle is, according to (2.2-13), given approximately by

$$\theta_B \simeq \frac{\lambda}{2n\Lambda} \tag{2.3-2}$$

where we use n for the index of refraction. It is customary to distinguish between the two regimes of diffraction by defining a dimensionless parameter Q:

$$Q = 2\pi\lambda L / n\Lambda^2 = 4\pi\theta_B/\psi \tag{2.3-3}$$

One defines the regime $Q > 1$ as the Bragg regime of optical diffraction. In this regime, multiple scattering is not permitted and only one order of diffraction of light is produced. Conversely, one defines the regime $Q < 1$ as the Raman–Nath regime of optical diffraction. In this regime, the angular spread of the grating wave vector is much larger than the Bragg angle, θ_B, and therefore many orders of diffraction can be observed. The incident light beam is split, after interacting with the grating, into several beams corresponding to different orders of diffraction. These orders are distinguished by the numbers 0, ± 1, ± 2, $\pm 3, \ldots, \pm m$ which correspond to wave vectors $\mathbf{k}$, $\mathbf{k} \pm \mathbf{K}$, $\mathbf{k} \pm 2\mathbf{K}, \ldots, \mathbf{k} \pm m\mathbf{K}$. The 0th beam is the extension of the incident beam, the beam $+m$ corresponds to the scattering with m Bragg diffractions while $-m$ corresponds to the scattering with m Bragg diffractions in the $-z$ direction.

To evaluate the diffraction efficiencies associated with each diffraction order, we consider a thin grating with a modulation in the index of refraction given by

$$\Delta n(x, y, z) = \begin{cases} n_1 \cos(\mathbf{K}\cdot\mathbf{r}), & 0 < x < L \\ 0, & \text{otherwise} \end{cases} \tag{2.3-4}$$

For the sake of simplicity in illustrating the basic ideas, we assume that the medium is isotropic and Δn is a scalar. An incident optical wave is represented by

$$\mathbf{E} = \mathbf{E}_0 \exp[i(\omega t - \mathbf{k} \cdot \mathbf{r})] \tag{2.3-5}$$

This wave is incident onto the thin sheet of grating at $x = 0$. In the Raman–Nath regime ($Q < 1$), the interaction length L is small enough so that this periodically perturbed sheet ($0 < x < L$) acts as a phase grating. In other words, the transmission of light through the perturbed region $0 < x < L$ only affects the phase of the plane wave. Thus, the transmitted wave can be written

$$\mathbf{E}_t = \mathbf{E}_0 \exp[-i\phi + i(\omega t - \mathbf{k} \cdot \mathbf{r})] \tag{2.3-6}$$

where ϕ is the phase shift which results from passing through the grating, and can be written

$$\phi = \int \frac{\omega}{c} \Delta n \, ds \tag{2.3-7}$$

where the integral is over the ray path within the region $0 < x < L$. Let θ be the angle of incidence (i.e., angle between the **k** vector and the x-axis), the path length in the perturbed region is $L/\cos\theta$. If L is small enough, the integral (2.3-7) can simply be written

$$\phi = \frac{\omega}{c} n_1 \frac{L}{\cos\theta} \cos(\mathbf{K} \cdot \mathbf{r}) \tag{2.3-8}$$

Substitution of (2.3-8) for ϕ in (2.3-6), leads to

$$\mathbf{E} = \mathbf{E}_0 \exp[i(\omega t - \mathbf{k} \cdot \mathbf{r}) - i\delta \cos(\mathbf{K} \cdot \mathbf{r})] \tag{2.3-9}$$

where δ is given by

$$\delta = \frac{\omega L n_1}{c \cos\theta} = \frac{2\pi L n_1}{\lambda \cos\theta} \tag{2.3-10}$$

and is referred to as the modulation index. We note that the transmitted field (2.3-9) is phase modulated. By using the following identity for Bessel's functions,

$$e^{-i\delta \cos x} = \sum_{m=-\infty}^{\infty} J_m(\delta)(-i)^m e^{-imx} \tag{2.3-11}$$

the transmitted field (2.3-9) can be written

$$\mathbf{E}_t = \mathbf{E}_0 \sum_{m=-\infty}^{\infty} J_m(\delta)(-i)^m \exp[i\omega t - i(\mathbf{k} + m\mathbf{K}) \cdot \mathbf{r}] \tag{2.3-12}$$

According to (2.3-12), the transmitted field is a linear superposition of plane waves with frequencies ω and wave vectors $(\mathbf{k} + m\mathbf{K})$ which are precisely the mth order diffracted beam discussed above.

It is important to note that Eq. (2.3-12) is actually valid only in the immediate vicinity of $x = L$ (i.e., near field). In the far field, the electric field is written

$$\mathbf{E}_t = \mathbf{E}_0 \sum_{m=\infty}^{\infty} J_m(\delta)(-i)^m \exp[i(\omega t - \mathbf{k}_m \cdot \mathbf{r})] \tag{2.3-13}$$

where $\mathbf{k}_m$ is the wave vector of the mth diffracted beam. The wave vector is determined by the following equations:

$$k_{mz} = k_z + mK_z \tag{2.3-14}$$

$$k_{mx} = \left[\left(\frac{\omega}{c}\right)^2 - k_{mz}^2\right]^{1/2} \tag{2.3-15}$$

where the subscripts x and z indicate x- and z-components of the corresponding vectors. Equation (2.3-14) is true because only the z-component of $\mathbf{K}$ is well defined. Equation (2.3-15) is a result of conservation of energy. The amplitude of the mth order diffracted beam is $J_m(\delta)(-i)^m$ according to (2.3-12). The diffraction efficiency for the mth order Raman–Nath diffraction is thus given by

$$\eta_m = J_m^2(\delta) = J_m^2\left(\frac{2\pi L n_1}{\lambda \cos\theta}\right) \tag{2.3-16}$$

In the absence of the modulation ($n_1 = 0$), all the light energy is found in the $m = 0$ order, i.e., $\eta_0 = 1$ and $\eta_m = 0$ for $m \neq 0$. The diffraction efficiency of the order $m = \pm 1$ is maximum when the modulation index, δ, is 1.85. The zeroth order is completely quenched when $\delta = 2.4$ since $J_0(2.4) = 0$. Using the identity

$$\sum_{m=-\infty}^{\infty} J_m^2(\delta) = J_0(0) = 1 \tag{2.3-17}$$

it follows that

$$\sum_{m=-\infty}^{\infty} \eta_m = 1 \tag{2.3-18}$$

In other words, the sum of the intensities of all the diffracted orders must be equal to the intensity of the incident beam so that the energy is conserved. Figure 2.13 illustrates the multiple diffraction order in the Raman–Nath regime.

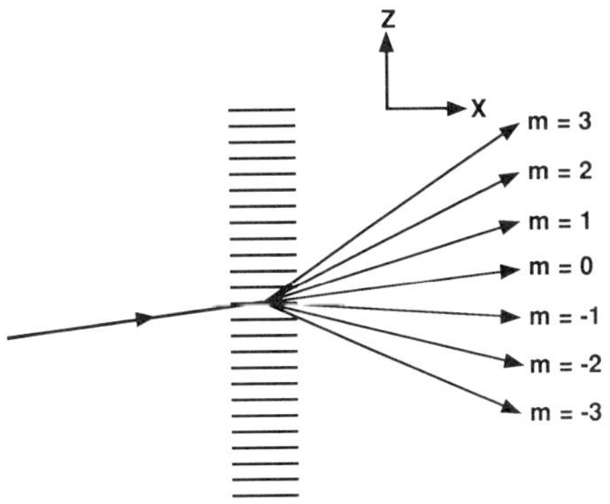

Figure 2.13. Raman–Nath diffraction.

2.4 BRAGG DIFFRACTION IN ANISOTROPIC MEDIA

We have seen in the previous sections that the diffraction of light by gratings can be pictured as an interaction process between the incident wave and the diffracted wave. The conservation of momentum requires that the three wave vectors associated with the photons and the grating form a triangle. In an isotropic medium, the refractive index associated with a light beam is independent of the direction of propagation and therefore $|\mathbf{k}'|$ is always the same as $|\mathbf{k}|$. Thus the triangle is an isosceles and its resolution leads to the Bragg diffraction condition (2.1-8).

In an anisotropic medium, the refractive index associated with a light beam is, in general, dependent on the direction of propagation as well as the polarization state. Since the diffracted light beam, in general, propagates in a different direction from the incident beam, the magnitude of the wave vectors is no longer the same. In some cases, there may even be a change of the polarization state between the incident and the diffracted beam. Let n' and n be the refractive indices associated with the diffracted and the incident beam, respectively. The triangle formed by $\mathbf{k}'$, $\mathbf{k}$, and $\mathbf{K}$ has sides equal, respectively, to $n'\omega/c$, $n\omega/c$, and K. Since n' and n are, in general, different, the triangle is not an isosceles. Let θ and θ' be the angles between the light beams and the wave front of the grating (Figure 2.14). The Bragg diffraction conditions are obtained from the triangle in Figure 2.14 and are given by

$$2k \sin \theta = K - (k'^2 - k^2)/K \tag{2.4-1}$$

and

$$2k' \sin \theta' = K + (k'^2 - k^2)/K \tag{2.4-2}$$

or equivalently,

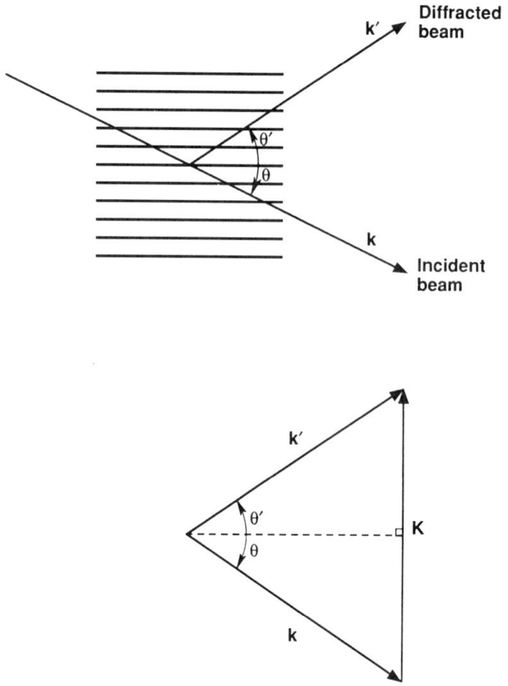

Figure 2.14. Bragg diffraction in an anisotropic medium. (a) Diffraction of a light beam from a grating in an anisotropic medium; (b) conservation of momentum.

$$2\Lambda \sin\theta = \frac{\lambda}{n} - \frac{\Lambda^2}{n\lambda}(n'^2 - n^2) \tag{2.4-3}$$

and

$$2\Lambda \sin\theta' = \frac{\lambda}{n'} + \frac{\Lambda^2}{n'\lambda}(n'^2 - n^2) \tag{2.4-4}$$

For $n' = n$, the Bragg diffraction condition (2.1-8) (with $N = 1$) is obtained again when $\theta' = \theta$. To illustrate the effect of anisotropy, we consider a special case of Bragg diffraction in a uniaxial crystal (e.g., $BaTiO_3$). We assume that both the light beams propagate in a plane perpendicular to the optic axis (c-axis) of the crystal. The incident light beam is linearly polarized parallel to the c-axis so that it propagates in the extraordinary mode of the crystal with a refractive index n_e. The diffracted light beam is assumed to be linearly polarized perpendicular to the c-axis and is in the ordinary mode of the crystal with a refractive index of n_0 (see Figure 2.15). The angles of incidence and diffraction for the beams are, according to (2.4-3) and (2.4-4),

$$\sin\theta = \frac{1}{2n_e}\left[\frac{\lambda}{\Lambda} - \frac{\Lambda}{\lambda}(n_o^2 - n_e^2)\right] \tag{2.4-5}$$

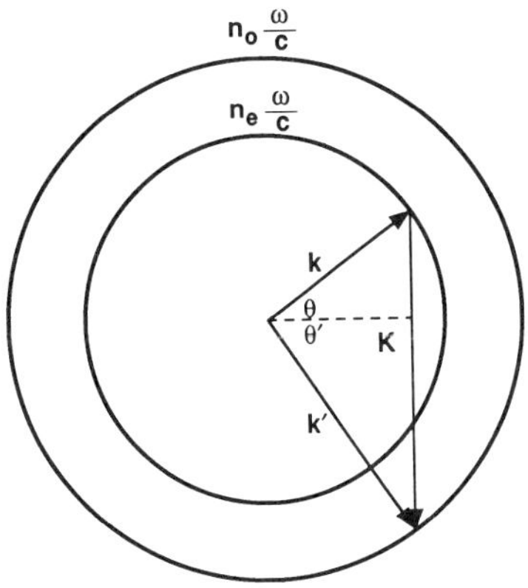

Figure 2.15. Anisotropic Bragg scattering in a negative uniaxial crystal (e.g., $BaTiO_3$). All three momentum vectors are in a plane perpendicular to the c-axis. The incident beam is an extraordinary wave while the diffracted beam is an ordinary wave.

and

$$\sin\theta' = \frac{1}{2n_o}\left[\frac{\lambda}{\Lambda} + \frac{\Lambda}{\lambda}(n_o^2 - n_e^2)\right] \qquad (2.4\text{-}6)$$

According to (2.4-5) and (2.4-6), the angle of incidence θ and the angle of diffraction θ' are both functions of (λ/Λ) for a given crystal when n_o and n_e are fixed. For $\lambda/\Lambda = |n_o \pm n_e|$, $|\theta'| = |\theta| = 90°$ the vectors **k**, **k**′, and **K** are

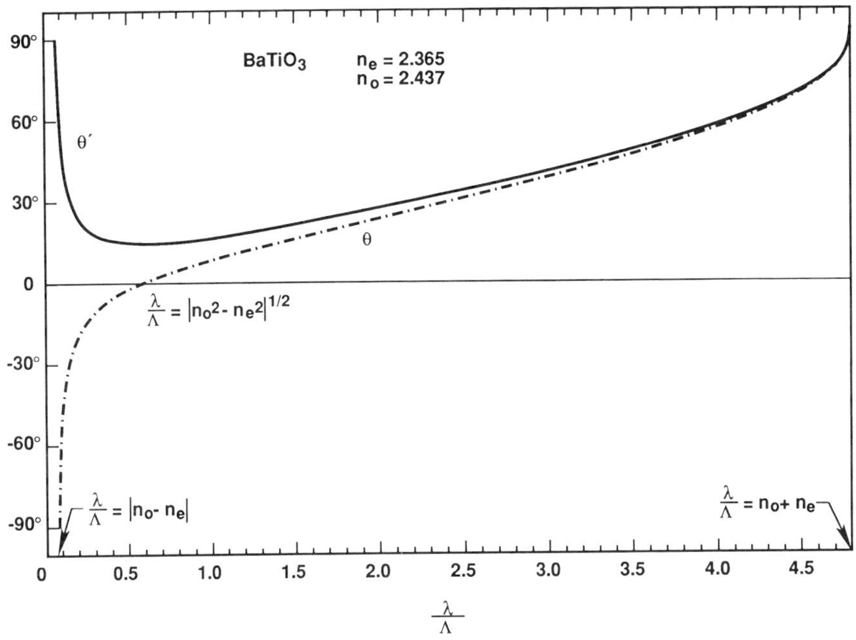

Figure 2.16. Angle of incidence θ and of diffraction θ' as a function of the ratio of the light wavelength and grating period in a $BaTiO_3$ crystal where we take $n_e = 2.365$ and $n_0 = 2.437$.

colinear. Bragg diffraction is only possible when $|n_o - n_e| \le \lambda/\Lambda \le |n_o + n_e|$, and the angles θ' and θ are real. In the regimes when $\lambda/\Lambda < |n_o - n_e|$ or $\lambda/\Lambda > |n_o + n_e|$, $\sin\theta'$ and $\sin\theta$ become greater than one and the diffraction is impossible. Figure 2.16 shows the dependence of θ and θ' on λ/Λ for a typical uniaxial crystal. The point $\lambda/\Lambda = |n_o - n_e|$ corresponds to a co-directional colinear Bragg coupling which occurs in Solc filters [1].

2.5 MULTIPLE GRATINGS AND FINITE GRATINGS

In the above discussion, we assumed that the index of refraction, as given by Eq. (2.2-1), varies sinusoidally in space and that the gratings are infinite in extent. In reality, all gratings are finite in size. For finite gratings, the index of refraction can no longer be purely sinusoidal, as given by Eq. (2.2-1). For example, consider a transmission grating with a finite aperture D with the index of refraction given by

$$n = \begin{cases} n_0 + n_1 \cos Kz\,, & |z| < \frac{1}{2}D \\ n_0\,, & \text{otherwise} \end{cases} \tag{2.5-1}$$

where n_0, n_1 are constants and K is the grating wave number. Such an index of refraction can be written in terms of a Fourier integral

$$n(z) = n_0 + n_1 \frac{D}{\pi}\left[\int_0^{\infty} \frac{\sin\left(\dfrac{kD}{2}\right)}{\left(\dfrac{kD}{2}\right)} \cos kz\, dk\right] \cos Kz \tag{2.5-2}$$

where the integral inside the square brackets is simply a Fourier representation of the rectangle function. This equation can also be written

$$n(z) = n_0 + n_1 \frac{D}{2\pi}\int_0^{\infty} \frac{\sin\left(\dfrac{kD}{2}\right)}{\left(\dfrac{kD}{2}\right)} [\cos(K+k)z + \cos(K-k)z]\, dk \tag{2.5-3}$$

As we can see from Eq. (2.5-3), a finite sinusoidal grating is a linear combination of an infinite number of purely sinusoidal gratings each with a well-defined grating wave number. In this section, we discuss Bragg diffraction from multiple gratings and finite gratings.

For the clarity of describing the concept, we consider the diffraction from a grating which consists of two grating wave vectors. Let the index of refraction be written

$$n = n_0 + n_1 \cos(\mathbf{K}_1 \cdot \mathbf{r}) + n_2 \cos(\mathbf{K}_2 \cdot \mathbf{r}) \tag{2.5-4}$$

where $\mathbf{K}_1$, $\mathbf{K}_2$ are the grating wave vectors and n_1, n_2 are constants. We also assume that such gratings exist between $x = 0$ and $x = L$. Two different beams may be generated when a beam is incident into the medium. The electric field of the three waves can be written

$$E = A_1 e^{-i\mathbf{k}_1\cdot\mathbf{r}} + A_2 e^{-i\mathbf{k}_2\cdot\mathbf{r}} + A_3 e^{-i\mathbf{k}_3\cdot\mathbf{r}} \tag{2.5-5}$$

where A_1 is the amplitude of the incident wave with wave vector $\mathbf{k}_1$, A_2 and A_3 are amplitudes of diffracted waves. Conservation of momentum (or Bragg condition) requires that these wave vectors satisfy the following equations (see Figure 2.17):

$$\mathbf{k}_2 = \mathbf{k}_1 + \mathbf{K}_1 \tag{2.5-6}$$

$$\mathbf{k}_3 = \mathbf{k}_1 + \mathbf{K}_2 \tag{2.5-7}$$

Using an argument similar to that leading to Eqs. (2.2-8), we obtain

$$\begin{aligned}\frac{dA_1}{dx} &= -i\kappa_{12}A_2 - i\kappa_{13}A_3\\ \frac{dA_2}{dz} &= -i\kappa_{21}A_1\\ \frac{dA_3}{dz} &= -i\kappa_{31}A_1\end{aligned} \tag{2.5-8}$$

where we assume that the momentum is conserved and

$$\kappa_{12} = \frac{\pi n_1}{\lambda \cos\theta_1}, \qquad \kappa_{21} = \frac{\pi n_1}{\lambda \cos\theta_2} \tag{2.5-9}$$

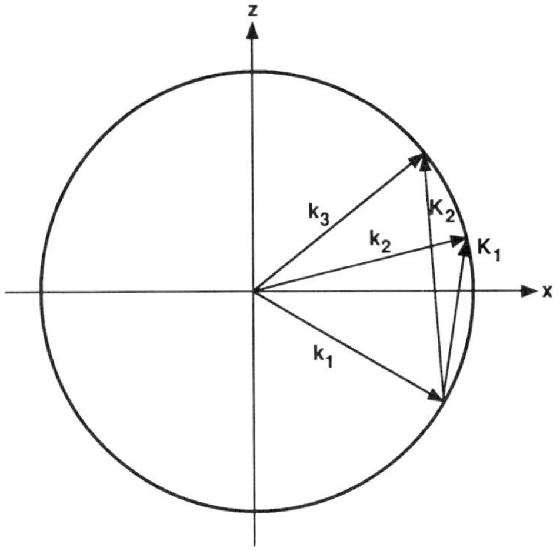

Figure 2.17. Bragg diffraction with two volume gratings.

$$\kappa_{13} = \frac{\pi n_2}{\lambda \cos \theta_1}, \qquad \kappa_{31} = \frac{\pi n_2}{\lambda \cos \theta_3} \tag{2.5-10}$$

The last two of Eqs. (2.5-8) describe the coupling from the incident wave A_1 to diffracted waves A_2 and A_3, respectively. The first of the Eqs. (2.5-8) describes the coupling back from waves A_2 and A_3 to wave A_1. These coupled equations can be easily solved. The solutions, subject to the boundary conditions $A_2(0) = A_3(0) = 0$, are

$$\begin{aligned} A_1(x) &= A_1(0) \cos \kappa x \\ A_2(x) &= -i \frac{\kappa_{21}}{\kappa} A_1(0) \sin \kappa x \\ A_3(x) &= -i \frac{\kappa_{31}}{\kappa} A_1(0) \sin \kappa x \end{aligned} \tag{2.5-11}$$

where

$$\kappa = (\kappa_{12}\kappa_{21} + \kappa_{13}\kappa_{31})^{1/2} \tag{2.5-12}$$

and $A_1(0)$ is the amplitude of the incident wave at $x = 0$. We note that these solutions as well as the coupled equation agree with conservation of power flux along x-direction, that is,

$$\cos \theta_1 |A_1|^2 + \cos \theta_2 |A_2|^2 + \cos \theta_3 |A_3|^2 = \text{constant} \tag{2.5-13}$$

Diffraction efficiencies of the gratings can be defined as

$$\eta_{12} = \frac{\cos \theta_2 |A_2(L)|^2}{\cos \theta_1 |A_1(0)|^2}, \qquad \eta_{13} = \frac{\cos \theta_3 |A_3(L)|^2}{\cos \theta_1 |A_1(0)|^2} \tag{2.5-14}$$

which can be written, according to Eqs. (2.5-11)

$$\eta_{12} = \frac{\kappa_{12}\kappa_{21}}{\kappa^2} \sin^2 \kappa L \tag{2.5-15}$$

$$\eta_{13} = \frac{\kappa_{13}\kappa_{31}}{\kappa^2} \sin^2 \kappa L \tag{2.5-16}$$

The results (Eqs. 2.5-15, 2.5-16) are very interesting. We note that the total diffraction efficiency is

$$\eta = \eta_{12} + \eta_{13} = \sin^2 \kappa L \tag{2.5-17}$$

and the diffraction efficiency of each of the grating components is only a fraction of η. For high diffraction efficiencies with $\eta \sim 1$, the maximum diffraction efficiencies for each of the gratings are

$$(\eta_{12})_{\max} = \frac{\kappa_{12}\kappa_{21}}{\kappa^2}, \qquad (\eta_{13})_{\max} = \frac{\kappa_{13}\kappa_{31}}{\kappa^2} \tag{2.5-18}$$

In the case of low diffraction efficiencies (i.e., $\eta \ll 1$ and $\kappa L \ll 1$), the diffraction efficiencies are

$$\eta_{12} \simeq \kappa_{12}\kappa_{21}L^2, \qquad \eta_{13} \simeq \kappa_{13}\kappa_{31}L^2 \tag{2.5-19}$$

According to Eq. (2.5-19), the diffraction efficiencies are independent of each other in the low diffraction efficiency regime. This is a result of negligible depletion for the incident beam. In the case of high diffraction efficiency, the total energy of the incident beam is shared by the two diffracted beams. The energy depletion and the sharing lead to the results given by (2.5-18). We must keep in mind that these results are obtained by assuming that all Bragg diffractions are perfectly phase-matched.

The results can be extended to multiple gratings that contain more than two components. Let the index of refraction be written

$$n = n_0 + \sum_{i=1}^{N} n_i \cos(\mathbf{K}_i \cdot \mathbf{r}) \tag{2.5-20}$$

where n_i are constants and $\mathbf{K}_i$ are the grating wave vectors. If all the gratings wave vectors are oriented such that an incident plane wave with wave vector $\mathbf{k}$ is scattered from all these gratings with perfect phase matching, then the wave vectors of these scattered waves are

$$\mathbf{k}_i = \mathbf{k} \pm \mathbf{K}_i, \quad i = 1, 2, \ldots, N \tag{2.5-21}$$

with

$$|\mathbf{k}_i| = |\mathbf{k}| \tag{2.5-22}$$

The last equation ensures that the frequency of the scattered wave is the same as that of the incident beam. Under these conditions, the diffraction efficiency of each of the grating components can be written

$$\eta_i = \frac{\kappa_i^2}{\kappa^2} \sin^2 \kappa L \tag{2.5-23}$$

where κ_i is the coupling constant for the ith grating defined as

$$\kappa_i^2 = \frac{(\pi n_i)^2}{\lambda^2 \cos\theta_i \cos\theta} \tag{2.5-24}$$

and

$$\kappa^2 = \sum_{i=1}^{N} \kappa_i^2 \tag{2.5-25}$$

In Eq. (2.5-24), θ is the angle for the incident beam and θ_i is that of the beam $\mathbf{k}_i$. If we further assume that the spread of the grating wave vectors is small so that $\cos\theta_i \simeq \cos\theta$; for all $i = 1, \ldots, N$ where θ is a constant, then Eqs. (2.5-23–25) can be written

$$\eta_i = \frac{n_i^2}{\sum_{j=1}^{N} n_j^2} \sin^2 \kappa L \tag{2.5-26}$$

with

$$\kappa^2 = \sum_{j=1}^{N} \frac{(\pi n_j)^2}{\lambda^2 \cos^2\theta} \tag{2.5-27}$$

According to Eq. (2.5-26), the diffraction efficiencies represent the spectral distribution of the grating. This is the basic principle of spectrum analyzer [2].

In the case of a finite transmission grating as shown in Figure 2.18, there are an infinite number of grating wave numbers. The spectral distribution of these wave numbers are given by Eq. (2.5-3) and are shown in Figure 2.18(b). Consider an incident plane wave which is Bragg-matched with the center wave number K. Such an incident plane wave will be scattered into an infinite number of plane waves. The distribution of energy among these scattered waves is governed by the grating strength for each of the grating components associated with the Bragg scattering as well as the phase-mismatch.

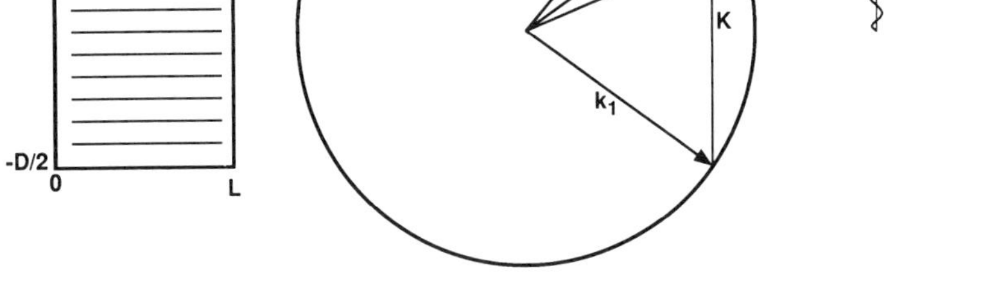

Figure 2.18. Bragg diffraction from a finite grating with a length L and an aperture D. The wave vector diagram shows the momentum mismatch $\Delta\alpha$, $\Delta\beta$ as well as a distribution of the grating wave vectors $\mathbf{K}$ due to the finite size of the grating.

Let the electric field of the incident and diffracted waves be written

$$E = A_1 \exp(-i\mathbf{k}_1 \cdot \mathbf{r}) + \int A_2(\beta_2) \exp(-i\mathbf{k}_2 \cdot \mathbf{r})\, d\beta_2 \qquad (2.5\text{-}28)$$

where

$$\mathbf{k}_1 = \alpha_1 \mathbf{x} + \beta_1 \mathbf{z}\,, \qquad \mathbf{k}_2 = \alpha_2 \mathbf{x} + \beta_2 \mathbf{z} \qquad (2.5\text{-}29)$$

with $\mathbf{x}$ and $\mathbf{z}$ being unit vectors along the axes and α_1, β_1, α_2, β_2 are components of the wave vectors. A_1 is the amplitude of the incident wave and $A_2(\beta)\, d\beta_2$ is the amplitude of the scattered wave with the z-component of the wave vector between $(\beta_2, \beta_2 + d\beta_2)$. Let the index of refraction be rewritten as

$$n(z) = n_0 + n_1 \frac{D}{2\pi} \int_0^{\infty} \frac{\sin\left(\frac{\beta D}{2}\right)}{\left(\frac{\beta D}{2}\right)} [\cos(K + \beta)z + \cos(K - \beta)z]\, d\beta \qquad (2.5\text{-}30)$$

or equivalently

$$n(z) = n_0 + n_1 \frac{D}{2\pi} \int_{-\infty}^{\infty} \frac{\sin\left(\frac{\beta D}{2}\right)}{\left(\frac{\beta D}{2}\right)} \cos(K + \beta)z\, d\beta \qquad (2.5\text{-}31)$$

The coupled equation for diffracted waves can be written

$$\frac{d}{dx} A_2 = -i\kappa_{21} \left[\frac{D}{2\pi} \frac{\sin\left(\frac{\Delta\beta\, D}{2}\right)}{\left(\frac{\Delta\beta\, D}{2}\right)} \right] A_1 e^{i\,\Delta\alpha\, x} \qquad (2.5\text{-}32)$$

where κ_{21} is given by Eq. (2.2-8) and

$$\Delta\beta = \beta_2 \pm K - \beta_1 \qquad (2.5\text{-}33)$$

$$\Delta\alpha = \alpha_2 - \alpha_1 \qquad (2.5\text{-}34)$$

We will assume that the depletion of energy from the incident beam A_1 is negligible so that Eq. (2.5-32) can be easily integrated. The result is

$$A_2 = -i\kappa_{21} \left[\frac{D}{2\pi} \frac{\sin\left(\frac{\Delta\beta\, D}{2}\right)}{\left(\frac{\Delta\beta\, D}{2}\right)} \right] \frac{e^{i\,\Delta\alpha\, L} - 1}{i\,\Delta\alpha} A_1 \qquad (2.5\text{-}35)$$

where L is the length of interaction. It is important to keep in mind that $\Delta\alpha$ is a function of β_2 (see Figure 2.18). The diffraction efficiency for the component $A_2(\beta_2)\,d\beta_2$ is given by

$$\frac{\cos\theta_2|A_2\,d\beta_2|^2}{\cos\theta_1|A_1|^2} = \kappa_{12}\kappa_{21}L^2\left(\frac{D}{2\pi}\,d\beta_2\right)^2 \frac{\sin^2\left(\frac{\Delta\beta\,D}{2}\right)}{\left(\frac{\Delta\beta\,D}{2}\right)^2} \frac{\sin^2\left(\frac{\Delta\alpha\,L}{2}\right)}{\left(\frac{\Delta\alpha\,L}{2}\right)^2} \tag{2.5-36}$$

We note that the diffraction efficiency drops as $\Delta\alpha$ or $\Delta\beta$ increases.

2.6 COUPLED MODE THEORY FOR POLARIZED WAVES

In the coupled mode analysis discussed in Section 2.2, we assume that both the index of refraction and the amplitude of the electric fields are scalars. The scalar coupled mode analysis is useful for introducing the concept of mode coupling. In practice, all amplitudes are vectors and the dielectric constant describing the index grating is, in general, a tensor. In this section, we treat the coupled mode analysis for polarized waves.

Let the periodic medium be described by

$$\epsilon' = \epsilon + \epsilon_1 \cos(\mathbf{K}\cdot\mathbf{r}) \tag{2.6-1}$$

where ϵ is the dielectric tensor in the absence of the periodic modulation and ϵ_1 is a tensor representing the amplitude of the dielectric modulation. $\mathbf{K}$ is the grating wave vector. The electric field of the incident and diffracted waves can be written

$$\mathbf{E} = A_1\mathbf{p}_1 \exp[i(\omega t - \mathbf{k}_1\cdot\mathbf{r})] + A_2\mathbf{p}_2 \exp[i(\omega t - \mathbf{k}_2\cdot\mathbf{r})] \tag{2.6-2}$$

where $\mathbf{p}_1$ and $\mathbf{p}_2$ are unit vectors representing the polarization states of the waves, respectively, and $\mathbf{k}_1, \mathbf{k}_2$ are wave vectors. We assume that $\mathbf{p}_1 \exp[i(\omega t - \mathbf{k}_1\cdot\mathbf{r})]$ and $\mathbf{p}_2 \exp[i(\omega t - \mathbf{k}_2\cdot\mathbf{r})]$ are normal modes of the medium when $\epsilon_1 = 0$. In other words, these two waves can propagate freely in the medium with $A_1\mathbf{p}_1$ and $A_2\mathbf{p}_2$ being constants. In the case of uniaxial crystals, $\mathbf{p}_1$ and $\mathbf{p}_2$ are either ordinary mode or extraordinary mode.

We will consider the case of transmission grating between $x = 0$ and $x = L$. The boundary condition requires that A_1 and A_2 are functions of x only. Let the xz-plane be the plane formed by $\mathbf{k}_1$ and $\mathbf{K}$. Conservation of momentum requires that $\mathbf{k}_2$ must also lie in this plane. The electric field can thus be written

$$\mathbf{E} = A_1(x)\mathbf{p}_1 \exp[i(\omega t - \alpha_1 x - \beta_1 z)] + A_2(x)\mathbf{p}_2 \exp[i(\omega t - \alpha_2 x - \beta_2 z)] \tag{2.6-3}$$

The electric field must satisfy the following wave equation:

$$(\nabla^2 + \omega^2\mu\epsilon')\mathbf{E} = 0 \tag{2.6-4}$$

where ϵ' is the dielectric tensor given by Eq. (2.6-1). Substitution of Eq. (2.6-1) for ϵ and Eq. (2.6-3) for $\mathbf{E}$ into Eq. (2.6-4) leads to

$$\sum_{j=1,2}\left[\left(\frac{d^2}{dx^2} - 2i\alpha_j \frac{d}{dx}\right)A_j\mathbf{p}_j\right]\exp[i(\omega t - \alpha_j x - \beta_j z)]$$
$$= -\omega^2\mu\epsilon_1 \cos(\mathbf{K}\cdot\mathbf{r}) \sum_{j=1,2} A_j\mathbf{p}_j \exp[i(\omega t - \alpha_j x - \beta_j z)] \tag{2.6-5}$$

Again, we neglect the second order derivatives. By following the procedure similar to that leading to Eqs. (2.2-8), we also obtain a set of coupled equations. They are given by

$$\frac{dA_1}{dx} = -i\kappa_{12}A_2 e^{-i\,\Delta\alpha\, x}, \qquad \frac{dA_2}{dx} = -i\kappa_{21}A_1 e^{i\,\Delta\alpha\, x} \tag{2.6-6}$$

with the coupling constants given by

$$\kappa_{12} = \frac{\pi}{2n_1\lambda\cos\theta_1\epsilon_0}\,\mathbf{p}_1\cdot\epsilon_1\mathbf{p}_2 \tag{2.6-7}$$

$$\kappa_{21} = \frac{\pi}{2n_2\lambda\cos\theta_2\epsilon_0}\,\mathbf{p}_2\cdot\epsilon_1\mathbf{p}_1 \tag{2.6-8}$$

and $\Delta\alpha$ is given by

$$\Delta\alpha = \alpha_2 - \alpha_1 \pm K_x \tag{2.6-9}$$

where K_x is the x-component of the grating wave vector $\mathbf{K}$, n_1 and n_2 are the indices of refraction associated with the normal modes $\mathbf{p}_1 \exp[i(\omega t - \mathbf{k}_1\cdot\mathbf{r})]$ and $\mathbf{p}_2 \exp[i(\omega t - \mathbf{k}_2\cdot\mathbf{r})]$, respectively. Since the medium is infinite along z-direction, the coupled equations are valid provided

$$\Delta\beta = \beta_2 - \beta_1 \pm K_z = 0 \tag{2.6-10}$$

where K_z is the z-component of the grating wave vector. These components of vectors are related to the angles θ_1 and θ_2 by

$$\alpha_1 = \frac{2\pi}{\lambda} n_1\cos\theta_1, \qquad \alpha_2 = \frac{2\pi}{\lambda} n_2\cos\theta_2$$
$$\beta_1 = \frac{2\pi}{\lambda} n_1\sin\theta_1, \qquad \beta_2 = \frac{2\pi}{\lambda} n_2\sin\theta_2 \tag{2.6-11}$$

According to Eqs. (2.6-6)–(2.6-8), we note that mode coupling for polarized waves is similar to that of scalar waves except that the coupling constant is proportional to the product

$$\mathbf{p}_1 \cdot \epsilon_1 \mathbf{p}_2 = \mathbf{p}_2 \cdot \epsilon_1 \mathbf{p}_1 \tag{2.6-12}$$

The equality is valid because ϵ_1 is a symmetric tensor. Equations (2.6-7) and (2.6-8) reduce to those of the scalar waves when ϵ_1 is a scalar and $\mathbf{p}_1 = \mathbf{p}_2$.

2.7 METHOD OF GRATING INTEGRAL

In the above discussion, we assume that the grating amplitude n_1 is a constant and the index of refraction is given by Eq. (2.2-1). In practice, gratings can be tapered and chirped. The tapering and chirping can be due to beam coupling in nonlinear media, or can be tailored intentionally for specific purposes. For these gratings, the index of refraction is written

$$n = n_0 + n_1(x) \cos[Kz + \phi(x)] \tag{2.7-1}$$

where $n_1(x)$ and $\phi(x)$ are slowly varying functions of x. The coupled equations can be written

$$\frac{d}{dx} A_1 = -i\kappa e^{-i\phi} A_2 \tag{2.7-2}$$

$$\frac{d}{dx} A_2 = -i\kappa e^{i\phi} A_1 \tag{2.7-3}$$

where κ is given by Eq. (2.2-16). We note that x is along the direction of propagation. Closed form solutions of the above equations can be obtained provided ϕ is a constant.

We now define a new variable, known as the grating integral,

$$u = \int_0^x \kappa(t)\, dt \tag{2.7-4}$$

where t is a dummy variable. By using this new variable u, the coupled equations (2.7-2) and (2.7-3) can be transformed into

$$\frac{d}{du} A_1 = -ie^{-i\phi} A_2 \tag{2.7-5}$$

$$\frac{d}{du} A_2 = -ie^{i\phi} A_1 \tag{2.7-6}$$

The general solutions are, assuming $\phi = \text{constant}$

$$A_1 = A_1(0) \cos u - ie^{-i\phi} A_2(0) \sin u \tag{2.7-7}$$

$$A_2 = A_2(0)\cos u - ie^{i\phi}A_1(0)\sin u \tag{2.7-8}$$

where $A_1(0)$ and $A_2(0)$ are the amplitude at $x = 0$ (or $u = 0$). We notice that the solutions are formally identical to Eq. (2.2-17). The diffraction efficiency, under the incidence condition $A_2(0) = 0$, is given by

$$\eta = \frac{|A_2(L)|^2}{|A_1(0)|^2} = \sin^2\left[\int_0^L \kappa(x)\, dx\right] \tag{2.7-9}$$

which reduces to Eq. (2.2-20) when κ is a constant. The integral

$$U = \int_0^L \kappa(x)\, dx \tag{2.7-10}$$

is a measure of the depth of volume gratings. The new variable defined by Eq. (2.7-4) is used later in the book to treat four-wave mixing in photorefractive media. The method of grating integral can also be employed for reflection grating.

PROBLEMS

2.1 **Square-Wave Periodic Medium.** Consider a periodic medium that consists of alternating layers of different materials. The index of refraction is written

$$n(z) = \begin{cases} n_1, & 0 < z < \dfrac{\Lambda}{2} \\ n_2, & \dfrac{\Lambda}{2} < z < \Lambda \end{cases}$$

where Λ is the period and $n(z) = n(z + \Lambda)$.

a. Using the notation in Eq. (2.1-4), derive expressions for n_0, n_1, and $p(z)$.

b. Consider a plate of such a medium with surfaces parallel to the z-axis. Let θ_i be the angle of incidence in air. Using Snell's law, show that the Bragg diffraction condition (2.1-7) can be written

$$2\Lambda \sin\theta_i = N\lambda$$

which is independent of the index of refraction n_0.

c. Let d be the thickness of such a plate with $d < \lambda$, find the amplitudes of Raman–Nath diffraction.

2.2 **Phase Shift Upon Bragg Diffraction.** Consider the case of phase-matched Bragg diffraction with $\kappa L < \pi/2$. The phase shift upon diffraction is described by Eq. (2.2-18) for transmission gratings and Eq. (2.2-52) for reflection gratings.

a. Show that the phase shift is $e^{-i\pi/2}$ provided κ is positive.

b. Show that the electric field of the waves at $x = L$ can be written

$$E = A_1(0) \exp[i(\omega t - \beta_1 z)] - i\rho A_1(0) \exp[i(\omega t - \beta_2 z)]$$

where ρ is a positive real number.

c. Derive the interference pattern by evaluating $|E|^2$. Show that the intensity can be written

$$I = I_0 \pm I_1 \sin Kz$$

where I_0 and I_1 are positive constants.

d. Show that the interference fringe pattern is shifted along the direction of grating wave vector by a quarter period relative to the grating. Thus, if $\beta_1 > 0$, then the shift is along $+z$ direction. The shift is along $-z$ direction if $\beta_1 < 0$.

2.3 **Sinc Function.** The diffraction efficiency of a transmission grating is given by Eq. (2.2-34), which is known as a sinc function. It can be written

$$\eta = \frac{\sin^2 \kappa L(1 + x^2)^{1/2}}{1 + x^2}, \quad \text{with } x = \frac{\Delta\alpha\ L}{2\kappa L} = \frac{\Delta\alpha}{2\kappa}$$

a. Plot η as a function of x with various values of κL (e.g., $\kappa L = \pi/4, \pi/2, 3\pi/4, \pi$).

b. Show that $\eta = 1$ occurs only when $x = 0$ and $\kappa L = \pi/2, 3\pi/2, \ldots$.

c. For $\kappa L = \pi/2$, show that η drops to $1/2$ at $x = 0.80$.

d. The diffraction efficiency η as a function of x consists of a series of side lobes. Find the location of these maxima and minima.

e. Show that the angular aperture of a transmission grating is proportional to the amplitude of index modulation n_1.

f. Discuss the situation when κL approaches zero.

2.4 **Reflection Gratings.** The diffraction efficiency of a reflection grating is given by Eq. (2.2-58).

a. Show that

$$\eta = \frac{(\kappa L)^2}{1 + (\kappa L)^2} \quad \text{at } \Delta\beta = 2\kappa$$

b. Show that minima ($\eta = 0$) occur at

$$\Delta\beta = 2(\kappa^2 + (m\pi)^2)^{1/2}, \quad m = 1, 2, 3, \ldots$$

c. Derive an expression for the spectral resolution of a reflection grating in terms of wavelength.

$$\left[\text{Answer: } \Delta\lambda_{1/2} = \frac{\kappa\Lambda}{\pi}\lambda\right]$$

2.5 **Phase Gratings.** Consider a thin plate of glass with a diffraction grating ruled on one side of the plate. Let x be an axis perpendicular to the surface, y and z be axes parallel to the surface. The surface ruling (or undulation) can be described by a height function

$$h = f(z)$$

where $f(z)$ is a periodic function.

a. Write down an expression for the index of refraction n as a function of x and z.

b. Let $f(z) = a \sin Kz$, find the amplitudes of Raman–Nath diffractions, for incident light parallel to the x-axis.

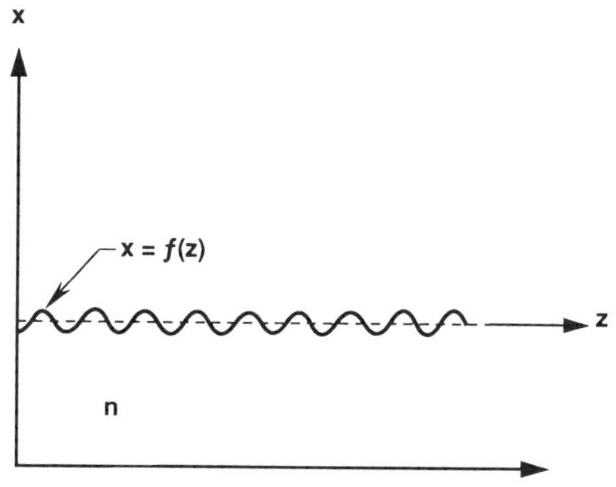

c. Consider an incident wave propagating in the xy plane with an angle of incidence θ. The wave vector can be written $\mathbf{k} = k(\cos\theta, \sin\theta, 0)$. Find the wave vectors of all diffracted beams.

2.6 **Amplitude Gratings.** An amplitude grating modulates the spatial intensity of a beam. It is made of a thin transparency with a typical amplitude transmittance given by

$$t(z) = \cos Kz$$

where K is a constant.

a. Let yz be the plane of the amplitude grating. An incident wave propagating along the x-axis is given by $E_1 = A \exp[i(\omega t - kx)]$. The transmitted wave can be written

$$E_t = tE_1$$

Show that there are only two diffracted beams at far field.

b. Let the incident wave be given by

$$E_1 = A \exp[i(\omega t - k \cos\theta x - k \sin\theta y)]$$

where θ is the angle of incidence. Find the diffracted beams at far field.

2.7 **Multiple Gratings.** Show that the result Eq. (2.5-26) is valid for multiple gratings with arbitrary phases, that is,

$$n = n_0 + \sum_{i=1}^{N} n_i \cos(\mathbf{K}_i \cdot \mathbf{r} + \phi_i)$$

2.8 **Reciprocity of Grating Diffraction.** Consider the grating described by Eq. (2.7-1). Let ρ and ρ' be the diffraction coefficients for incidence at $x = 0$ and $x = L$, respectively.

a. Use the method of grating integral and derive expressions for ρ and ρ'

b. Show that

$$\rho = \rho'$$

c. Use the argument of time reversal symmetry and show that

$$t'\rho^* + t^*\rho' = 0\,, \qquad t^*t' + \rho^*\rho' = 1$$

where t and t' are transmission coefficients.

2.9 **Local Diffraction Coefficient.** For the grating defined by Eq. (2.7-1), we define a local diffraction coefficient as

$$\rho = \frac{A_2}{A_1}$$

a. Show that

$$\frac{d\rho}{dx} = -i\kappa e^{i\phi} + i\kappa\rho^2 e^{-i\phi}$$

b. If ϕ is a constant, we can define

$$\rho = \Gamma e^{i\phi}$$

Show that

$$\frac{d\Gamma}{dx} = -i\kappa(1 - \Gamma^2)$$

c. Show that

$$\Gamma = -i \tan\left(\int_0^x \kappa(t)\, dt\right)$$

2.10 Weak Gratings. Show that in the limit when κL approaches zero, the diffraction efficiency of a grating (either transmission or reflection) can be written

$$\eta = (\kappa L)^2 \frac{\sin^2\left(\dfrac{\Delta k}{2} L\right)}{\left(\dfrac{\Delta k}{2} L\right)^2}$$

where Δk is the momentum mismatch.

REFERENCES

1. See, for example, P. Yeh, Optical Waves in Layered Media (Wiley, New York, 1988).
2. See, for example, A. Yariv and P. Yeh, Optical Waves in Crystals (Wiley, New York, 1984).

3

Photorefractive Effect

The photorefractive effect is a phenomenon in which the local index of refraction of a medium is changed by the illumination of a beam of light with spatial variation of the intensity. Such an effect was first discovered in 1966 when researchers were studying the transmission of laser beams through electro-optic crystals. It was found that the presence of laser beams inside some electro-optic crystals leads to an index inhomogeneity which distorts the wave front of the transmitted laser beam. Such an effect was referred to as "optical damage" [1]. The photorefractive effect has since been observed in many electro-optic crystals including $BaTiO_3$, $KNbO_3$, $LiNbO_3$, $LiTaO_3$, $Sr_{1-x}Ba_xNb_2O_6$ (SBN), $Ba_{2-x}Sr_xK_{1-y}Na_yNb_5O_{15}$ (KNSBN), $Bi_{12}SiO_{20}$ (BSO), $Bi_{12}GeO_{20}$ (BGO), GaAs, InP, CdTe, etc. In this chapter, we first introduce a simple physical model which describes most of the observed photorefractive phenomena. We then solve for the induced index grating and discuss the speed of hologram formation. At the end of the chapter, we also discuss moving gratings in photorefractive media.

3.1 BAND TRANSPORT MODEL

We consider a simple model which is very useful for explaining most of the observed photorefractive phenomena. In the model, photorefractive media are assumed to contain certain types of impurities or imperfections. For the sake of simplicity in explaining the concept, we assume that all donor impurities are identical and have exactly the same energy state somewhere in the middle of the bandgap (see Figure 3.1). These donor impurities can be ionized by absorbing photons. As a result of the ionization, electrons are generated in the conduction band leaving empty states behind. Such ionized impurities are capable of capturing electrons. Let the density of donor impurity be N_D of which N_D^i are ionized. The rate of electron generation is $(sI + \beta)(N_D - N_D^i)$, whereas the rate of trap capture is $\gamma_R N N_D^i$, where N is the electron density, s is the cross section for photoexcitation, I is the light intensity, β is the rate of thermal generation of electrons and γ_R is the electron-ionized trap recombination rate. The rate equation for N_D^i, neglecting the thermal generation (i.e., $\beta \ll sI$), can thus be written [2, 3]

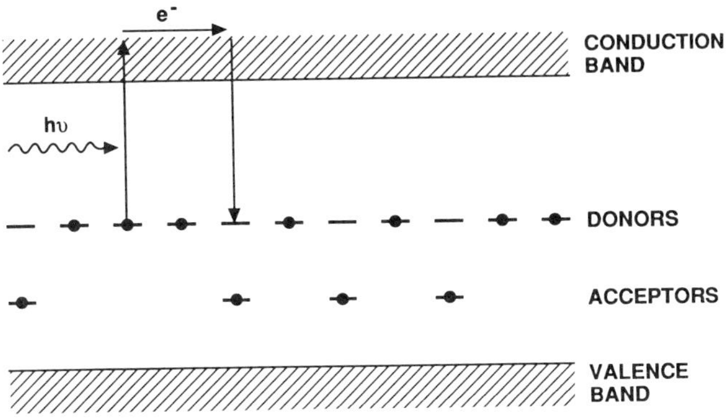

Figure 3.1. Model of the photorefractive effect.

$$\frac{\partial N_D^i}{\partial t} = sI(N_D - N_D^i) - \gamma_R N N_D^i \tag{3.1-1}$$

For every impurity ionized, there is an electron generated. On the other hand, an electron is eliminated when a recapture occurs which fills an empty impurity. The rate of generation of electrons is the same as that of the ionized impurities except that the electrons are mobile whereas the impurities are stationary. This is essential for photorefractive effect. The transport of electrons may affect the electron density. The rate equation for electron density can thus be written

$$\frac{\partial N}{\partial t} - \frac{\partial N_D^i}{\partial t} = \frac{1}{q} \nabla \cdot \mathbf{j} \tag{3.1-2}$$

where $\mathbf{j}$ is the current density and $-q$ is the electron charge ($q = 1.602 \times 10^{-19}$ C). The presence of these charge carriers (electrons) leads to a space-charge field which in turn affects the transport of the carriers. The current density consists of contributions from the drift of charge carriers due to the electric field and the diffusion due to the gradient of carrier density and thus can be written

$$\mathbf{j} = qN\mu\mathbf{E} + k_B T \mu \nabla N \tag{3.1-3}$$

where μ is the mobility tensor, $\mathbf{E}$ is the electric field, $k_B T$ is the product of the Boltzmann constant and temperature. The electric field obeys the following Poisson equation:

$$\nabla \cdot \epsilon \mathbf{E} = \rho(\mathbf{r}) = -q(N + N_A - N_D^i) \tag{3.1-4}$$

where ϵ is the dielectric tensor, $\rho(\mathbf{r})$ is the charge density and N_A is the density of the acceptor impurity. The presence of acceptor impurities is essential to maintain the charge neutrality. In the absence of light illumination, the charge neutrality is written

$$(N + N_A - N_D^i) = 0 \tag{3.1-5}$$

In the case when the electron density is small, we have $N_D^i = N_A$ in the absence of light. In other words, the density of ionized donor impurities equals the density of acceptor impurities. In Eq. (3.1-3), $qN\mu\mathbf{E}$ represents the current density due to drift, and $k_B T\mu\nabla N$ represents the current density due to diffusion. We have neglected the current due to photovoltaic effect which is discussed in Section 3.7.

As indicated in Figure 3.1, there are donor impurities and acceptor impurities. The density of donors is often much larger than that of the acceptor impurities. Here we assume that all acceptors are also identical. In the case when there is no electrons in the conduction band and no holes in the valence band, the density of ionized donor impurities is identical to that of the acceptor impurities. The neutral donor impurities are capable of donating electrons for photoexcitation and the ionized ones are capable of capturing these photoelectrons. In our model, the acceptor impurities are for the purpose of charge neutrality only. They do not directly participate in the photorefractive effect.

We now consider the incidence of two laser beams into a photorefractive medium. Let the electric field be written

$$\mathbf{E} = \mathbf{E}_a \exp(i\omega t - i\mathbf{k}_a \cdot \mathbf{r}) + \mathbf{E}_b \exp(i\omega t - i\mathbf{k}_b \cdot \mathbf{r}) \tag{3.1-6}$$

where we assume that the frequency of the beams are identical, $\mathbf{E}_a$ and $\mathbf{E}_b$ are the amplitudes and $\mathbf{k}_a$ and $\mathbf{k}_b$ are the wave vectors. If the polarization states of these two waves are not orthogonal, they will form an interference pattern with an intensity given by

$$I(\mathbf{r}) = I_0 + \text{Re}\{I_1 e^{-i\mathbf{K}\cdot\mathbf{r}}\} \tag{3.1-7}$$

where

$$I_0 = |\mathbf{E}_a|^2 + |\mathbf{E}_b|^2 \tag{3.1-8}$$

$$I_1 = 2\mathbf{E}_b \cdot \mathbf{E}_a^* \tag{3.1-9}$$

and $\mathbf{K}$ is the grating wave vector defined as

$$\mathbf{K} = \mathbf{k}_b - \mathbf{k}_a \tag{3.1-10}$$

The magnitude of the grating wave vector is related to the period (Λ) of the interference pattern by

$$K = 2\pi/\Lambda \tag{3.1-11}$$

Maximum fringe visibility occurs when the two waves are polarized along the same direction and have the same magnitude (i.e., $\mathbf{E}_a = \mathbf{E}_b$). The intensity for this case is given by

$$I = I_0(1 + \cos \mathbf{K} \cdot \mathbf{r}) \tag{3.1-12}$$

It represents an array of interference fringes. Dark fringes occur at $\mathbf{K} \cdot \mathbf{r} = (2m + 1)\pi$ where $I = 0$.

Consider now the illumination of such an intensity inside a photorefractive medium. In the bright regions near $\cos \mathbf{K} \cdot \mathbf{r} = 1$ and $I = 2I_0$, photoionized charges are generated by the absorption of photons. These charge carriers can diffuse away from the bright regions leaving behind positively charged ionized donor impurities. If these charge carriers are trapped in the dark regions, they will remain there because there is no light to reexcite them. This leads to a charge separation as depicted in Figure 3.2. As a result of the illumination with periodic intensity in the photorefractive medium, the dark regions are negatively charged and the bright regions are positively charged (see Figure 3.2). The buildup of space-charge separation continues until the diffusion current is counterbalanced by the drift current. The fundamental component of the space-charge density can thus be written

$$\rho = \rho_0 \cos \mathbf{K} \cdot \mathbf{r} \tag{3.1-13}$$

where ρ_0 is a constant. An integration of the Poisson Eq. (3.1-4) leads to

$$\mathbf{E} = \rho_0 \frac{\mathbf{K}}{\mathbf{K} \cdot \epsilon \mathbf{K}} \sin(\mathbf{K} \cdot \mathbf{r}) \tag{3.1-14}$$

We note that the space-charge field is shifted in space by $\pi/2$ relative to the intensity pattern. This space-charge field will induce a change in the index of refraction via Pockels' effect. Figure 3.2 illustrates the spatial variation of the light intensity, space-charge density, space-charge field and the induced index change.

In summary, photorefractive effect basically consists of five fundamental processes which occur in electro-optic crystals: (1) photoionization of impurities and the generation of charge carriers; (2) transport of these charge carriers; (3) trapping of charge carriers and the formation of space-charge density; (4) formation of the photoinduced space-charge electric field; (5) formation of index grating via the linear electro-optic effect (the Pockels effect). The photorefractive effect is a macroscopic phenomenon and requires the generation and transport of a large number of charge carriers (typically, of the order of 10^{15} cm^{-3}).

Given an illumination intensity of Eq. (3.1-7), we need to solve for the space-charge field as a function of both $\mathbf{r}$ and t using Eqs. (3.1-1)–(3.1-4).

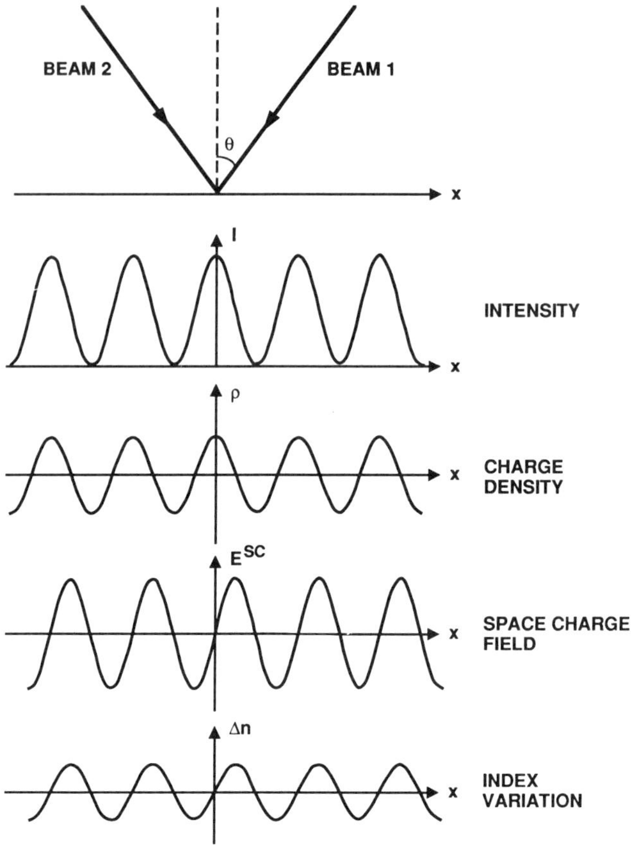

Figure 3.2. Photorefractive index gratings produced by a pair of interfering beams. The periodic variation of the intensity in the crystal creates a space-charge separation which mimics the intensity pattern. The space-charge field induces an index grating in the medium via Pockels' effect.

Closed form general solutions are, however, not available. For the moment, let us examine the situation of steady state. By setting the time derivatives to zero, we obtain the following steady state equations, according to Eqs. (3.1-1)–(3.1-4):

$$\begin{aligned} sI(N_D - N_D^i) - \gamma_R N N_D^i &= 0 \\ \nabla \cdot \mathbf{j} &= 0 \\ \mathbf{j} &= qN\mu\mathbf{E} + k_B T \mu \nabla N \\ \nabla \cdot \epsilon \mathbf{E} = \rho(\mathbf{r}) &= -q(N + N_A - N_D^i) \end{aligned} \qquad (3.1\text{-}15)$$

Using these equations, we can solve for **E** in terms of the intensity $I(\mathbf{r})$.

Closed form solutions are available for the case when the intensity modulation is small (i.e., $|I_1| \ll I_0$). Under such an assumption, we can ignore the higher spatial harmonic terms and the steady state solutions can be written

$$N(\mathbf{r}) = N_0 + \mathrm{Re}\{N_1 e^{-i\mathbf{K}\cdot\mathbf{r}}\} \tag{3.1-16}$$

$$N_D^i(\mathbf{r}) = N_{D0}^i + \mathrm{Re}\{N_{D1}^i e^{-i\mathbf{K}\cdot\mathbf{r}}\} \tag{3.1-17}$$

$$\mathbf{j}(\mathbf{r}) = \mathbf{j}_0 + \mathrm{Re}\{\mathbf{j}_1 e^{-i\mathbf{K}\cdot\mathbf{r}}\} \tag{3.1-18}$$

$$\mathbf{E}(\mathbf{r}) = \mathbf{E}_0 + \mathrm{Re}\{\mathbf{E}_1 e^{-i\mathbf{K}\cdot\mathbf{r}}\} \tag{3.1-19}$$

where N_0, N_{D0}^i, $\mathbf{j}_0$, $\mathbf{E}_0$, N_1, N_{D1}^i, $\mathbf{j}_1$, $\mathbf{E}_1$ are constants. The problem at hand is to solve for these constants in terms of I_0, I_1. In what follows, we will solve for these physical quantities. We will first consider a limiting case when $I_1 = 0$. This occurs when we have either uniform illumination of a single beam or two beams with orthogonal polarization ($I_1 = 2\mathbf{E}_b \cdot \mathbf{E}_a^* = 0$). We will then consider the case of illumination with periodic intensity variation.

a. *Uniform illumination* ($I_1 = 0$). In this case, there is no spatial variation of all the physical quantities. The charge carrier densities are related by

$$sI(N_D - N_D^i) = \gamma_R N N_D^i \tag{3.1-20}$$

$$(N + N_A - N_D^i) = 0 \tag{3.1-21}$$

These two equations can be used to solve for the two unknowns N and N_D^i. From Eq. (3.1-20), we obtain the density for ionized donor impurities

$$N_D^i = \frac{N_D sI}{(sI + \gamma_R N)} \tag{3.1-22}$$

This is the density for the ionized donor impurities under a uniform illumination. Substituting Eq. (3.1-22) for N_D^i in Eq. (3.1-21), we obtain the following quadratic equation for N,

$$\gamma_R N^2 + (sI + \gamma_R N_A)N + sI(N_A - N_D) = 0$$

This equation can be solved for N. If we further assume that $N \ll N_A$ and $sI \ll \gamma_R N_A$, then the solution for N is

$$N = \frac{N_D - N_A}{\gamma_R N_A} sI \tag{3.1-23}$$

This is the electron density under the illumination of a uniform intensity I.

The density for ionized donor impurities N_D^i can also be written

$$N_D^i = N_A + \frac{N_D - N_A}{\gamma_R N_A} sI \tag{3.1-24}$$

b. *Periodic illumination* ($I_1 \neq 0$). We consider the case of periodic illumination with $I_1 \neq 0$. This occurs when two coherent beams intersect inside the medium. Substituting Eqs. (3.1-16)–(3.1-19) and neglecting higher order terms, we obtain

$$sI_1(N_D - N_{D0}^i) + sI_0(-N_{D1}^i) - \gamma_R N_1 N_{D0}^i - \gamma_R N_0 N_{D1}^i = 0 \tag{3.1-25}$$

$$sI_0(N_D - N_{D0}^i) - \gamma_R N_0 N_{D0}^i = 0 \tag{3.1-26}$$

$$\mathbf{K} \cdot (qN_1 \mu \mathbf{E}_0 + qN_0 \mu \mathbf{E}_1 - ik_B T \mu \mathbf{K} N_1) = 0 \tag{3.1-27}$$

$$-i\mathbf{K} \cdot \epsilon \mathbf{E}_1 = -q(N_1 - N_{D1}^i) \tag{3.1-28}$$

$$N_0 + N_A - N_{D0}^i = 0 \tag{3.1-29}$$

In these equations, we recall that N_A, N_D, and $\mathbf{E}_0$ are constants. N_0 and N_{D0}^i can be obtained by solving Eqs. (3.1-26) and (3.1-29). The results are the same as Eqs. (3.1-22) and (3.1-23). The other three equations can be employed to solve for N_1, N_{D1}^i, and E_1.

The curl Maxwell Eq. (1.1-1) for the steady state case, $\nabla \times \mathbf{E} = 0$ dictates that $\mathbf{K} \times \mathbf{E}_1 = 0$ which means $\mathbf{E}_1$ is parallel to $\mathbf{K}$. After a few steps of algebraic manipulation and approximations, we obtain the amplitude of the space-charge field

$$E_1 = \frac{iK \dfrac{k_B T}{q} - \dfrac{\mathbf{K} \cdot \mu \mathbf{E}_0}{K\langle \mu \rangle}}{1 + \dfrac{K^2}{k_D^2} + i\dfrac{q\mathbf{K} \cdot \mu \mathbf{E}_0}{k_B T k_D^2 \langle \mu \rangle}} \frac{I_1}{I_0} \tag{3.1-30}$$

where

$$k_D^2 = \frac{q^2}{\langle \epsilon \rangle k_B T} \frac{N_A}{N_D} (N_D - N_A) \tag{3.1-31}$$

and $\langle \epsilon \rangle$, $\langle \mu \rangle$ are effective dielectric constant and effective mobility, respectively

$$\langle \epsilon \rangle = \frac{\mathbf{K} \cdot \epsilon \mathbf{K}}{K^2} \tag{3.1-32}$$

$$\langle \mu \rangle = \frac{\mathbf{K} \cdot \mu \mathbf{K}}{K^2} \tag{3.1-33}$$

In arriving at Eq. (3.1-30), we also assumed that (1) $\gamma_R N_A \gg sI_0$, (2) $N_D sI_0 \ll \gamma_R N_A^2$ in addition to the small modulation approximation, $I_1 \ll I_0$.

In the case when the dc field $\mathbf{E}_0$ is also parallel to the grating vector $\mathbf{K}$, Eq. (3.1-30) reduces to

$$E_1 = \frac{iK\dfrac{k_B T}{q} - \dfrac{\mathbf{K}\cdot\mathbf{E}_0}{K}}{1 + \dfrac{K^2}{k_D^2} + i\dfrac{q\mathbf{K}\cdot\mathbf{E}_0}{k_B T k_D^2}} \frac{I_1}{I_0} \tag{3.1-34}$$

This expression is further reduced for the case when $\mathbf{E}_0 = 0$ (known as the case of pure diffusion). The space-charge field is given by

$$E_1 = \frac{iK\dfrac{k_B T}{q}}{1 + \dfrac{K^2}{k_D^2}} \frac{I_1}{I_0} \tag{3.1-35}$$

These two equations for the amplitude of the space-charge field are key results of the band transport model. Here we recall that $\mathbf{K}$ is the grating wave vector, k_D is given by Eq. (3.1-31) and is known as the Debye wave number. In the theory of electrodynamics in plasmas, the electrons move in such a way as to screen out the Coulomb field of a test charge in a distance of $\Lambda_D = 2\pi/k_D$. Such a distance, known as the Debye screening radius, is determined by the balance between the thermal kinetic energy and the electrostatic energy of these charge carriers. In the case when $N_D \gg N_A$, k_D can be written, according to Eq. (3.1-31)

$$k_D^2 = \frac{q^2 N_A}{\langle\epsilon\rangle k_B T} = \frac{q^2 N_{D0}^i}{\langle\epsilon\rangle k_B T} \tag{3.1-36}$$

where we recall that N_{D0}^i is the density of ionized donor impurities which are positively charged and are capable of capturing electrons.

3.2 SPACE-CHARGE FIELD

We now discuss the space-charge field in two separate categories.

a. $E_0 = 0$ (*No Applied Field*). In the absence of an applied electric field, the photoelectrons are transported by diffusion. The diffusion is a result of the gradient of the electron density. When the electrons are generated by illumination light, a current is produced by the diffusion and charges are separated. The static electric field produced by the buildup of the space charges will move these electrons in the opposite direction. At steady state, there is no net current and the electric field is given by, according to Eq. (3.1-3) and assuming $N_0 = 0$,

$$E = iK \frac{k_B T}{q} \equiv iE_d \tag{3.2-1}$$

This is known as the diffusion field. We note that the diffusion field E_d is independent of dopant densities.

Consider now the electric field that can be produced by a given charge density N^i_{D0}. If we assume a charge separation with a space-charge density of $qN^i_{D0} \exp(-i\mathbf{K}z)$, the amplitude of the space-charge field is

$$E = i \frac{qN^i_{D0}}{K\langle \epsilon \rangle} = i \frac{qN_A}{\langle \epsilon \rangle K} \equiv iE_q \tag{3.2-2}$$

where we assume $N_0 \ll N_A$. E_q is the maximum achievable electric field (the so-called saturation field) for a given charge density of N^i_{D0}.

Using these two parameters, the amplitude of the space-charge field, as given by Eq. (3.1-35), can be written

$$E_1 = \frac{iK \dfrac{k_B T}{q}}{1 + \dfrac{K^2}{k_D^2}} \frac{I_1}{I_0} = \frac{iE_d}{1 + \dfrac{E_d}{E_q}} \frac{I_1}{I_0} \tag{3.2-3}$$

We now examine the dependence of E_1 on the grating period $\Lambda = 2\pi/K$, density of donors/acceptors N_A. First, we notice that the space-charge field is proportional to I_1/I_0. The factor i in Eq. (3.2-3) represents a spatial phase shift of $\pi/2$. This means that the electric field is spatially shifted by 1/4 of a period relative to the interference fringe pattern. This phase shift is illustrated in Figure 3.2. Such a spatial phase shift plays a key role in the energy exchange in two-wave mixing. To examine the dependence on grating period Λ (or grating wave number K), we plot the space-charge field E_1 as a function of Λ. According to Eq. (3.2-3), E_1 is always less than the smaller of E_d and E_q. Plotting as functions of grating period Λ, E_q is a straight line and E_d is a hyperbola (Figure 3.3). We note that BSO crystal has a much higher saturation field E_q by virtue of its smaller dielectric constant $\langle \epsilon \rangle = 56$. The space-charge field E_1 is always under these two curves even if $I_1/I_0 = 1$. We note that the space-charge field increases linearly as a function of Λ for small Λ and reaches a maximum at $\Lambda = \Lambda_D$ when $E_d = E_q$. For grating periods $\Lambda \gg \Lambda_D$, the space-charge field decreases like $1/\Lambda$. The maximum field is given by

$$E_{1\,\max} = \frac{1}{2} \left(\frac{N_A k_B T}{\langle \epsilon \rangle} \right)^{1/2} \frac{I_1}{I_0} \tag{3.2-4}$$

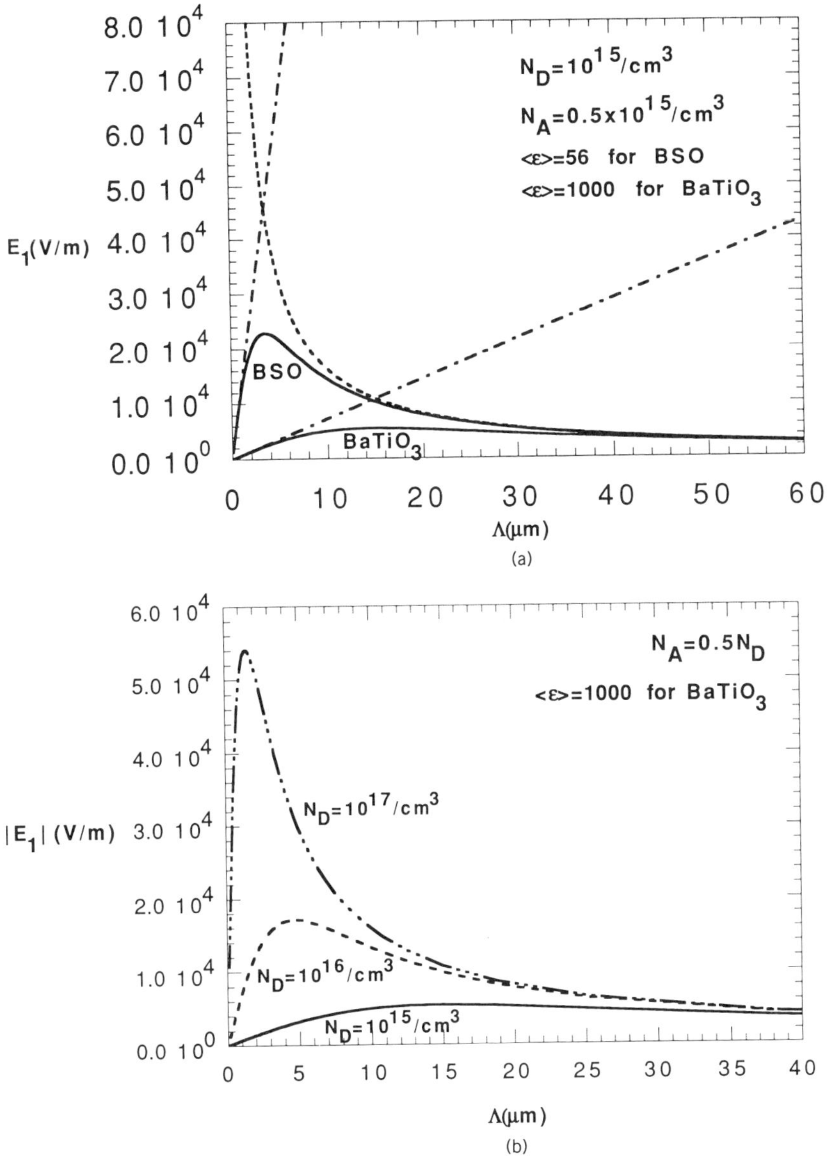

Figure 3.3. (a) Space-charge field E_1 versus grating period Λ for $BaTiO_3$ and BSO. (b) Space-charge field E_1 versus grating period Λ for various dopant densities.

We note that both the magnitude and the location of the maximum depend on dopant density N_A (or N^i_{D0}). The increase in the maximum space-charge field is significant in the regime where the grating periods are small.

According to Eqs. (3.2-1) and (3.2-2), we find that the diffusion field E_d is proportional to K, whereas the saturation field E_q is inversely proportional to K. Thus for larger grating spacing (small K), the space-charge field is dominated by diffusion (i.e., $E_d \ll E_q$) and we have

$$E_1 \simeq iE_d \frac{I_1}{I_0} \tag{3.2-5}$$

On the other hand, for small grating spacing (large K), the space-charge field is given by

$$E_1 \simeq iE_q \frac{I_1}{I_0} \tag{3.2-6}$$

We recall that E_1 is always less than the smaller of E_d and E_q. At $\Lambda = \Lambda_D$ (or $K = k_D$), $E_d = E_q$, E_1 reaches the maximum.

To get an idea about these physical parameters, we consider a typical example of a photorefractive crystal with $N_A = 10^{16}\ \text{cm}^{-3}$, $\langle \epsilon \rangle = 10\epsilon_0$ and a temperature of $T = 300\ \text{K}$. The maximum space-charge field $E_{1\,\text{max}} = 0.34 \times 10^6\ (\text{V/m})\ I_1/I_0$, according to Eq. (3.2-4).

b. $E_0 \neq 0$ (*With Applied Field*). In the case when there is an applied electric field, the amplitude of the space-charge field is given by Eq. (3.1-34). Using Eqs. (3.2-1) and (3.2-2), we can rewrite Eq. (3.1-34) as

$$E_1 = \frac{iE_d - E_0}{1 + \dfrac{E_d}{E_q} + i\dfrac{E_0}{E_q}} \frac{I_1}{I_0} \tag{3.2-7}$$

where we assume that the applied electric field is parallel to the grating wave vector. In this case the space-charge field E_1 is no longer limited by the smaller of E_d and E_q. To appreciate the effect of the applied field, the space-charge field is written as

$$E_1 = \frac{iE_d}{1 + \dfrac{E_d}{E_q}} \left[\frac{1 + i\left(\dfrac{E_0}{E_d}\right)}{1 + i\left(\dfrac{E_0}{E_d + E_q}\right)} \right] \frac{I_1}{I_0} \tag{3.2-8}$$

The first term on the right-hand side of the equation is simply the space-charge field in the absence of applied field. The term inside the square brackets represents a scaling factor due to the presence of a DC electric field E_0. This scaling factor is a complex number. Thus the applied electric field not only changes the magnitude of the space-charge field, but also alters its spatial phase.

By examining the scaling factor in Eq. (3.2-8), we note that the effect due to applied field E_0 is insignificant when $E_q \ll E_d$ which occurs, according to Eqs. (3.2-1) and (3.1-2), at large wave numbers (or smaller grating periods)

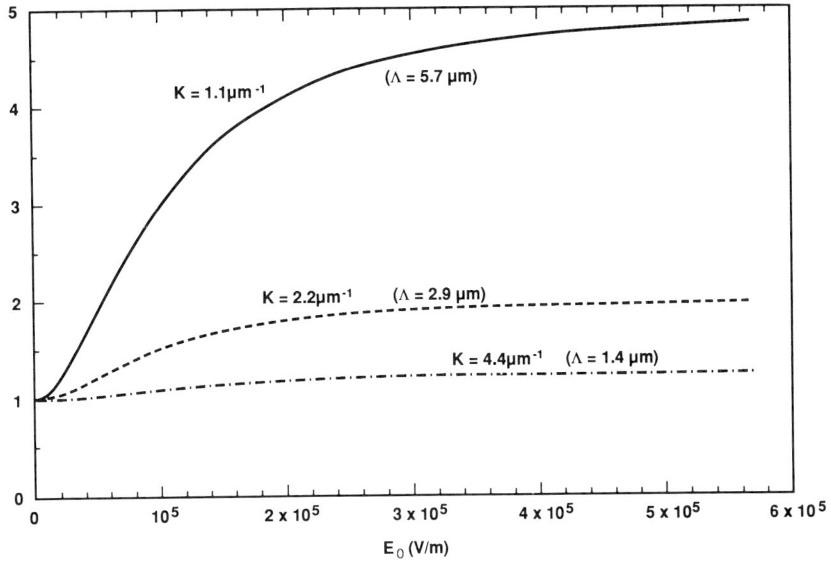

Figure 3.4. The scaling factor due to applied field E_0 in Eq. (3.2-8) is plotted as a function of E_0 for various values of grating periods Λ.

$$k_D^2 \ll K^2 \tag{3.2-9}$$

This can be seen in Figure 3.4, which plots the magnitude of the scaling factor as a function of the applied field E_0 for various values of K. Generally speaking, the externally applied field E_0 has a large effect on the space-charge field when $E_q \gg E_d$ and $E_0 \geq E_d$.

Before we consider an example, let us examine these two parameters E_d and E_q. In terms of the grating period Λ, they can be written

$$E_d = \frac{2\pi}{\Lambda} \frac{k_B T}{q} \tag{3.2-10}$$

and

$$E_q = \frac{qN_A}{2\pi\langle \epsilon \rangle} \Lambda \tag{3.2-11}$$

respectively. We notice that the diffusion field E_d is inversely proportional to the grating period, whereas the saturation field E_q is directly proportional to Λ.

Example: $BaTiO_3$ Consider the example of $BaTiO_3$ crystal with $N_A = 0.5 \times 10^{15}/\text{cm}^3$, $\Lambda = 5\ \mu\text{m}$, $\langle \epsilon \rangle = 1000\epsilon_0$, and a temperature of $T = 300$ K.

Using $k_B T = 26\,\text{meV}$ at room temperature, the diffusion field is, according to Eq. (3.2-10)

$$E_d = 33\,\text{kV/m}$$

The saturation field E_q is, according to Eq. (3.2-11)

$$E_q = 7.2\,\text{kV/m}$$

We note that E_q is much smaller than the diffusion field and the externally applied field will not increase the space-charge field. The space-charge field for the case when $I_1/I_0 = 1$, $E_0 = 0$, is given by $E_1 = 6\,\text{kV/cm}$, according to Eq. (3.28).

Example: BSO Consider the case of BSO crystal with the same acceptor density $N_A = 0.5 \times 10^{15}/\text{cm}^3$. The dielectric constant of BSO is $\epsilon = 56\epsilon_0$. With the same grating period Λ and the same temperature, the saturation field is

$$E_q = 126\,\text{kV/m}$$

We note that E_q is much larger than the diffusion field and the externally applied field can increase the space-charge field. This is left as an exercise for the reader (see also Problem 3.1).

3.3 PHOTOREFRACTIVE VOLUME GRATINGS

All photorefractive materials are electro-optic crystals. The presence of an electric field in these media will introduce a change in the index of refraction via the linear electro-optic effect (Pockels' effect):

$$\Delta\left(\frac{1}{n^2}\right) = r_{ijk} E_k^{sc} \tag{3.3-1}$$

where r_{ijk} is the electro-optic coefficients and E_k^{sc} is kth component of the space-charge field. Summation over $k = 1, 2, 3$ (or x, y, z) in Eq. (3.3-1) is assumed. The space-charge field is written

$$\mathbf{E}^{sc} = \text{Re}\{\mathbf{E}_1 e^{-i\mathbf{K}\cdot\mathbf{r}}\} \tag{3.3-2}$$

with $\mathbf{E}_1$ given by Eqs. (3.1-34) and (3.1-35), as well as various forms of expression in the last section. According to Eqs. (3.3-1) and (3.3-2), a volume index grating is formed when a photorefractive crystal is illuminated by two beams of coherent light.

To illustrate the use of these two equations, we consider in the following several interesting examples.

3.3.1 Photorefractive Grating in $LiNbO_3$

Referring to Figure 3.5, we consider a two-beam interference such that a photorefractive grating is formed with its grating wave vector parallel to the c-axis of the crystal. Let the space-charge field be written

$$\mathbf{E}^{sc} = \begin{bmatrix} 0 \\ 0 \\ 1 \end{bmatrix} E^{sc} \cos Kz \tag{3.3-3}$$

where the z-axis is taken along the c-axis of the crystal and E^{sc} is the amplitude of the space-charge field. The crystal has a point group symmetry of $3m$ and its electro-optic coefficients are in the form (see Table 1.1)

$$\begin{bmatrix} 0 & -r_{22} & r_{13} \\ 0 & r_{22} & r_{13} \\ 0 & 0 & r_{33} \\ 0 & r_{51} & 0 \\ r_{51} & 0 & 0 \\ -r_{22} & 0 & 0 \end{bmatrix} \tag{3.3-4}$$

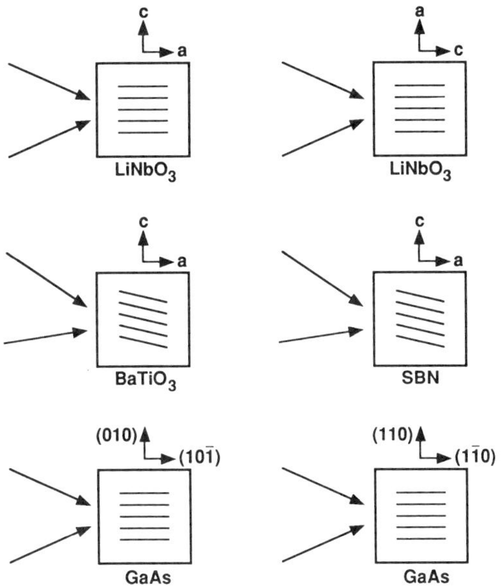

Figure 3.5. Photorefractive index gratings in various crystals.

The volume index grating is thus written, according to Eq. (3.3-1)

$$\Delta\left(\frac{1}{n^2}\right)_{ij} = \begin{bmatrix} r_{13} & 0 & 0 \\ 0 & r_{13} & 0 \\ 0 & 0 & r_{33} \end{bmatrix} E^{sc}\cos(Kz) \qquad (3.3\text{-}5)$$

If, instead, the grating wave vector is parallel to the x-axis, the space-charge field is written

$$\mathbf{E}^{sc} = \begin{bmatrix} 1 \\ 0 \\ 0 \end{bmatrix} E^{sc}\cos(Kx) \qquad (3.3\text{-}6)$$

The index grating in this case becomes

$$\Delta\left(\frac{1}{n^2}\right)_{ij} = \begin{bmatrix} 0 & -r_{22} & r_{51} \\ -r_{22} & 0 & 0 \\ r_{51} & 0 & 0 \end{bmatrix} E^{sc}\cos(Kx) \qquad (3.3\text{-}7)$$

We note that the index gratings as given by Eqs. (3.3-5) and (3.5-7) are, in general, tensors that contain off-diagonal elements. This means that the effective amplitude of the index grating seen by an incident wave may depend on the polarization state of the light.

Let $\mathbf{p}$ be the polarization state of the light and the scattering does not involve a change of polarization. The index grating which is responsible for the scattering is

$$\Delta\left(\frac{1}{n^2}\right) = \left\langle \mathbf{p} \middle| \Delta\left(\frac{1}{n^2}\right) \middle| \mathbf{p} \right\rangle \qquad (3.3\text{-}8)$$

which is the scalar product of $\mathbf{p}$ and $\Delta(1/n^2)\mathbf{p}$ and is physically the expectation value of the tensor $\Delta(1/n^2)_{ij}$. Equation (3.3-8) can also be written

$$\Delta n = -\frac{1}{2} n^3 \left\langle \mathbf{p} \middle| \Delta\left(\frac{1}{n^2}\right) \middle| \mathbf{p} \right\rangle \qquad (3.3\text{-}9)$$

where n is the index of refraction associated with the polarization state $\mathbf{p}$. Thus, if the light is polarized along the y-axis and the grating is described by Eq. (3.3-5), then the index grating can be written

$$\Delta n = -\tfrac{1}{2} n^3 r_{13} E^{sc}\cos(Kz) \qquad (3.3\text{-}10)$$

In the case when there is a change of polarization state, the index grating should be expressed in terms of $\Delta\epsilon$ (as in Eq. (1.4-2)) for the purpose of studying the mode coupling. The change in the dielectric constant is written

$$\Delta\epsilon = -\frac{1}{\epsilon_0}\,\epsilon\,\Delta\eta\,\epsilon \qquad (3.3\text{-}11)$$

where $\Delta\eta = \Delta(1/n^2)_{ij}$ is the impermeability tensor. In the principal coordinate in which ϵ is diagonal, the change can be written

$$\Delta\epsilon_{ij} = -\epsilon_0 n_i^2 n_j^2 \, \Delta\left(\frac{1}{n^2}\right)_{ij} \tag{3.3-12}$$

where n_i, n_j are the principal indices of refraction. The coupling constants for Bragg scattering with a change of polarization state are given by Eqs. (2.6-7) and (2.6-8).

Example: Photorefractive Index Grating in $LiNbO_3$ Consider the photorefractive index grating in a $LiNbO_3$ crystal. Assuming a grating period of $\Lambda = 1.85\ \mu m$ along the direction of the c-axis, the space-charge field is approximately $E_1 = 0.86$ kV/cm for a modulation depth of $I_1/I_0 = 0.27$ [10]. The index grating for the extraordinary wave is given by

$$\Delta n = -\tfrac{1}{2} n^3 r_{33} E_1 \cos(Kz)$$

Using $r_{33} = 31 \times 10^{-12}$ m/V and $n = 2.2$ from Table 1.2, we obtain an index modulation of

$$\Delta n = -1.4 \times 10^{-5} \cos(Kz)$$

3.3.2 Photorefractive Gratings in $BaTiO_3$, SBN

Both $BaTiO_3$ and SBN crystals have a point group symmetry of $4mm$ at room temperature. Referring to Figure 3.5, we consider a two-beam interference inside a photorefractive crystal with symmetry of $4mm$. The grating wave vector is in the xz-plane with an angle of θ measured from the c-axis. Let the space-charge field be written

$$\mathbf{E}^{sc} = \begin{pmatrix} \sin\theta \\ 0 \\ \cos\theta \end{pmatrix} E^{sc} \cos(\mathbf{K}\cdot\mathbf{r}) \tag{3.3-13}$$

where $\mathbf{K}$ is the grating wave vector and E^{sc} is the amplitude of the space-charge field. We recall that the space-charge field is parallel to the grating wave vector $\mathbf{K}$.

Using the results obtained in the examples in Section 1.4, we obtain the following expression for the index grating

$$\Delta\left(\frac{1}{n^2}\right)_{ij} = \begin{bmatrix} r_{13}\cos\theta & 0 & r_{42}\sin\theta \\ 0 & r_{13}\cos\theta & 0 \\ r_{42}\sin\theta & 0 & r_{33}\cos\theta \end{bmatrix} E^{sc} \cos(\mathbf{K}\cdot\mathbf{r}) \tag{3.3-14}$$

3.3.3 Photorefractive Gratings in Cubic Crystals

Cubic crystals such as GaAs and BSO have point group symmetry of $\bar{4}3m$ and 23, respectively. Their electro-optic coefficients are in the form

$$\begin{bmatrix} 0 & 0 & 0 \\ 0 & 0 & 0 \\ 0 & 0 & 0 \\ r_{41} & 0 & 0 \\ 0 & r_{41} & 0 \\ 0 & 0 & r_{41} \end{bmatrix} \tag{3.3-15}$$

Let an arbitrary space-charge field be written

$$\mathbf{E}^{sc} = \begin{bmatrix} E_x \\ E_y \\ E_z \end{bmatrix} \cos(\mathbf{K} \cdot \mathbf{r}) \tag{3.3-16}$$

The change in the impermeability tensor due to the space-charge field is

$$\Delta\left(\frac{1}{n^2}\right)_{ij} = \begin{bmatrix} 0 & E_z & E_y \\ E_z & 0 & E_x \\ E_y & E_x & 0 \end{bmatrix} r_{41} \cos(\mathbf{K} \cdot \mathbf{r}) \tag{3.3-17}$$

Let the space-charge field $\mathbf{E}^{sc}$ and the grating wave vector $\mathbf{K}$ be parallel to the (0 1 0) direction and the polarization state of the light be along the (1 0 1) direction. The index grating which is responsible for the Bragg scattering can be obtained by using Eq. (3.3-9) and is given by

$$\Delta n = -\tfrac{1}{2} n^3 r_{41} E^{sc} \cos(\mathbf{K} \cdot \mathbf{r}) \tag{3.3-18}$$

If the crystal is oriented such that the space-charge field is parallel to the (1 0 1) direction, then the index grating Eq. (3.3-17) can be written

$$\Delta\left(\frac{1}{n^2}\right)_{ij} = \frac{1}{\sqrt{2}} \begin{bmatrix} 0 & 1 & 0 \\ 1 & 0 & 1 \\ 0 & 1 & 0 \end{bmatrix} E^{sc} r_{41} \cos(\mathbf{K} \cdot \mathbf{r}) \tag{3.3-19}$$

or

$$\Delta\epsilon = -\epsilon_0 \frac{1}{\sqrt{2}} \begin{bmatrix} 0 & 1 & 0 \\ 1 & 0 & 1 \\ 0 & 1 & 0 \end{bmatrix} n^4 E^{sc} r_{41} \cos(\mathbf{K} \cdot \mathbf{r}) \tag{3.3-20}$$

By examining the matrix elements in Eqs. (3.3-19) and (3.3-20), we notice that an incident light with a polarization state $\mathbf{p}$ which is parallel to $(1\,0\,\bar{1})$ will not be scattered by the grating because $\Delta\epsilon\,\mathbf{p} = 0$. On the other hand, if the incident polarization state $\mathbf{p}$ is parallel to (0 1 0), the scattered polarization state $\mathbf{p}'$ must have a finite component along the (1 0 1) direction. The dynamic condition for a scattering to occur is

$$\mathbf{p}' \cdot \Delta\epsilon\,\mathbf{p} \neq 0 \tag{3.3-21}$$

where $\mathbf{p}'$ is the polarization state of the scattered wave.

3.4 SPEED OF GRATING FORMATION

The photorefractive effect is a macroscopic phenomenon. The formation of a grating with significant diffraction efficiency requires the excitation and transport of a large number of charge carriers [4]. Based on the model described in Section 3.1, the generation of charge carriers (e.g., electrons) is a prerequisite of the photorefractive effect. Because of the need to absorb photons and to create charge carriers, the speed of the photorefractive effect is fundamentally limited by the rate of photon arrival (i.e., the photon flux).

Instead of finding the transient solutions by solving the coupled differential equations (3.1-1)–(3.1-4), we examine the minimum time required for the formation of an index grating in photorefractive crystals from the point of view of energy deposition needed for the photoexcitation. This approach will yield a fundamental limit for the speed of the photorefractive effect [5].

Assuming that the separation of a pair of charge particles requires the absorption of at least one photon, we can calculate the energy required to form a given volume index grating. To illustrate this, let us consider the photorefractive effect in $BaTiO_3$. Generally speaking, an efficient beam coupling would require a charge carrier density of approximately 10^{16} cm^{-3}. Such a charge separation would require the absorption of at least 10^{16} photons in a volume of 1 cm^3. Using a power of 1 W in the visible spectrum, the photon flux is approximately 10^{19}/s. Thus, assuming a photoexcitation absorption of 10%, it takes at least 10 ms just to deposit enough photons to create the charge separation.

To investigate this energy deposition time quantitatively, let us consider a simple model in which the charge density is a sinusoidal function given by

$$\rho = \rho_0 \cos(Kx) \tag{3.4-1}$$

where ρ_0 is a constant and K is the grating wave number ($K = 2\pi/\Lambda$, Λ is the grating period) and is related to the wave vectors $\mathbf{k}_1$ and $\mathbf{k}_2$ of the two beams by

$$|\mathbf{k}_2 - \mathbf{k}_1| = K = 2k \sin(\theta) \tag{3.4-2}$$

where 2θ is the angle between the beams and k is the wave number of light. In order to create a space-charge density given by Eq. (3.4-1), the number of charge particle pairs separated per unit volume is given by

$$N = \frac{\rho_0}{\pi q} \tag{3.4-3}$$

where q is the electronic charge.

Assuming a quantum efficiency of η, the number of photoexcitations per unit volume required is given by

$$N_p = \frac{N}{\eta} \tag{3.4-4}$$

Here, we note that only a fraction of the photoexcitation leads to a useful charge separation for the creation of a periodic space-charge field.

The space-charge field generated by the charge density function (3.4-1) is governed by the Poisson equation and can be written

$$E = \frac{\rho_0}{\epsilon K} \sin(Kx) \tag{3.4-5}$$

where ϵ is the relevant dielectric constant. In anisotropic crystals, this dielectric constant depends on the direction of the electric field relative to the crystal orientation. The space-charge field will induce a volume index grating via the Pockels effect. Such an index grating can be written as

$$\Delta n = n_1 \sin(Kx) \tag{3.4-6}$$

where the depth of index modulation n_1 is related to the electro-optic coefficient and can be written

$$n_1 = \frac{1}{2} n^3 r \frac{\rho_0}{\epsilon K} \tag{3.4-7}$$

where n is the relevant refractive index and r is the effective electro-optic coefficient. Here, we lumped all the directional cosines into the effective electro-optic coefficient r.

The coupling constant γ for Bragg scattering is related to the depth of index modulation n_1 by

$$\gamma = \frac{2\pi}{\lambda} n_1 \tag{3.4-8}$$

and can be written as

$$\gamma = \frac{\Lambda}{2\lambda} n^3 r \frac{\rho_0}{\epsilon} \tag{3.4-9}$$

where we have used $K = 2\pi/\Lambda$. The parameter γ defined above is an intensity coupling constant. It is different from κ defined in Eq. (2.2-16) by a factor of 2. In Eq. (3.4-8) we assume that the Bragg angle is small so that $\cos\theta_B \simeq 1$. The intensity coupling constant γ is an important measure of the photorefractive effect and plays an important role in two-wave mixing.

We now calculate the time required to generate N_p photoexcitations in a unit volume. Consider a photorefractive medium with an area of illumination A and a length of interaction L. Let P be the total power of illumination and α_p be the photoexcitation absorption coefficient, the time required for the energy deposition is given by

$$t = \frac{h\nu}{I\alpha_p} N_p \tag{3.4-10}$$

where $h\nu$ is the photon energy and $I = P/A$ is the intensity of light. In arriving at Eq. (3.4-10), we assume that $\alpha_p L \ll 1$. Using Eqs. (3.4-3), (3.4-4), and (3.4-9), such a time can be written

$$t = \left(\frac{h\nu}{q}\right)\left(\frac{\lambda}{\Lambda}\right)\left(\frac{\gamma}{\alpha_p}\right)\frac{2}{\pi\eta}\,\frac{\epsilon}{In^3 r} \tag{3.4-11}$$

where we recall that I is the intensity of light. Note that the time constant is directly proportional to the coupling constant γ and is inversely proportional to the light intensity. Equation (3.4-11) is the expression for the minimum time required for the formation of an index grating which provides a coupling constant γ.

The minimum time t, as given by Eq. (3.4-11), is different from the time constant that is described in Section 3.6. As we will see in that section, the growth of the index grating in the transient regime is approximately given by

$$\gamma = \gamma_s(1 - e^{-t/\tau}) \tag{3.4-12}$$

where γ_s is the saturated coupling constant and τ is the exponential time constant. In other words, at $t = \tau$, a coupling constant of $\gamma_s(1 - e^{-1})$ is obtained. According to Eq. (3.4-12), the coupling constant γ, in the time regime $t \ll \tau$, is given by

$$\gamma = \gamma_s \frac{t}{\tau} \tag{3.4-13}$$

According to this result, the minimum time t is related to the exponential time constant by

$$t = \frac{\gamma}{\gamma_s}\tau \tag{3.4-14}$$

A figure-of-merit for photorefractive material is defined as

$$Q = \frac{n^3 r}{\epsilon} \tag{3.4-15}$$

Table 3.1 lists such parameters for some photorefractive materials. Using such a parameter, the photon-limited time for the index grating formation becomes

$$t = \left(\frac{h\nu}{q}\right)\left(\frac{\lambda}{\Lambda}\right)\left(\frac{\gamma}{\alpha_p}\right)\frac{2}{\pi\eta IQ} \tag{3.4-16}$$

Here, we note that this photon-limited time is inversely proportional to the material's figure-of-merit and is proportional to the coupling constant γ.

Table 3.1 Figure-of-Merit for Some Photorefractive Materials [5]

Materials	λ (μm)	r (pm/V)	n	ϵ/ϵ_0	Q^a (pm/Vϵ_0)	Q (MKS)
$BaTiO_3$	0.5	$r_{42} = 1640$	$n_e = 2.4$	$\epsilon_1 = 3600$	6.3	0.71
SBN	0.5	$r_{33} = 1340$	$n_e = 2.3$	$\epsilon_3 = 3400$	4.8	0.54
GaAs	1.1	$r_{12} = 1.43$	$n_e = 3.4$	$\epsilon_3 = 12.3$	4.7	0.53
BSO	0.6	$r_{41} = 5$	$n = 2.54$	$\epsilon = 56$	1.5	0.17
$LiNbO_3$	0.6	$r_{33} = 31$	$n_e = 2.2$	$\epsilon_3 = 32$	10.3	1.16
$LiTaO_3$	0.6	$r_{33} = 31$	$n_e = 2.2$	$\epsilon_3 = 45$	7.3	0.83
$KNbO_3$	0.6	$r_{42} = 380$	$n = 2.3$	$\epsilon_3 = 240$	19.3	2.2
GaP	0.56	$r_{41} = 1.07$	$n = 3.45$	$\epsilon = 12$	3.7	0.41

[a]The figure-of-merit Q depends on the configuration of interaction.

Consider now an example of this minimum time for the formation of an index grating which yields a coupling constant of 10 cm^{-1}. For materials such as $BaTiO_3$, SBN, BSO and GaAs, the figure-of-merit Q is of the order of 1 (see Table 3.1) in MKS units (M^2/FV). In many of the experiments reported recently, $h\nu$ is approximately 2 eV, λ/Λ is of the order 0.1. We further assume that the photoexcitation absorption coefficient is 1 cm^{-1} and that the quantum efficiency η is 1. Using these parameters and a light intensity of 1 W/cm^2, Eq. (3.4-13) yields a minimum time of 0.15 ms. This is a minimum time required for the formation of an index grating which can provide a two-wave mixing coupling constant of 10 cm^{-1}. By virtue of its photoexcitation nature, photorefractive effect is slow at low intensity because of the finite time required to absorb the photons. Table 3.2 shows the

Table 3.2 Comparison of Measured Time Constants and the Fundamental Limit

Materials	λ (μm)	Λ (μm)	α_p (cm^{-1})	γ (cm^{-1})	τ (s)[a]	t (s)[b]
GaAs	1.06	1.0	1.2	0.4	80×10^{-6}	45×10^{-6}
GaAs:Cr	1.06	1.1	4.0	0.6	53×10^{-6}	31×10^{-6}
$BaTiO_3$	0.515	1.3	1.0	20.0	1.3	2×10^{-3}
BSO	0.568	23.0	0.13	10.0	15×10^{-3}	2×10^{-3}
GaP	0.633	1.0	2.0	0.3	3×10^{-4}	3×10^{-5}
SBN (undoped)	0.5145	2.0	0.1	7.0	8.4	5×10^{-3}
SBN:60:Ce	0.5145	1.5	0.6	12.0	2.85	2×10^{-3}
SBN:75:Ce	0.5145	3.0	0.4	11.4	30.0	1×10^{-3}
SBN:60:Cr	0.5145	2.95	0.45	4.1	0.078	5×10^{-4}
SBN:75:Cr	0.5145	9.83	0.6	9.1	2.13	2×10^{-4}
SBN:60:Rh (0.015%)	0.5145	1.07	0.54	4.4	1.2	1×10^{-3}
SBN:75:Rh	0.5145	5.90	0.41	5.0	2.70	3×10^{-4}

[a]τ and t are time constants at incident intensity 1 W/cm^2.
[b]t is the calculated time constant by using Eq. (3.4-16) and assuming a quantum efficiency of 1.

comparison between the measured time constants with the calculated minimum time from Eq. (3.4-16).

3.5 MOVING GRATINGS

Consider the intersection of two laser beams with different frequencies inside a photorefractive medium. Let the frequencies be ω_a and ω_b. The intensity of the interference pattern can be written

$$I = I_0 + \mathrm{Re}\{I_1 e^{i(\Omega t - \mathbf{K}\cdot\mathbf{r})}\} \tag{3.5-1}$$

where I_0 and I_1 are given by Eqs. (3.1-8) and (3.1-9) and

$$\Omega = \omega_b - \omega_a \tag{3.5-2}$$

$$\mathbf{K} = \mathbf{K}_b - \mathbf{K}_a \tag{3.5-3}$$

The intensity as described by Eq. (3.5-1) represents a traveling fringe pattern which moves along the direction of **K** with a speed of

$$v = \frac{\Omega}{|\mathbf{K}|} \tag{3.5-4}$$

We have mentioned in the last section that the formation of a volume grating with a finite index modulation requires a finite time to generate and trap charge carriers. The motion of the fringe pattern will cause a similar motion for the space-charge density. Because of the finite formation time, the space-charge density and the index grating may not be able to catch up with the motion of the intensity pattern if the intensity pattern moves too fast. In this section, we investigate the formation of a moving index grating inside photorefractive media.

By using a similar assumption that $I_1 \ll I_0$, we may write steady state solutions as

$$N = N_0 + \mathrm{Re}\{N_1 e^{i(\Omega t - \mathbf{K}\cdot\mathbf{r})}\} \tag{3.5-5}$$

$$N_D^i = N_{D0}^i + \mathrm{Re}\{N_{D1}^i e^{i(\Omega t - \mathbf{K}\cdot\mathbf{r})}\} \tag{3.5-6}$$

$$\mathbf{j} = \mathbf{j}_0 + \mathrm{Re}\{\mathbf{j}_1 e^{i(\Omega t - \mathbf{K}\cdot\mathbf{r})}\} \tag{3.5-7}$$

$$\mathbf{E} = \mathbf{E}_0 + \mathrm{Re}\{\mathbf{E}_1 e^{i(\Omega t - \mathbf{K}\cdot\mathbf{r})}\} \tag{3.5-8}$$

Substituting these equations into Eqs. (3.1-1)–(3.1-4), we obtain

$$i\Omega N_{D1}^i = sI_1(N_D - N_{D0}^i) + sI_0(-N_{D1}^i) - \gamma_R N_1 N_{D0}^i - \gamma_R N_0 N_{D1}^i \tag{3.5-9}$$

$$i\Omega N_1 - i\Omega N_{D1}^i = -i\mathbf{K}\cdot\mu\mathbf{E}_0 N_1 - iN_0\mathbf{K}\cdot\mu\mathbf{E}_1 - \frac{k_B T}{q} N_1\mathbf{K}\cdot\mu\mathbf{K} \tag{3.5-10}$$

$$-i\mathbf{K}\cdot\epsilon\mathbf{E}_1 = -q(N_1 - N_{D1}^i) \tag{3.5-11}$$

$$0 = sI_0(N_D - N_{D0}^i) - \gamma_R N_0 N_{D0}^i \tag{3.5-12}$$

$$0 = N_0 + N_A - N_{D0}^i \tag{3.5-13}$$

The last two equations are identical to Eqs. (3.1-20) and (3.1-21) and the solutions are given by Eqs. (3.1-22) and (3.1-24). Using Eq. (3.5-11), we can eliminate $\mathbf{E}_1$ from Eq. (3.5-10). This leads to two linear equations for N_1 and N_{D1}^i. Solving these equations for N_1 and N_{D1}^i, we obtain the space-charge field E_1, according to Eq. (3.5-11)

$$E_1 = \frac{iE_d}{1 + \dfrac{E_d}{E_q}} \; \frac{1 + \dfrac{iE_0}{E_d}}{1 + \dfrac{iE_0}{E_d + E_q}} \; \frac{1}{1 + i\Omega t_0 \dfrac{E_d + E_\mu + iE_0}{E_d + E_q + iE_0}} \left(\frac{I_1}{I_0}\right) \tag{3.5-14}$$

where the diffusion and saturation fields E_d, E_q are given by Eqs. (3.2-1) and (3.2-2), E_μ is another characteristic field (the so-called drift field) given by

$$E_\mu = \frac{\gamma_R N_A}{\langle \mu \rangle K} \tag{3.5-15}$$

and t_0 is a characteristic time constant of the medium given by

$$t_0 = \frac{N_A}{N_D s I_0} \tag{3.5-16}$$

In arriving at the result (3.5-14), we assume that the applied field $\mathbf{E}_0$ is parallel to the grating wave vector $\mathbf{K}$. In addition, we also assume that $N_0 \ll N_A \ll N_D$ as well as $N_D s I_0 \ll \gamma_R N_A^2$.

Examining the result, we notice that the motion of the intensity pattern not only decreases the amplitude of the space-charge field, but also introduces a phase shift between the grating and the intensity pattern. We consider the case when there is no applied field (i.e., $E_0 = 0$). The amplitude of the space-charge field becomes

$$E_1 = \frac{iE_d}{1 + \dfrac{E_d}{E_q}} \; \frac{1}{1 + i\Omega\tau} \; \frac{I_1}{I_0} \tag{3.5-17}$$

where τ is the photorefractive grating decay constant given by

$$\tau = t_0 \frac{E_d + E_\mu}{E_d + E_q} \tag{3.5-18}$$

Such an amplitude can also be written

$$E_1 = |E_1| e^{i\phi} \tag{3.5-19}$$

with

$$|E_1| = \frac{E_d}{1 + \dfrac{E_d}{E_q}} \left(\frac{1}{1 + (\Omega\tau)^2} \right)^{1/2} \left| \frac{I_1}{I_0} \right| \tag{3.5-20}$$

$$\phi = \frac{\pi}{2} - \tan^{-1}(\Omega t) \tag{3.5-21}$$

We note that as a result of the motion, an additional phase shift of $-\tan^{-1}(\Omega\tau)$ is introduced between the intensity pattern and the space-charge field. The amplitude of the space-charge field is also dropped by a factor of $[1 + (\Omega\tau)^2]^{-1/2}$.

3.6 TRANSIENT SOLUTIONS OF SPACE-CHARGE FIELDS

To understand the formation of a photorefractive grating, we need to solve for the transient solutions in the band transport equations. In the previous sections, we have discussed the steady state space-charge field and the fundamental limit of grating formation. The steady state solution (Eq. (3.1-30)) describes the amplitude and phase of the grating. The fundamental speed limit (Eq. (3.4-11)) describes the minimum time required for a grating to grow to a certain strength. However, to see how the space-charge field builds up with time before reaching the steady state, we need to consider not only the rate of photoexcitation but also other processes, such as the rate of recombination and the speed of charge transportation. All these mechanisms are included in the band transport model described by Eqs. (3.1-1)–(3.1-4).

In this section, we solve the time-dependent Eqs. (3.1-1)–(3.1-4) for the case of low modulation depth (i.e., $|I_1| \ll I_0$). In this case, the nonlinear coupled differential equations can be linearized in a similar way as the linearization used in Section 3.1. These linearized first order differential equations can then be solved with standard methods to obtain transient solutions.

We start by linearizing the band transport equations. Similar to the steady state linearization (3.1-16)–(3.1-19), the time-dependent solutions can be written as

$$N(\mathbf{r}, t) = N_0(t) + \mathrm{Re}\{N_1(t) e^{-i\mathbf{K}\cdot\mathbf{r}}\} \tag{3.6-1}$$

$$N_D^i(\mathbf{r}, t) = N_{D0}^i(t) + \mathrm{Re}\{N_{D1}^i(t) e^{-i\mathbf{K}\cdot\mathbf{r}}\} \tag{3.6-2}$$

$$\mathbf{j}(\mathbf{r}, t) = \mathbf{j}_0(t) + \mathrm{Re}\{\mathbf{j}_1(t)e^{-i\mathbf{K}\cdot\mathbf{r}}\} \tag{3.6-3}$$

$$\mathbf{E}(\mathbf{r}, t) = \mathbf{E}_0(t) + \mathrm{Re}\{\mathbf{E}_1(t)e^{-i\mathbf{K}\cdot\mathbf{r}}\} \tag{3.6-4}$$

where we have assumed that the intensity modulation is small, that is, $|I_1| \ll I_0$, so that the higher harmonic terms can be neglected. For simplicity, we will further assume that the grating wave vector, the electric field and the current are all in the z-direction. Therefore, in the following discussions, we will treat $E(z, t)$ and $j(z, t)$ as scalars. By substituting Eqs. (3.6-1)–(3.6-4) in Eqs. (3.1-1)–(3.1-4) and separating zeroth order terms and first order terms, we can derive the following set of equations:

$$\frac{dN^i_{D0}}{dt} = (N_D - N^i_{D0})sI_0 - \gamma_R N^i_{D0} N_0 \tag{3.6-5}$$

$$\frac{dN_0}{dt} = \frac{dN^i_{D0}}{dt} \tag{3.6-6}$$

$$j_0 = q\mu N_0 E_0 \tag{3.6-7}$$

$$0 = N^i_{D0} - N_0 - N_A \tag{3.6-8}$$

and

$$\frac{d(N^i_{D1})}{dt} = -(\gamma_R N_0 + sI_0)N^i_{D1} - \gamma_R N^i_{D0} N_1 + (N_D - N^i_{D0})sI_1 \tag{3.6-9}$$

$$\frac{d(N_1)}{dt} = \frac{d(N^i_{D1})}{dt} - \frac{iK}{q} j_1 \tag{3.6-10}$$

$$j_1 = q\mu(E_0 N_1 + N_0 E_1) - iKk_B T\mu N_1 \tag{3.6-11}$$

$$-iK\epsilon E_1 = q(N^i_{D1} - N_1) \tag{3.6-12}$$

where ϵ is the effective dielectric constant $\langle \epsilon \rangle$, and μ is the effective mobility $\langle \mu \rangle$. We notice that the first order equations are linear in terms of N^i_{D1}, N_1, j_1, and E_1. In what follows, we first solve the zeroth order equations, then the first order ones.

The zeroth order equations describe the time dependence of carrier and ionized trap densities upon uniform illumination. When there is an applied dc field E_0, $E_0(t)$ in Eq. (3.6-7) is a constant, that is,

$$E_0(t) = E_0 \tag{3.6-13}$$

By making the approximations $sI_0 \ll \gamma_R N_A$ (i.e., electron recombination rate is much larger than generation rate) and $N_0 \ll N_A$, we obtain the solutions

$$N^i_{D0}(t) = N_A \tag{3.6-14}$$

$$N_0(t) = N_0(1 - e^{-t/\tau_R}) \tag{3.6-15}$$

where

$$N_0 = \frac{(N_D - N_A)sI_0}{\gamma_R N_A} \tag{3.6-16}$$

and

$$\tau_R = \frac{1}{\gamma_R N_A} \tag{3.6-17}$$

The recombination time constant τ_R is less than 100 ns in most of the photorefractive crystals. Thus, electron density $N_0(t)$ reaches its steady state (N_0 in Eq. (3.6-16)) in a very short time compared with the time constant for grating formation. Therefore, we can use the steady state solutions, Eqs. (3.6-14) and (3.6-16), to evaluate the coefficients in Eqs. (3.6-9)–(3.6-12).

Now we solve the first order equations for the formation of the photorefractive grating. Substituting Eq. (3.6-11) and Eq. (3.6-12) in Eq. (3.6-10), we obtain

$$\frac{d(N_1 - N^i_{D1})}{dt} = \left(-i\mu E_0 K - \frac{k_B T\mu K^2}{q} - \frac{q\mu N_0}{\epsilon}\right)N_1 + \frac{q\mu N_0}{\epsilon} N^i_{D1} \tag{3.6-18}$$

To solve the space-charge field $E_1(t)$, we first solve Eq. (3.6-9) and Eq. (3.6-18) to obtain $N_1 - N^i_{D1}$, then substitute it into Eq. (3.6-12). If we assume that

$$sI_0 \ll \gamma_R N_A \tag{3.6-19}$$

and

$$q\mu N_0/\epsilon \ll \gamma_R N_A \tag{3.6-20}$$

the final solution for the space-charge field can be written as

$$E_1(t) = E_1[1 - e^{-t/\tau}] \tag{3.6-21}$$

where E_1 is the steady state ($t = \infty$) space-charge field given by

$$E_1 = \frac{I_1}{I_0} \frac{i\dfrac{k_B TK}{q} - E_0}{1 + \dfrac{K^2}{k_D^2} + i\dfrac{qKE_0}{k_B Tk_D^2}} \tag{3.6-22}$$

and the complex time constant is

$$\tau = \frac{\epsilon}{q\mu N_0} \frac{1 + \dfrac{k_B T\mu K^2}{q\gamma_R N_A} + \dfrac{i\mu K E_0}{\gamma_R N_A}}{1 + \dfrac{K^2}{k_D^2} + i\dfrac{qKE_0}{k_B T k_D^2}} \tag{3.6-23}$$

In the above expressions, we recall that k_D is the Debye wave number given by

$$k_D^2 = \frac{q^2 N_A(N_D - N_A)}{k_B T\epsilon N_D} \tag{3.6-24}$$

Equations (3.6-22) and (3.6-23) can also be written in the following form:

$$E_1 = \frac{iE_d - E_0}{1 + \dfrac{E_d}{E_q} + i\dfrac{E_0}{E_q}} \frac{I_1}{I_0} \tag{3.6-25}$$

and

$$\tau = t_0 \frac{E_d + E_\mu + iE_0}{E_d + E_q + iE_0} \tag{3.6-26}$$

where E_0 is the applied field and we recall that the characteristic fields are

$$E_d = \frac{k_B TK}{q} \tag{3.6-27}$$

$$E_q = \frac{qN_A(N_D - N_A)}{\epsilon K N_D} \tag{3.6-28}$$

$$E_\mu = \frac{\gamma_R N_A}{\mu K} \tag{3.6-29}$$

and the characteristic time constant t_0 is given by

$$t_0 = \frac{N_A}{N_D s I_0} \tag{3.6-30}$$

From Eqs. (3.6-25) and (3.6-26), we notice that E_1 is the same as Eq. (3.2-7), and the complex time constant τ is the same as that appeared in Eq. (3.5-14).

The real part of Eq. (3.6-21) is the transient solution of the space-charge field. This solution can be explicitly written as

$$E_{sc}(z, t) = E_{sc}[-\cos(Kz + \phi) + \exp(-t/\tau_g)\cos(Kz + \omega_g t + \phi)] \tag{3.6-31}$$

In the above equation E_{sc} is the amplitude of the steady state space-charge field,

$$E_{sc} = \frac{I_1}{I_0} E_q \left[\frac{E_0^2 + E_d^2}{E_0^2 + (E_d + E_q)^2} \right]^{1/2} \tag{3.6-32}$$

ϕ is the phase shift between the steady state space-charge field and the intensity fringes,

$$\phi = \tan^{-1} \frac{(E_d + E_q)E_d + E_0^2}{E_0 E_q} \tag{3.6-33}$$

τ_g is the time constant for the space-charge field to reach its steady state,

$$\tau_g = \tau_{di} \frac{\left(1 + \frac{E_d}{E_\mu}\right)^2 + \left(\frac{E_0}{E_\mu}\right)^2}{\left(1 + \frac{E_d}{E_q}\right)\left(1 + \frac{E_d}{E_\mu}\right) + \frac{E_0}{E_\mu}\frac{E_0}{E_q}} \tag{3.6-34}$$

and ω_g results from the imaginary part of the complex time constant,

$$\omega_g = \frac{1}{\tau_{di}} \frac{\frac{E_0}{E_q} - \frac{E_0}{E_\mu}}{\left(1 + \frac{E_d}{E_\mu}\right)^2 + \left(\frac{E_0}{E_\mu}\right)^2} \tag{3.6-35}$$

where the dielectric relaxation time is given by

$$\tau_{di} = \frac{\epsilon}{q\mu N_0} \tag{3.6-36}$$

From the above solutions, we can see the effect of an applied field. In the case of no applied field, that is, $E_0 = 0$, the constant τ in Eq. (3.6-26) is a real number. Therefore, $E_1(t)$ in Eq. (3.6-21) grows exponentially, as shown in Figure 3.6. In the case of a finite applied field, τ is complex. $E_1(t)$ will oscillate as well as exponentially build up. Figure 3.7 shows the oscillation due to the complex time constant. The oscillation frequency is given by Eq. (3.6-35). The applied field will also change the real part of the time constant, as indicated by Eq. (3.6-34).

In the above discussion, we have solved for the transient solution to the space-charge field during the writing process. We can also derive the transient solution for the erasing process. During the erasure, the crystal with a recorded grating is illuminated by a uniform light with intensity I_0. The solution is similar to Eq. (3.6-21), except that the initial condition is

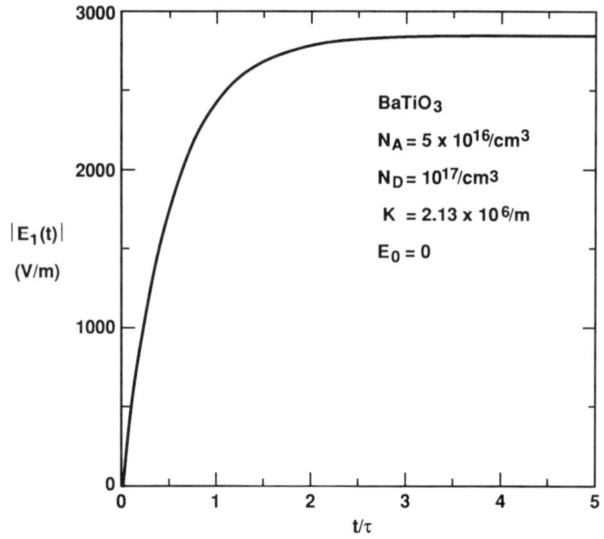

Figure 3.6. Space-charge field E_1 as a function of time.

$E_1(0) = E_1$ and the steady state is $E_1(t = \infty) = 0$. The space-charge field during erasing can be written as

$$E_1(t) = E_1 e^{-t/\tau} \tag{3.6-37}$$

where τ is given by Eq. (3.6-23).

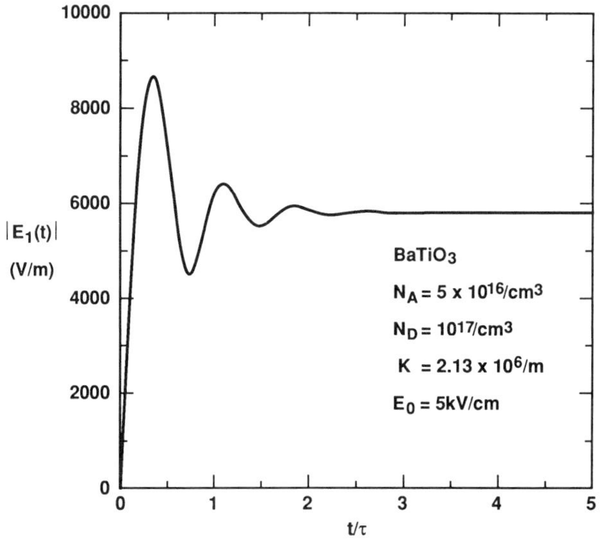

Figure 3.7. The temporal growth of the space-charge field when the time constant is complex.

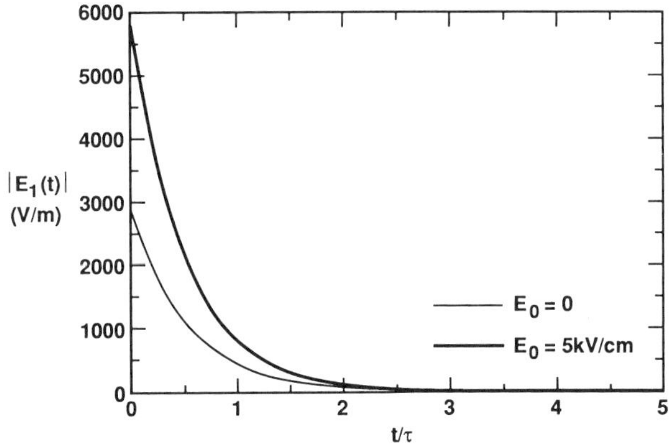

Figure 3.8. The decay of $|E_1|$ during erasure.

The applied field will not cause oscillation in the erasing process. Since the imaginary part of the time constant adds only a phase to $E_1(t)$ and only the magnitude of E_1 affects the diffraction efficiency, $|E_1(t)|$ will exponentially decay during the erasure. The time constant for erasure is the real part of τ, that is, τ_g given in Eq. (3.6-34). Figure 3.8 shows the decay of $|E_1(t)|$ during erasure.

3.7 PHOTOVOLTAIC EFFECT

In some photorefractive crystals, for example, $LiNbO_3$, $BaTiO_3$ and $LiTaO_3$, uniform illumination of light can generate open circuit voltage or equivalently short circuit current [6]. This effect is known as the photovoltaic effect (or photogalvanic effect).

Photovoltaic effect arises from the asymmetry of crystals. Phenomenologically, it is described by the following expansion:

$$j_i = \sigma_{ij} E_j + \alpha_{ijk} E_j E_k \tag{3.7-1}$$

where j is the photo-induced current density, σ is the conductivity tensor, E is the electric field and α is the photovoltaic tensor. In the above expression, we have assumed summation over the repeated indices. In the case of linearly polarized light, α is symmetric over the last two indices (i.e., $\alpha_{ijk} = \alpha_{ikj}$). Similar expressions can be found for elliptically polarized light [7]. In this book, we will only introduce the basic concept of photovoltaic effect for linearly polarized light. Under spatial inversion, j_i changes sign but $E_j E_k$ does not. Therefore, α_{ijk} will be nonzero only in crystals without a center of symmetry. Similar to electro-optic tensor, the photovoltaic tensor can also be written in terms of contracted notation:

$$\begin{aligned} \alpha_{i1} &= \alpha_{i11} \\ \alpha_{i2} &= \alpha_{i22} \\ a_{i3} &= \alpha_{i33} \\ \alpha_{i4} &= \alpha_{i23} = \alpha_{i32} \\ \alpha_{i5} &= \alpha_{i13} = \alpha_{i31} \\ \alpha_{i6} &= \alpha_{i12} = \alpha_{i21} \end{aligned} \tag{3.7-2}$$

The form of photovoltaic tensor for all the noncentrosymmetric crystal classes is the same as that of the nonlinear optical tensor [8].

For example, in $LiNbO_3$, the nonzero terms of the photovoltaic tensor are α_{33}, $\alpha_{31} = \alpha_{32}$, $\alpha_{21} = -\alpha_{22} = \alpha_{16}$, and $\alpha_{24} = \alpha_{15}$. The experimental data [9] shows that α_{31} and α_{33} are very close. Therefore, when a linearly polarized light propagates along the y-direction, the current density is approximately independent of the polarization. When the light is propagating along other directions so that the electric field component along the y-direction is not zero, there might be photovoltaic current along x- and y-directions. However, they are much smaller than the current along the z-direction, since α_{22} and α_{15} are much smaller than α_{31} and α_{33}.

With the presence of the photovoltaic effect, Eq. (3.1-3) of the band transport model needs to be modified. Equations (3.1-1)–(3.1-4) are now written as

$$\frac{\partial N_D^i}{\partial t} = (N_D - N_D^i)(sI + \beta) - \gamma_R N_D^i N \tag{3.7-3}$$

$$\frac{\partial N}{\partial t} = \frac{\partial N_D^i}{\partial t} + \frac{1}{q}\frac{\partial j}{\partial z} \tag{3.7-4}$$

$$j = q\mu NE + k_B T\mu \frac{\partial N}{\partial z} + pI \tag{3.7-5}$$

$$\epsilon \frac{\partial E}{\partial z} = q(N_D^i - N - N_A) \tag{3.7-6}$$

where pI is the photovoltaic current density and p is the equivalent photovoltaic constant. If the crystal is uniformly illuminated and open circuited, there will be a constant electric field in the crystal at steady state. The drift current due to this field will cancel the photovoltaic current. This constant field is often called the photovoltaic field

$$E_{ph} = -\frac{p\gamma_R N_A}{q\mu s(N_D - N_A)} \tag{3.7-7}$$

When two writing beams are used to write a grating in the crystal, we can solve Eqs. (3.7-4)–(3.7-7) under low modulation depth approximation.

Following similar steps used in the last section, we obtain the transient solution of space-charge field

$$E_{sc}(z,t) = \frac{I_1}{I_0} E_{sc}[-\cos(Kz+\phi) + \exp(-t/\tau_g)\cos(Kz+\omega_g t+\phi)], \tag{3.7-8}$$

where the amplitude E_{sc} is given by

$$E_{sc} = E_q \left[\frac{(E_0 - E_{ph})^2 + E_d^2}{E_0^2 + (E_d + E_q)^2} \right]^{1/2} \tag{3.7-9}$$

and the phase of the grating is given by

$$\tan\phi = \frac{(E_d + E_q)E_d + E_0(E_0 - E_{ph})}{(E_0 - E_{ph})E_q - E_{ph}E_d} \tag{3.7-10}$$

The photorefractive time constant is given by

$$\tau_g = \tau_{di} \frac{\left(1 + \frac{E_d}{E_\mu}\right)^2 + \left(\frac{E_0}{E_\mu}\right)^2}{\left(1 + \frac{E_d}{E_q}\right)\left(1 + \frac{E_d}{E_\mu}\right) + \frac{E_0}{E_\mu}\frac{E_0}{E_q}} \tag{3.7-11}$$

and the oscillation frequency of the grating is

$$\omega_g = \frac{1}{\tau_{di}} \frac{\frac{E_0}{E_q} - \frac{E_0}{E_\mu}}{\left(1 + \frac{E_d}{E_\mu}\right)^2 + \left(\frac{E_0}{E_\mu}\right)^2} \tag{3.7-12}$$

where

$$\tau_{di} = \frac{\epsilon}{q\mu N_0} \tag{3.7-13}$$

$$E_d = \frac{k_B T K}{q} \tag{3.7-14}$$

$$E_q = \frac{qN_A(N_D - N_A)}{\epsilon K N_D} \tag{3.7-15}$$

$$E_\mu = \frac{\gamma_R N_A}{\mu K} \tag{3.7-16}$$

and E_0 is the applied field. We note from the above expressions that the photovoltaic effect cannot be thought of as an equivalent applied field since the above expressions for the amplitude, phase, time constant, and oscillatory frequency of the grating do not depend on a simple linear combination of E_0 and E_{ph}. In the limiting case of no photovoltaic effect (i.e., $p = 0$), these solutions reduce to previous results. Bulk photovoltaic effect in lithium niobate ($LiNbO_3$) has been measured. The experimental data of the photovoltaic coefficients varies from samples to samples by an order of magnitude. For visible light with $\lambda = 500$ nm, the values for Fe:$LiNbO_3$ are in the range of $p = 10^{-9}$ A/W. Typical numbers for the photovoltaic field in Fe:$LiNbO_3$ are in the range of $E_{ph} = 10$ kV/cm.

PROBLEMS

3.1 Space-Charge Field

a. Show that the amplitude of the space-charge field E_1 is independent of the intensity of the beams. It only depends on the contrast of the fringe visibility.

b. Equation (3.2-8) can be written

$$E_1 = \frac{iE_d}{1 + \dfrac{E_d}{E_q}} Se^{i\phi} \frac{I_1}{I_0}$$

where $Se^{i\phi}$ is a scale factor due to the applied field. Show that

$$S^2 = \frac{1 + \left(\dfrac{E_0}{E_d}\right)^2}{1 + \left(\dfrac{E_0}{E_d + E_q}\right)^2}$$

and

$$\phi = \tan^{-1}\left(\frac{E_0}{E_d}\right) - \tan^{-1}\left(\frac{E_0}{E_d + E_q}\right) = \tan^{-1} \frac{E_0 E_q}{E_0^2 + E_d(E_d + E_q)}$$

c. Show that $S^2 \simeq 1$ when $E_q \ll E_d$.

d. Show that

$$S^2 = \frac{(E_d^2 + E_0^2)E_q^2}{(E_q^2 + E_0^2)E_d^2}, \quad \text{when } E_q \gg E_d$$

e. In the limit when $E_0 \gg (E_q + E_d)$, the scaling factor becomes

$$S = 1 + \frac{E_q}{E_d}$$

f. Show that $E_1 = iE_q I_1/I_0$, in the limit when $E_0 = \infty$.

3.2 **Photorefractive Absorption.** The absorption coefficient α_p in Eq. (3.4-10) is related to the photoexcitation cross section introduced in Eq. (3.1-1).

a. Using $I = I_0 \exp(-\alpha_p L)$, show that the rate of carrier generation is $I_0\alpha_p/h\nu$ (i.e., the number of electrons per unit time per unit volume), where I_0 is incident intensity.

b. Show that

$$\alpha_p = s(N_D - N_D^i)h\nu$$

3.3 **Moving Gratings and Causality.** Suppose that the formation of gratings is similar to that of the charging of a capacitor. Then the grating amplitude as a function of time can be written

$$\Delta n(t) = \Delta n_0(1 - e^{-t/\tau})$$

where τ is the characteristic time constant.

a. Show that for an infinitesimal time period of dt, the grating is given by

$$\Delta n(t) = \frac{dt}{\tau}\,\Delta n_0$$

b. In the case of moving intensity pattern, the induced index grating can be written as an integral

$$\Delta n(t) = \int_{-\infty}^{t} \frac{dt'}{\tau}\,\Delta n_0 \cos(Kx - \Omega t')\exp\left[-\frac{(t-t')}{\tau}\right]$$

where K and Ω are related to the moving intensity pattern by $I = I_0 + I_1\cos(Kx - \Omega t)$. The integral represents the linear superposition of all the gratings form during $(t', t' + dt')$ for t' from $-\infty$ to t. All these gratings are scaled by an exponential factor $\exp[-(t-t')/\tau]$ representing the grating decay.

c. Carry out the integration in (**b**) and show that

$$\Delta n(t) = \frac{\Delta n_0}{2}\left[\frac{e^{i(\Omega t - Kx)}}{1 + i\Omega\tau} + \frac{e^{-i(\Omega t - Kx)}}{1 - i\Omega\tau}\right]$$

d. Show that

$$\Delta n(t) = \frac{\Delta n_0}{1 + (\Omega\tau)^2} [\cos(\Omega t - Kx) + \Omega\tau \sin(\Omega t - Kx)]$$

We note that the amplitude is decreased and there is a finite phase shift between the intensity and the index grating.

e. Show that the phase shift is given by

$$\phi = -\tan^{-1}(\Omega\tau)$$

3.4 Photorefractive Gratings in $BaTiO_3$. Consider a photo-induced grating produced by intersecting two coherent laser beams with the same polarization state in a $BaTiO_3$ crystal. Let the grating wave vector be along the a-axis of the crystal.

a. Describe the index grating.

b. Show that, although the index grating is proportional to r_{42} (the largest EO coefficient), the index grating cannot be read by either one of the writing beams provided they are pure mode of propagation (i.e., either ordinary or extraordinary mode).

c. Show that r_{42} is not utilized if the grating wave vector is parallel to the c-axis.

REFERENCES

1. A. Ashkin, G. D. Boyd, J. M. Dziedzic, R. G. Smith, A. A. Ballman, J. J. Levinstein, and K. Nassau, "Optically Induced Refractive Index Inhomogeneities in $LiNbO_3$ and $LiTaO_3$," Appl. Phys. Lett. **9**, 72–74 (1966).
2. V. L. Vinetskii, N. V. Kukhtarev, S. G. Odulov, and M. S. Soskin, "Dynamic Self-Diffraction of Coherent Light Beams," Sov. Phys. Usp. **22**, 742–756 (1979).
3. N. V. Kukhatarev, V. B. Markov, S. G. Odulov, M. S. Soskin, and V. L. Vinetskii, "Holographic Storage in Electro-Optic Crystals. Beam Coupling and Light Amplification," Ferroelectrics **22**, 961–964 (1979).
4. See, for example, A. M. Glass, "The Photorefractive Effect," Opt. Eng. 17, 470–480 (1978).
5. P. Yeh, "Fundamental Limit of the Speed of Photorefractive Effect and its Impact on Device Applications and Material Research," Appl. Opt. **26**, 602–605 (1987).
6. A. M. Glass, D. Von der Linde, and T. J. Negran, "High-Voltage Bulk Photovoltaic Effect and the Photorefractive Process in $LiNbO_3$," Appl. Phys. Lett. **25(4)**, 233 (1974).
7. V. I. Belinicher and B. I. Sturman, "The Photogalvanic Effect in Media Lacking a Center of Symmetry." Sov. Phys. Usp. **23(3)**, 199–223 (1980).

8. See, for example, A. Yariv and P. Yeh, Optical Waves in Crystals (Wiley, New York, 1984).
9. V. M. Fridkin and R. M. Magomadov, "Anomalous Photovoltaic Effect in $LiNbO_3$:Fe in Polarized Light," JETP Lett. **30(11)**, 686 (1979).
10. C. Gu, J. Hong, H. Li, D. Psaltis, and P. Yeh, "Dynamics of Grating Formation in Photovoltaic Media," J. Appl. Phys. **69**, 1167–1172 (1991).

ADDITIONAL REFERENCES FOR FURTHER READING

1. P. Günter, Ed., Electric-Optic and Photorefractive Materials (Springer-Verlag, Berlin, 1987).
2. P. Günter and J.-P. Huignard, Eds., Photorefractive Materials and Their Applications (Springer-Verlag, Berlin, 1988).

4

Wave Mixing in Photorefractive Media

When two beams of coherent electromagnetic radiation intersect inside a photorefractive medium, the periodic variation of the intensity due to interference will induce a volume index grating. The grating wave vector is given by $\mathbf{K} = \pm(\mathbf{k}_2 - \mathbf{k}_1)$, where $\mathbf{k}_1$ and $\mathbf{k}_2$ are wave vectors of the beams. The presence of such an index grating will affect the propagation of these two beams. In fact, these waves are strongly diffracted by the index grating because Bragg scatterings are perfectly phase-matched. Thus Beam 1 is scattered by the index grating and the diffracted beam is propagating along the direction of Beam 2. Similarly Beam 2 is scattered by the same grating and the diffracted beam is propagating along the direction of Beam 1 (see Figure 4.1). This leads to energy coupling and the scattering is known as self-diffraction [1]. In this chapter, we treat the coupling between waves inside photorefractive media.

4.1 DEGENERATE TWO-WAVE MIXING

We first consider the interaction of two laser beams inside a photorefractive medium (see Figure 4.2). If the two beams are of the same frequency, a stationary interference pattern is formed. Let the electric field of the two waves be written

$$E_j = A_j \exp[i(\omega t - \mathbf{k}_j \cdot \mathbf{r})], \quad j = 1, 2 \tag{4.1-1}$$

where A_1, A_2 are the wave amplitudes, ω is the angular frequency, and $\mathbf{k}_1$, $\mathbf{k}_2$ are the wave vectors. For simplicity, we assume that the medium is isotropic and both beams are polarized perpendicular to the plane of incidence (i.e., s-polarized).

Within a factor of proportionality, the intensity of the electromagnetic radiation can be written

$$I = |E|^2 = |E_1 + E_2|^2 \tag{4.1-2}$$

Using Eq. (4.1-1) for the electric field, the intensity can be written

$$I = |A_1|^2 + |A_2|^2 + A_1^* A_2 e^{-i\mathbf{K}\cdot\mathbf{r}} + A_1 A_2^* e^{i\mathbf{K}\cdot\mathbf{r}} \tag{4.1-3}$$

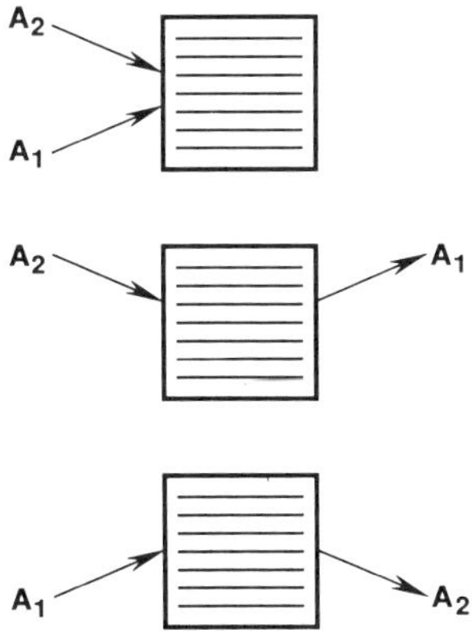

Figure 4.1. Bragg scattering due to volume gratings in photorefractive media. Top: A volume grating is induced by a pair of beams (A_1, A_2). Middle: Beam A_2 is diffracted by the volume grating producing Beam A_1. Bottom: Beam A_1 is diffracted by the volume grating producing Beam A_2.

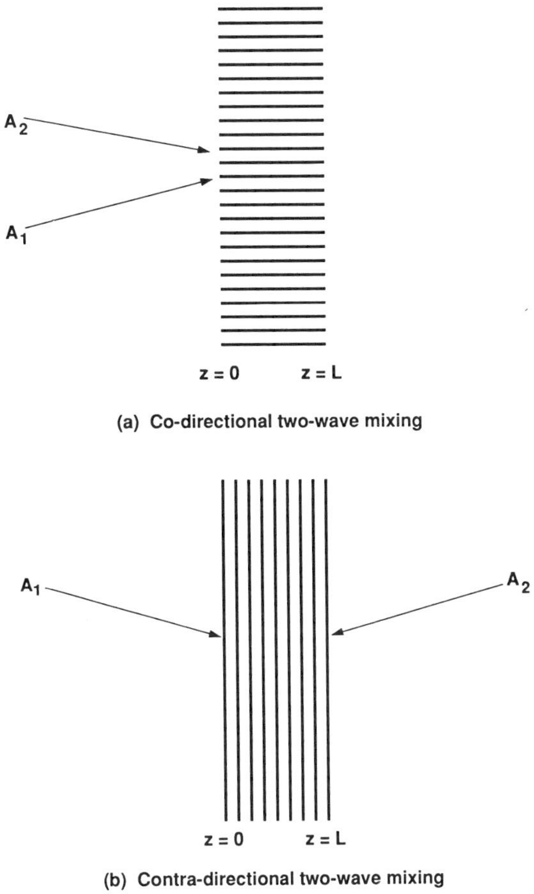

Figure 4.2. Two-wave mixing in photorefractive media: (a) co-directional two-wave mixing; (b) contra-directional two-wave mixing.

where

$$\mathbf{K} = \mathbf{k}_2 - \mathbf{k}_1 \tag{4.1-4}$$

The magnitude of the vector **K** is $2\pi/\Lambda$, where Λ is the period of the fringe pattern. The intensity (Eq. (4.1-3)) represents a spatial variation of optical energy inside the photorefractive medium. According to the model described in the last chapter, such an intensity pattern will generate and redistribute charge carriers. As a result, a space-charge field is created in the medium. This field induces an index volume grating via the Pockels effect. In general, the index grating will have a spatial phase shift relative to the interface pattern. The index of refraction including the fundamental component of the intensity-induced gratings can be written

$$n = n_0 + \left[\frac{n_1}{2} \exp(i\phi) \frac{A_1^* A_2}{I_0} \exp(-i\mathbf{K}\cdot\mathbf{r}) + \text{c.c.} \right] \tag{4.1-5}$$

where c.c. represents complex conjugation and

$$I_0 = I_1 + I_2 \equiv |A_1|^2 + |A_2|^2 \tag{4.1-6}$$

n_0 is the index of refraction when no light is present, ϕ is real and n_1 is a real and positive number. Here again, for the sake of simplicity, we assume a scalar grating. The phase ϕ indicates the degree to which the index grating is shifted spatially with respect to the light interference pattern. In photorefractive media that operate by diffusion only (i.e., no external static field), for example, $BaTiO_3$, the magnitude of ϕ is $\pi/2$ with its sign depending on the direction of the c-axis. **K** is the grating wave vector and I_0 is the sum of the intensities. The parameter n_1 depends on the grating spacing and its direction, as well as on the applied electric field and the material properties of the crystal, for example, the electro-optic coefficient. Expressions for $n_1 e^{i\phi}$ can be derived using the results obtained in the last chapter.

The finite spatial phase shift between the interference pattern and the induced volume index grating has been known for some time. The presence of such a phase shift allows the possibility of nonreciprocal steady-state transfer of energy between the beams. To investigate the coupling, we substitute Eq. (4.1-5) for the index of refraction and $E = E_1 + E_2$ for the electric field into the following wave equation:

$$\nabla^2 E + \frac{\omega^2}{c^2} n^2 E = 0 \tag{4.1-7}$$

where c is the velocity of light.

We assume that both waves propagate in the xz plane. Generally speaking, if the beams are of finite extent (i.e., comparable to the intersection of the beams), the amplitudes may depend on both x and z. Here we assume, for the sake of simplicity, that the transverse dimensions of the beams are of infinite extent so that the boundary condition requires that the wave amplitudes A_1 and A_2 be functions of z only (see Figure 4.2). We will solve for the steady state so that A_1 and A_2 are also taken to be time-independent.

Using the slowly varying amplitude approximation, that is,

$$\left|\frac{d^2}{dz^2} A_j\right| \ll \left|\beta_j \frac{d}{dz} A_j\right|, \quad j=1,2 \tag{4.1-8}$$

we obtain

$$2i\beta_1 \frac{d}{dz} A_1 = \frac{\omega^2 n_0 n_1}{c^2 I_0} e^{-i\phi} A_2^* A_2 A_1 \tag{4.1-9}$$

$$2i\beta_2 \frac{d}{dz} A_2 = \frac{\omega^2 n_0 n_1}{c^2 I_0} e^{i\phi} A_1^* A_1 A_2 \tag{4.1-10}$$

where β_1 and β_2 are the z-component of the wave vectors $\mathbf{k}_1$ and $\mathbf{k}_2$ inside the medium, respectively. The energy coupling depends on the relative sign of β_1 and β_2. Thus, two-wave mixing is divided into the following two categories.

4.1.1 Co-Directional Two-Wave Mixing ($\beta_1\beta_2 > 0$)

Referring to Figure 4.2(a), we consider the case when the two laser beams enter the medium from the same side at $z=0$. Without loss of generality, we assume that

$$\beta_1 = \beta_2 = k \cos\theta = \frac{2\pi}{\lambda} n_0 \cos\theta \tag{4.1-11}$$

where 2θ is the angle between the beams inside the medium, and n_0 is the index of refraction of the medium.

Substituting Eq. (4.1-11) for β_1 and β_2 in Eqs. (4.1-9) and (4.1-10), and using $(\omega/c) = 2\pi/\lambda$, we obtain

$$\begin{aligned} \frac{d}{dz} A_1 &= -\frac{1}{2I_0} \Gamma |A_2|^2 A_1 - \frac{\alpha}{2} A_1 \\ \frac{d}{dz} A_2 &= \frac{1}{2I_0} \Gamma^* |A_1|^2 A_2 - \frac{\alpha}{2} A_2 \end{aligned} \tag{4.1-12}$$

where we have added terms that account for the attenuation with α as the bulk absorption coefficient, and Γ is the complex coupling constant

$$\Gamma = i \frac{2\pi n_1}{\lambda \cos\theta} e^{-i\phi} \tag{4.1-13}$$

In the event of asymmetrical incidence (i.e., $\beta_1 \neq \beta_2$), a similar set of equations with a common coupling constant Γ can be derived by a change of variables.

We now write

$$A_1 = \sqrt{I_1} \exp(-i\psi_1), \qquad A_2 = \sqrt{I_2} \exp(-i\psi_2) \tag{4.1-14}$$

where ψ_1 and ψ_2 are phases of the complex amplitudes A_1 and A_2, respectively. Using Eqs. (4.1-14) and (4.1-6), the coupled Eqs. (4.1-12) can be written as

$$\frac{d}{dz} I_1 = -\gamma \frac{I_1 I_2}{I_1 + I_2} - \alpha I_1 \tag{4.1-15}$$

$$\frac{d}{dz} I_2 = \gamma \frac{I_1 I_2}{I_1 + I_2} - \alpha I_2 \tag{4.1-16}$$

and

$$\frac{d}{dz} \psi_1 = \beta \frac{I_2}{I_1 + I_2} \tag{4.1-17}$$

$$\frac{d}{dz} \psi_2 = \beta \frac{I_1}{I_1 + I_2} \tag{4.1-18}$$

where γ and β are related to Γ with

$$\Gamma = \gamma + 2i\beta \tag{4.1-19}$$

$$\gamma = \frac{2\pi n_1}{\lambda \cos\theta} \sin\phi \quad \text{and} \quad \beta = \frac{\pi n_1}{\lambda \cos\theta} \cos\phi \tag{4.1-20}$$

By examining the coupled equations, we note that in the absence of material absorption ($\alpha = 0$), I_2 is an increasing function of z provided γ is positive. This indicates that the energy is flowing from beam 1 to beam 2. The direction of energy flow is determined by the sign of γ which depends on the orientation of the crystal axes.

The solutions for the intensities $I_1(z)$ and $I_2(z)$ are (see Problem 4.1)

$$I_1(z) = I_1(0) \frac{1 + m^{-1}}{1 + m^{-1} e^{\gamma z}} e^{-\alpha z} \tag{4.1-21}$$

$$I_2(z) = I_2(0) \frac{1 + m}{1 + m e^{-\gamma z}} e^{-\alpha z} \tag{4.1-22}$$

where m is the input intensity ratio

$$m = \frac{I_1(0)}{I_2(0)} \tag{4.1-23}$$

Figure 4.3 plots the beam intensities as functions of z. Note that in the absence of absorption ($\alpha = 0$), $I_2(z)$ is an increasing function of z and $I_1(z)$ is a decreasing function of z, provided γ is positive (see Figure 4-3(a)). We notice that beam 2 is taking energy from beam 1 which decays exponentially. In the presence of material absorption (i.e., $\alpha > 0$), beam 2 can still be amplified provided the gain due to beam coupling is large enough to overcome the loss (see Figure 4-3(b)). For small input signals with $I_2(0) \ll I_1(0)$, Eq. (4.1-22) can be written $I_2(z) = I_2(0)\exp[(\gamma - \alpha)z]$, provided $m\exp(-\gamma z) \gg 1$. Thus beam amplification requires that $\alpha < \gamma$. In addition, optimum gain occurs at an appropriate interaction length as shown in Figure 4.3(b). Beyond this distance, the beam intensity $I_2(z)$ starts to decrease. For $m\exp(-\gamma L) \ll 1$, $I_2(L)$ becomes $[I_1(0) + I_2(0)]\exp(-\alpha L)$. Physically, beam 2 takes all the energy and then decays exponentially due to material absorption.

With $I_1(z)$, $I_2(z)$ known, the phases ψ_1 and ψ_2 can be obtained by a direct integration of Eqs. (4.1-17) and (4.1-18). Substituting Eqs. (4.1-21) and (4.1-22) into Eqs. (4.1-17) and (4.1-18) for I_1 and I_2, respectively, we obtain

$$\psi_2(z) - \psi_2(0) = \int_0^z \frac{\beta\, dz'}{1 + m^{-1}e^{\gamma z'}} \tag{4.1-24}$$

Note that this photorefractive phase shift is independent of the absorption coefficient α. Carrying out the integration in Eq. (4.1-24), we obtain

$$\psi_2(z) - \psi_2(0) = \frac{\beta}{\gamma}\ln\left(\frac{1+m}{1+me^{-\gamma z}}\right) \tag{4.1-25}$$

From Eqs. (4.1-17) and (4.1-18), we note that

$$\frac{d}{dz}(\psi_1 + \psi_2) = \beta \tag{4.1-26}$$

Thus, $\psi_1(z)$ can be written

$$\begin{aligned}\psi_1(z) - \psi_1(0) &= \beta z - [\psi_2(z) - \psi_2(0)] \\ &= -\frac{\beta}{\gamma}\ln\left(\frac{m+1}{m+e^{\gamma z}}\right)\end{aligned} \tag{4.1-27}$$

If we refer to A_2 as the signal beam, then a useful parameter is the gain

$$g = \frac{I_2(L)}{I_2(0)} = \frac{1+m}{1+me^{-\gamma L}}\, e^{-\alpha L} \tag{4.1-28}$$

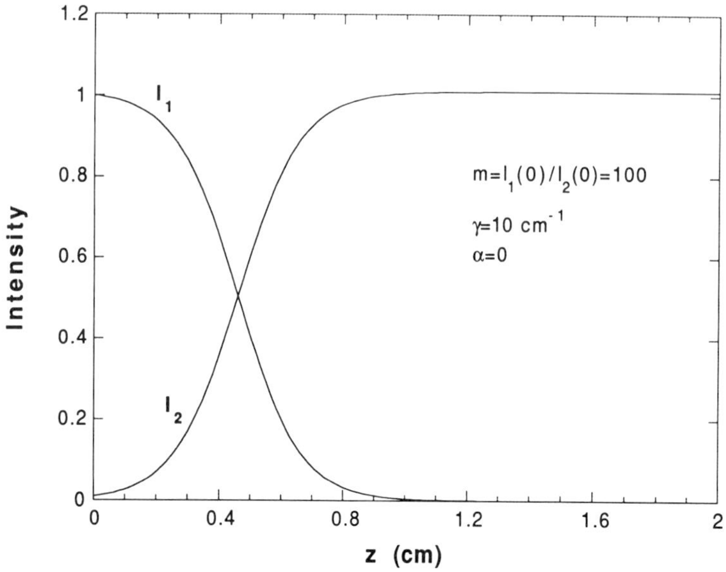

(a)

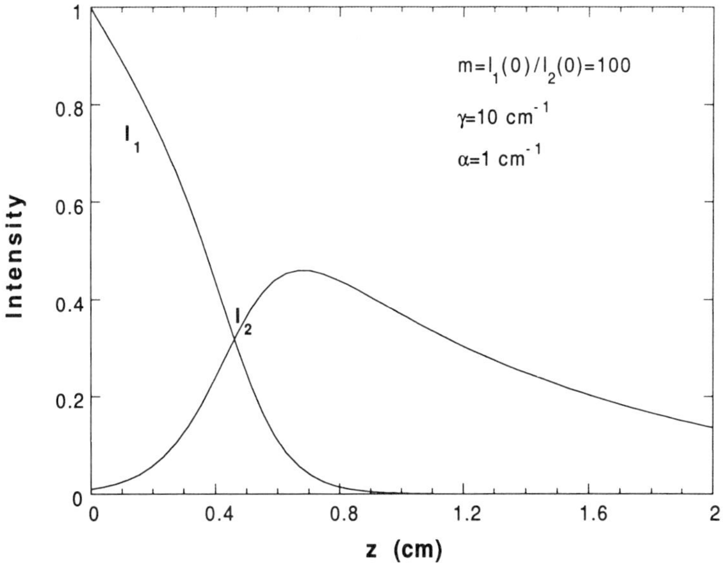

(b)

Figure 4.3. Beam intensities I_1 and I_2 as functions of z in a co-directional two-wave mixing. (a) Lossless with $\alpha = 0$. (b) Finite loss with $\alpha = 1\ \text{cm}^{-1}$.

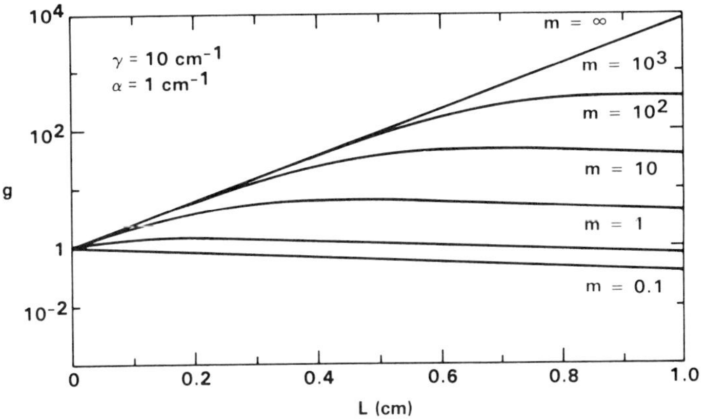

Figure 4.4. Signal gain g versus interaction length L with various beam intensity ratios m for $\gamma = 10\ \text{cm}^{-1}$ and $\alpha = 1\ \text{cm}^{-1}$.

where we recall that m is the intensity ratio at input face ($z = 0$). Figure 4.4 plots the gain as a function of the length of interaction L for various values of m.

Example: Beam Amplification Using $\boldsymbol{BaTiO_3}$ Consider the amplification of weak signal beam with intensity $I_2(0) = 1.0 \times 10^{-6}\ \text{W/cm}^2$ by using a photorefractive $BaTiO_3$ crystal. Let the pump beam intensity be $I_1(0) = 1.0 \times 10^{-3}\ \text{W/cm}^2$ and the coupling strength be $\gamma L = 10$. The intensity gain, according to Eq. (4.1-28) is $g = 958$ assuming $\alpha = 0$. If the signal beam intensity is $I_2(0) = 1.0 \times 10^{-8}\ \text{W/cm}^2$, the gain becomes $g = 9.6 \times 10^4$.

4.1.2 Contra-Directional Two-Wave Mixing

We now consider the case when the two beams enter the medium from opposite faces, as shown in Figure 4.2. In the co-directional two-wave mixing, the sum of the beam power is a constant of integration provided the medium is lossless, whereas in the contra-directional two-wave mixing, the difference of the beam power (i.e., net Poynting power flow) is a constant. In addition, the coupled-mode equation which governs the wave amplitudes is also different from that of the co-directional coupling. This leads to qualitative differences in the energy exchange between the two waves in two cases.

Let

$$\beta_1 = -\beta_2 = k \cos\theta = \frac{2\pi}{\lambda} n_0 \cos\theta \qquad (4.1\text{-}29)$$

where θ is the angle between each of the beams and the z-axis. Substitution of Eq. (4.1-29) for β_1 and β_2 in Eqs. (4.1-9) and (4.1-10) yields a similar set of coupled equations. Using Eq. (4.1-14), such a set of coupled equations becomes

$$\frac{d}{dz} I_1 = -\gamma \frac{I_1 I_2}{I_1 + I_2} - \alpha I_1 \tag{4.1-30}$$

$$\frac{d}{dz} I_2 = -\gamma \frac{I_1 I_2}{I_1 + I_2} + \alpha I_2 \tag{4.1-31}$$

and

$$\frac{d}{dz} \psi_1 = \beta \frac{I_2}{I_1 + I_2} \tag{4.1-32}$$

$$\frac{d}{dz} \psi_2 = -\beta \frac{I_1}{I_1 + I_2} \tag{4.1-33}$$

where β and γ are phase and intensity coupling constants and are given by

$$\gamma = \frac{2\pi n_1}{\lambda \cos \theta} \sin \phi \tag{4.1-34}$$

$$\beta = \frac{\pi n_1}{\lambda \cos \theta} \cos \phi \tag{4.1-35}$$

Comparing with Eqs. (4.1-15)–(4.1-18), we notice the sign difference in these equations for Beam 2.

The solutions for Eqs. (4.1-30)–(4.1-33) can be obtained in closed form for the case when $\alpha = 0$ (i.e., lossless). In the lossless case, we note that the Poynting power flow along $+z$ is conserved, that is,

$$\frac{d}{dz} (I_1 - I_2) = 0 \tag{4.1-36}$$

and the solution of Eqs. (4.1-30) and (4.1-31) with $\alpha = 0$ is [2]

$$I_1(z) = -C + [C^2 + B \exp(-\gamma z)]^{1/2} \tag{4.1-37}$$

$$I_2(z) = C + [C^2 + B \exp(-\gamma z)]^{1/2} \tag{4.1-38}$$

where B and C are constants and are related to the boundary condition. B and C can be expressed in terms of any two of the four boundary values $I_1(0)$, $I_2(0)$, $I_1(L)$, and $I_2(L)$, where L is the length of interaction. In terms of $I_1(0)$ and $I_2(0)$, B and C are given by

$$B = I_1(0) I_2(0) \tag{4.1-39}$$

$$C = [I_2(0) - I_1(0)]/2 \tag{4.1-40}$$

In practice, it is convenient to express B and C in terms of the incident intensities $I_1(0)$ and $I_2(L)$. In this case, B and C become

$$B = I_1(0)I_2(L)\frac{I_1(0) + I_2(L)}{I_2(L) + I_1(0)\exp(-\gamma L)} \tag{4.1-41}$$

$$C = \frac{1}{2}\frac{I_2^2(L) - I_1^2(0)\exp(-\gamma L)}{I_2(L) + I_1(0)\exp(-\gamma L)} \tag{4.1-42}$$

According to Eqs. (4.1-30) and (4.1-31), both $I_1(z)$ and $I_2(z)$ are decreasing functions of z, provided γ is positive and $\alpha = 0$. The transmittance for both waves, according to Eqs. (4.1-37)–(4.1-42), are

$$t_1 = \frac{I_1(L)}{I_1(0)} = \frac{1 + m^{-1}}{1 + m^{-1}\exp(\gamma L)} \tag{4.1-43}$$

$$t_2 \equiv \frac{I_2(0)}{I_2(L)} = \frac{1 + m}{1 + m\exp(-\gamma L)} \tag{4.1-44}$$

where m is the incident intensity ratio $m \equiv I_1(0)/I_2(L)$. Note that $t_1 < 1$ and $t_2 > 1$ for positive γ. The sign of γ depends on the direction of the c-axis. It is interesting to note that these expressions for transmittance are formally identical to those of the co-directional coupling, even though the spatial variations of $I_1(z)$ and $I_2(z)$ with respect to z are very different. Note that t_1 and t_2 are related by $t_2 = t_1\exp(\gamma L)$.

The relative phase of the two waves is obtained by solving Eqs. (4.1-32) and (4.1-33) in cooperation with Eqs. (4.1-37) and (4.1-38), and is given by

$$\psi_2 - \psi_1 = -\beta z + \text{constant} \tag{4.1-45}$$

where β is the phase coupling constant. The relative phase varies linearly with z, and thus leads to a change in the grating wave vector by β along the z-direction (i.e., the grating wave vector becomes $\mathbf{K} - \beta\hat{z}$).

The nonreciprocal transmittance of photorefractive media may have important applications in many optical systems. It is known that in linear optical media, the transmittance of a layered structure (including absorbing material) is independent of the side of incidence (the so-called left and right incidence theorem) [3]. Right now, with the photorefractive material available, it is possible to make a "one-way" window which favors transmission from one side only.

The solutions of Eqs. (4.1-37) and (4.1-38) did not take into account the effect of the bulk absorption of light. The attenuation due to finite absorption coefficient is reflected by the $-\alpha I_1$ term on the right-hand side of the first equation in Eqs. (4.1-37) and (4.1-38), and the $+\alpha I_2$ term on the right-hand side of the other equation. With these two additional terms accounting for bulk absorption, closed-form solutions are not easily available. However, Eqs. (4.1-30) and (4.1-31) can still be integrated numerically. It is found that a very good approximate solution is

$$I_1^{\alpha}(z) = I_1^{\alpha=0}(z)\exp(-\alpha z) \tag{4.1-46}$$

$$I_2^{\alpha}(z) = I_2^{\alpha=0}(z)\exp[\alpha(z-L)] \tag{4.1-47}$$

The approximation is legitimate provided $\alpha \ll |\gamma|$. Figure 4.5 illustrates the intensity variation in the medium.

Example Consider a contra-directional two-wave mixing in a medium with $\gamma = -10\ \text{cm}^{-1}$, $\alpha = 1.6\ \text{cm}^{-1}$, and $L = 2.5$ mm. If the loss were neglected (i.e., $\alpha = 0$), the transmittance would be $t_1 = 1.81$ and $t_2 = 0.15$. With $\alpha = 1.6\ \text{cm}^{-1}$, the transmittances become $t_1 = 1.27$ and $t_2 = 0.11$ according to a numerical integration. The approximate solution (Eqs. (4.1-46) and (4.1-47)) would lead to $t_1 = 1.21$ and $t_2 = 0.10$. Note that even with the presence of absorption, the transmittance can still be greater than unity.

By using the approximate solution (Eqs. (4.1-46) and (4.1-47)), the transmittance become

$$t_1 = \frac{1+m^{-1}}{1+m^{-1}\exp(\gamma L)}\exp(-\alpha L) \tag{4.1-48}$$

$$t_2 = \frac{1+m}{1+m\exp(-\gamma L)}\exp(-\alpha L) \tag{4.1-49}$$

There are two extreme cases worth mentioning. In the case when $I_2(L) \gg I_1(0)$ $m \ll 1$, the transmittance become $t_1 \simeq \exp[(-\gamma-\alpha)L]$ and $t_2 \simeq \exp(-\alpha L)$, whereas in the case when $I_1(0) \gg I_2(L)$, $m \gg 1$, the transmittances are $t_1 \simeq \exp(-\alpha L)$ and $t_2 \simeq \exp[(\gamma-\alpha)L]$.

Coupling Constant γ. In the above derivation of two-wave mixing, the grating is expressed in terms of its index of refraction as given by Eq. (4.1-5). The same grating can also be expressed in terms of the dielectric constant as

$$\epsilon' = \epsilon + \frac{1}{2}\left[\epsilon_1 e^{i\phi}\frac{A_1^* A_2}{I_0}\exp(-i\mathbf{K}\cdot\mathbf{r}) + \text{c.c.}\right] \tag{4.1-50}$$

where

$$\epsilon = \epsilon_0 n_0^2 \tag{4.1-51}$$

$$\epsilon_1 = 2\epsilon_0 n_0 n_1 \tag{4.1-52}$$

Using these parameters, the coupling constant can be written

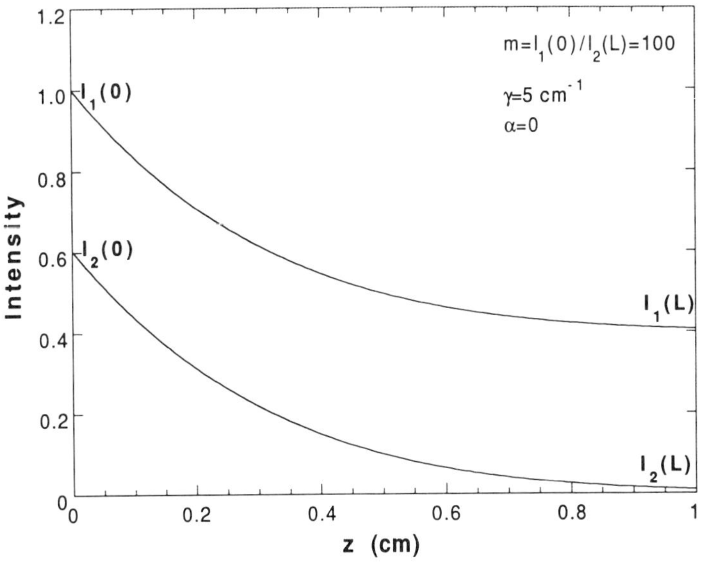

(a)

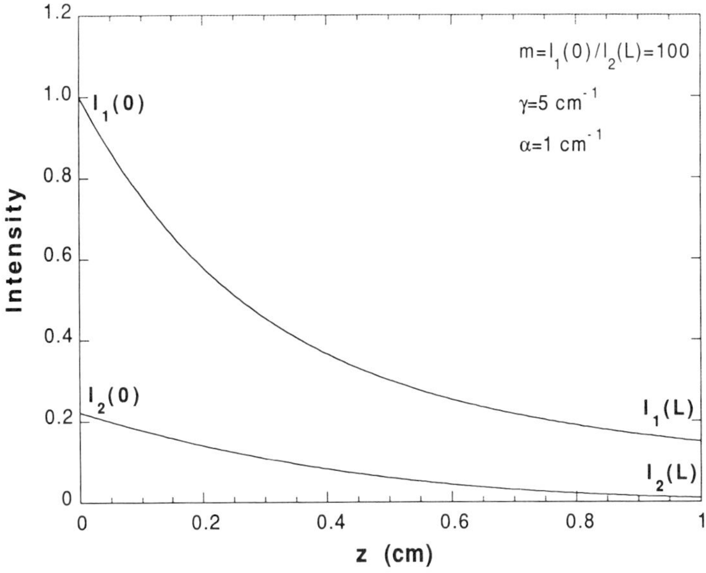

(b)

Figure 4.5. Beam intensities I_1 and I_2 as functions of z in a contra-directional two-wave mixing. (a) Lossless with $\alpha = 0$. (b) Finite loss with $\alpha = 1\ \mathrm{cm}^{-1}$.

$$\gamma = \frac{\pi \epsilon_1}{n_0 \epsilon_0 \lambda \cos \theta} \tag{4.1-53}$$

Equation (4.1-50) is a more general expression for the volume grating. The parameters ϵ and ϵ_1 can be tensors describing anisotropic media and nonscalar gratings. If we assume that the two waves are polarized and they travel at nearly the same phase velocity c/n, then the coupling constant is written

$$\gamma = \frac{\pi}{n_0 \epsilon_0 \lambda \cos \theta} \mathbf{p}_1 \cdot \epsilon_1 \mathbf{p}_2 \tag{4.1-54}$$

where $\mathbf{p}_1$ and $\mathbf{p}_2$ are unit vectors describing the polarization states of the waves.

The amplitude of the dielectric perturbation ϵ_1 is related to the amplitude of the space-charge field and the electro-optic effect. Let $\Delta\epsilon$ represent the second term of Eq. (4.1-50) and the space-charge field is represented by Eq. (3.3-2), then $\Delta\epsilon$ can be written, according to Eq. (3.3-12)

$$\Delta \epsilon_{ij} = -\epsilon_0 n_i^2 n_j^2 r_{ijk} E_k^{sc} \tag{4.1-55}$$

where summation over dummy index k is assumed. By comparing Eqs. (4.1-50), (4.1-55) and (3.3-2), we obtain

$$(\epsilon_1)_{ij} e^{i\phi} \frac{A_1^* A_2}{I_0} = -\epsilon_0 n_i^2 n_j^2 r_{ijk} E_{1k} \tag{4.1-56}$$

where $(\epsilon_1)_{ij}$ is the (ij)th component of the tensor ϵ_1 and E_{1k} is the kth component of the amplitude of the space-charge field.

We define a saturated space-charge field $\mathbf{E}^s$ as

$$\mathbf{E}_1 = \mathbf{E}^s \frac{I_1}{I_0} \tag{4.1-57}$$

then

$$(\epsilon_1)_{ij} e^{i\phi} = -2 \sum_k \epsilon_0 n_i^2 n_j^2 r_{ijk} E_k^s \tag{4.1-58}$$

where E_k^s is the kth component of $\mathbf{E}^s$.

Consider now the case of $BaTiO_3$ or SBN photorefractive crystals (point group symmetry $4mm$). Let α be the angle between the grating wave vector and the c-axis of the crystal (see Figure 4.6). According to Eqs. (3.3-12) and (3.3-14), ϵ_1 can be written

$$\epsilon_1 e^{i\phi} = -2\epsilon_0 \begin{pmatrix} n_o^4 r_{13} \cos \alpha & 0 & n_o^2 n_e^2 r_{42} \sin \alpha \\ 0 & n_o^4 r_{13} \cos \alpha & 0 \\ n_o^2 n_e^2 r_{42} \sin \alpha & 0 & n_e^4 r_{33} \cos \alpha \end{pmatrix} E^s \tag{4.1-59}$$

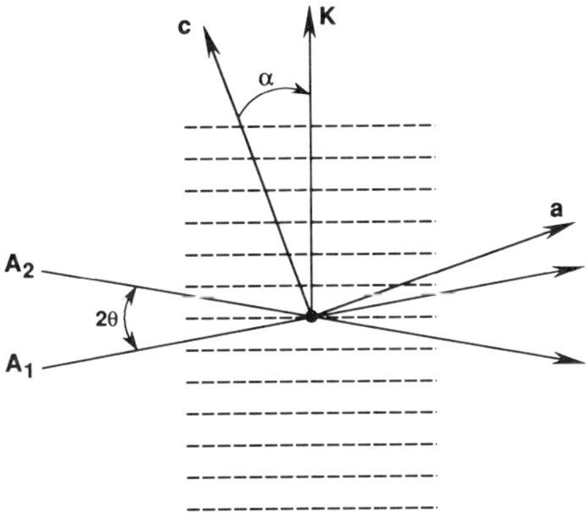

Figure 4.6. A volume grating is induced in a photorefractive medium. The grating wave vector K is in the ac-plane of the crystal coordinate and at an angle α with respect to the c-axis of the crystal.

where E^s is the magnitude of $\mathbf{E}^s$. The saturated space-charge field for the case of pure diffusion is given by, according to Eq. (3.1-35)

$$E^s = \frac{iK \dfrac{k_B T}{q}}{1 + \dfrac{K^2}{k_D^2}} \tag{4.1-60}$$

We note that E^s depends on the magnitude of K which is related to the angle between the beams by

$$K = 2k \sin\theta \tag{4.1-61}$$

assuming that both beams have the same wave number k. In addition, E^s also depends on the direction of $\mathbf{K}$ via k_D^2. According to Eq. (3.1-31), k_D^2 is inversely proportional to $\langle \epsilon \rangle$ which is given by

$$\langle \epsilon \rangle = \epsilon_{\parallel} \cos^2\alpha + \epsilon_{\perp} \sin^2\alpha \tag{4.1-62}$$

where $\epsilon_{\parallel}$ and $\epsilon_{\perp}$ are the dielectric constants for an electric field parallel and perpendicular to the c-axis, respectively. Here it is important to remember that $\langle \epsilon \rangle$ given by Eq. (4.1-62) is the dielectric constant of the medium at $\omega \simeq 0$ (DC field), whereas the ϵ' or ϵ given by other equations are for ω at optical frequencies.

The coupling constant γ, as given by Eq. (4.1-54) is proportional to E^s and thus depends on angles θ and α. In addition, it is also dependent on the polarization state of the waves. Let the ac-plane be the plane of incidence (i.e., the plane formed by the a-axis and the c-axis of the crystal). If both beams are ordinary waves, the polarization states in the abc-coordinate are given by

$$\mathbf{p}_1 = \mathbf{p}_2 = (0, 1, 0) \tag{4.1-63}$$

The coupling constant γ, as given by Eq. (4.1-54), can be written, according to Eq. (4.1-59)

$$\gamma = \frac{-2\pi}{\lambda \cos\theta} n_o^3 r_{13} E^s \cos\alpha \tag{4.1-64}$$

where n_o is the ordinary index of refraction.

If both beams are extraordinary waves, the polarization states in the *abc*-coordinate are given by

$$\mathbf{p}_1 = (\cos\theta_1, 0, \sin\theta_1) \tag{4.1-65}$$

$$\mathbf{p}_2 = (\cos\theta_2, 0, \sin\theta_2) \tag{4.1-66}$$

where θ_1, θ_2 are angles of polarization vector measured from the a-axis and are given by

$$\theta_1 = \frac{\pi}{2} + \theta - \alpha \tag{4.1-67}$$

$$\theta_2 = \frac{\pi}{2} - \theta - \alpha \tag{4.1-68}$$

The coupling constant γ, in this case, is given by

$$\gamma = \frac{-2\pi}{n\lambda \cos\theta} [\cos\theta_1 \cos\theta_2 n_o^4 r_{13} \cos\alpha + n_o^2 n_e^2 r_{42} \sin\alpha \sin 2\alpha$$
$$+ \sin\theta_1 \sin\theta_2 n_e^4 r_{33} \cos\alpha] E^s \tag{4.1-69}$$

where n is given by

$$\frac{1}{n^2} = \frac{\cos^2\alpha}{n_e^2} + \frac{\sin^2\alpha}{n_o^2} \tag{4.1-70}$$

We assumed that 2θ is small and the beams travel at approximately the same speed.

In $BaTiO_3$ crystals where r_{42} is the dominant electro-optic coefficient, the coupling constant is given approximately as

$$\gamma = \frac{-2\pi}{n\lambda \cos\theta} n_o^2 n_e^2 r_{42} E^s \sin\alpha \sin 2\alpha \tag{4.1-71}$$

Example: Coupling Constant in BaTiO$_3$ Consider two-wave mixing in a $BaTiO_3$ crystal with a grating vector which is 45° with respect to the c-axis.

Let the amplitude of the space-charge field be $E^s = 10^4$ V/m. Take $n_o \simeq n_e \simeq n = 2.4$, $\theta = 10°$, $\lambda = 0.5$ μm, and $r_{42} = 1640 \times 10^{-12}$ m/V, Eq. (4.1-71) yields a coupling constant of $\gamma = 20\ \text{cm}^{-1}$.

In SBN crystals, r_{33} is the dominant one; the coupling constant is then given approximately as

$$\gamma = \frac{-2\pi}{n\lambda \cos\theta} n_e^4 r_{33} \sin\theta_1 \sin\theta_2 \cos\alpha\ E^s \tag{4.1-72}$$

In the case of $2k$-gratings (i.e., gratings with wave number ($K = 2k$)), we have

$$\mathbf{p}_1 = \mathbf{p}_2 = (-\cos\alpha, 0, \sin\alpha) \tag{4.1-73}$$

and a similar derivation leads to a coupling constant γ of

$$\gamma = \frac{-2\pi}{n\lambda} [n_o^4 r_{13} \cos^3\alpha - n_o^2 n_e^2 r_{42} \sin 2\alpha \sin\alpha + n_e^4 r_{33} \sin^2\alpha \cos\alpha] E^s \tag{4.1-74}$$

where E^s is evaluated at $K = 2k$. In most photorefractive crystals, $k \gg k_D$, the space-charge field E^s is therefore relatively small.

The coupling constant γ depends on the angles of the beams as well as their polarization states. It is important to note that γ is proportional to the linear electro-optic coefficients which change sign when the crystal orientation is reversed.

Photorefractive Beam Fanning. When a laser beam passes through a photorefractive crystal, significant scattering often occurs. The scattered light appears to be asymmetrical with respect to the beam except for propagation along the c-axis. For laser beams of moderate power, the scattered light appears to grow in time and eventually reaches a steady state scattering pattern. This is known as photorefractive beam fanning.

The beam fanning originates from the energy coupling between the incident beam and the scattered beam. Scattering due to surface roughness, index inhomogeneity, impurities in solids, etc., occurs when a beam of light passes through a photorefractive crystal. If the incident beam has a finite transverse cross section, the scattered beam will have an overlap with the incident beam. Since the scattered beam has a large number of spatial components, the overlap leads to the formation of a large number of photo-induced gratings. Depending on the orientation of the crystal, some of the spatial components may be amplified by the incident beam via the energy transfer in two-beam coupling. The amplified scattering can be quite significant as many of the photorefractive crystals exhibit strong coupling strength (i.e., $\gamma L \gg 1$), even though the initial scattering is quite small.

Beam fanning is inevitable in high gain media, even if the material is near perfect. Nature phenomena, such as Rayleigh scattering, Brillouin scattering, are always present in optical media. Even if the initial scattering is as small as 10^{-8}, the amplified scattering can be very significant, considering photorefractive gain strength of $\gamma L = 20$ or more. Photorefractive beam fanning plays an important role in the initiation of many phase conjugators and resonators, even though the fanning itself can be a source of noise in many experimental measurements.

4.2 NONDEGENERATE TWO-WAVE MIXING

When the frequencies of the two laser beams are different, the interference fringe pattern is no longer stationary. A volume index grating can still be induced provided the fringe pattern does not move too fast. The amplitude of the index modulation decreases as the speed of the fringe pattern increases. This is related to the finite time needed for the formation of index grating in the photorefractive medium.

Let ω_1 and ω_2 be the frequency of the two beams. The electric field of these two beams can be written

$$E_j = A_j \exp[i(\omega_j t - \mathbf{k}_j \cdot r)] , \quad j = 1, 2 \tag{4.2-1}$$

where $\mathbf{k}_1$ and $\mathbf{k}_2$ are the wave vectors, and A_1, A_2 are the wave amplitudes. The intensity of the electromagnetic radiation, similar to that given by Eq. (4.1-3), can be written

$$I = |A_1|^2 + |A_2|^2 + A_1^* A_2 e^{i(\Omega t - \mathbf{K}\cdot\mathbf{r})} + A_1 A_2^* e^{-i(\Omega t - \mathbf{K}\cdot\mathbf{r})} \tag{4.2-2}$$

where

$$\Omega = \omega_2 - \omega_1 \tag{4.2-3}$$

$$\mathbf{K} = \mathbf{k}_2 - \mathbf{k}_1 \tag{4.2-4}$$

Such an intensity distribution represents a traveling fringe pattern at a speed of

$$v = \frac{\Omega}{K} = \frac{\Omega\Lambda}{2\pi} \tag{4.2-5}$$

where Λ is the period of the fringe pattern.

The index of refraction including the fundamental component of the intensity-induced grating can be written

$$n = n_0 + \frac{n_1}{2} \left\{ \exp(i\phi) \frac{A_1^* A_2}{I_0} \exp[i(\Omega t - \mathbf{K}\cdot\mathbf{r})] + \text{c.c.} \right\} \tag{4.2-6}$$

where

$$I_0 = I_1 + I_2 \equiv |A_1|^2 + |A_2|^2 \qquad (4.2\text{-}7)$$

ϕ is real and n_1 is a real and positive number. Here again, for the sake of simplicity, we assume a scalar grating. The phase ϕ indicates the degree to which the index grating is shifted spatially with respect to the light interference pattern. According to the results obtained in the last chapter, ϕ and n_1 can be written respectively, as

$$\phi = \phi_0 - \tan^{-1}(\Omega\tau) \qquad (4.2\text{-}8)$$

and

$$n_1 = \frac{n_{10}}{(1 + \Omega^2\tau^2)^{1/2}} \qquad (4.2\text{-}9)$$

where τ is the decay time constant of the photorefractive grating, n_{10} is the amplitude of the photoinduced index change at $\Omega = 0$, and ϕ_0 is a constant phase shift related to the nonlocal response of the crystal under fringe illumination. Both parameters n_{10} and ϕ_0 depend on the grating spacing $2\pi/K$ and its direction, as well as on applied electric field and the material properties of the crystal, for example, the electro-optic coefficients. Expressions for n_{10} and ϕ_0 can be found in the last chapter. In photorefractive media, for example, $BaTiO_3$, that operate by diffusion only (i.e., no external static field), the magnitude of ϕ_0 is $\pi/2$ with its sign depending on the orientation of the c-axis (note that these crystals are acentric).

Following the procedure similar to the one used in the previous section, coupled equations for the intensities $I_1(z)$, $I_2(z)$ and the phases $\psi_1(z)$, $\psi_2(z)$ are obtained. They are formally identical to those of the degenerate case, that is, Eqs. (4.1-15)–(4.1-18) for co-directional coupling and Eqs. (4.1-30)–(4.1-33) for contra-directional coupling. The intensity coupling constant γ, however, is now a function of the frequency detuning Ω.

For crystals such as $BaTiO_3$ that operate by diffusion only, the coupling constant can be written, according to Eqs. (4.1-19), (4.2-8), and (4.2-9), as

$$\gamma = \gamma_0/[1 + (\Omega\tau)^2] \qquad (4.2\text{-}10)$$

where γ_0 is the coupling constant for the case of degenerate two-wave mixing (i.e., $\Omega = \omega_1 - \omega_2 = 0$) and is given by

$$\gamma_0 = \frac{2\pi n_{10}}{\lambda \cos\theta} \qquad (4.2\text{-}11)$$

In arriving at Eq. (4.2-10), we have used $\pi/2$ for ϕ_0 in Eq. (4.2-8).

The two-wave mixing gain can be written

$$g \equiv \frac{I_2(L)}{I_2(0)} = \frac{1+m}{1+me^{-\gamma L}}\, e^{-\alpha L} \tag{4.2-12}$$

where we recall that m is the input beam ratio $m = I_1(0)/I_2(0)$ and L is the length of interaction.

Figure 4-7 shows the signal gain (Eq. (4.2-12)) as a function of the frequency detuning $\Omega\tau$ for various values of m. We note that for the case of pure diffusion, signal gain decreases as $\Omega\tau$ increases. This is true for both co-directional and contra-directional coupling. When $\Omega\tau \gg 1$, the intensity coupling constant γ decreases significantly. The time constant τ, as discussed in Chapter 3, depends on materials as well as the intensity of the laser beams.

In cases when an external electric field is applied in the photorefractive medium to enhance the space-charge field and the index modulation, the phase ϕ_0 is no longer $\pi/2$. Under these conditions, nondegenerate two-wave mixing is useful to achieve optimum beam coupling. This is the situation in BSO crystals. By properly choosing the frequency detuning, it is possible to achieve a spatial phase shift of $\phi = \pi/2$ which leads to maximum beam coupling.

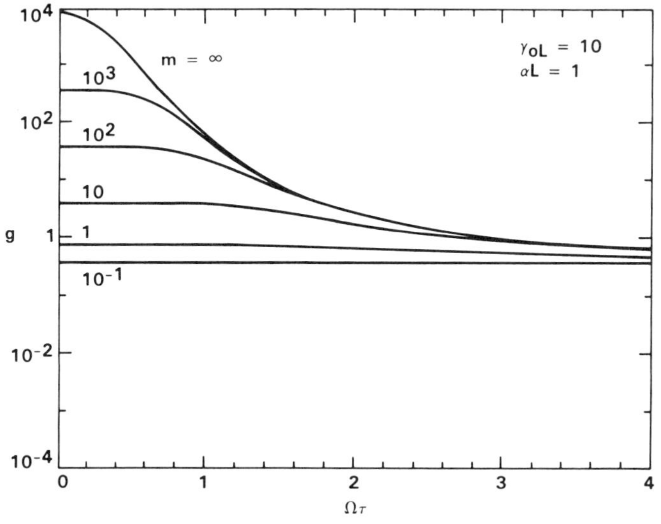

Figure 4.7. Signal gain as a function of frequency detuning $\Omega\tau$ in a co-directional two-wave mixing for various beam intensity ratios m.

4.3 ANISOTROPIC TWO-WAVE MIXING

In the above discussion of two-wave mixing, we assume that the two waves are of the same polarization state. Most of the photorefractive media are optically anisotropic. In these media, each beam of light can have two independent polarization states. These two modes propagate at different speeds. In what follows, we consider the coupling of these polarized waves in anisotropic photorefractive media.

Referring to Figure 4.8(a), we consider the intersection of two beams of light in a photorefractive medium. Because of the optical anisotropy, each beam of light consists of two independent modes of propagation. The electric field of these waves can be written

$$\mathbf{E} = A_1\mathbf{p}_1 \exp(-i\mathbf{k}_{a1}\cdot\mathbf{r}) + A_2\mathbf{p}_2 \exp(-i\mathbf{k}_{a2}\cdot\mathbf{r}) + B_1\mathbf{q}_1 \exp(-i\mathbf{k}_{b1}\cdot\mathbf{r}) + B_2\mathbf{q}_2 \exp(-i\mathbf{k}_{b2}\cdot\mathbf{r}) \qquad (4.3\text{-}1)$$

where A_1, A_2, B_1, B_2 are complex amplitudes of the waves, $\mathbf{p}_1$, $\mathbf{p}_2$, $\mathbf{q}_1$, $\mathbf{q}_2$ are unit vectors representing the polarization states of the modes of propagation and $\mathbf{k}_{a1}$, $\mathbf{k}_{a2}$, $\mathbf{k}_{b1}$, $\mathbf{k}_{b2}$ are wave vectors. The polarization states and magnitude of the wave vectors depend on the orientation of the crystal axes.

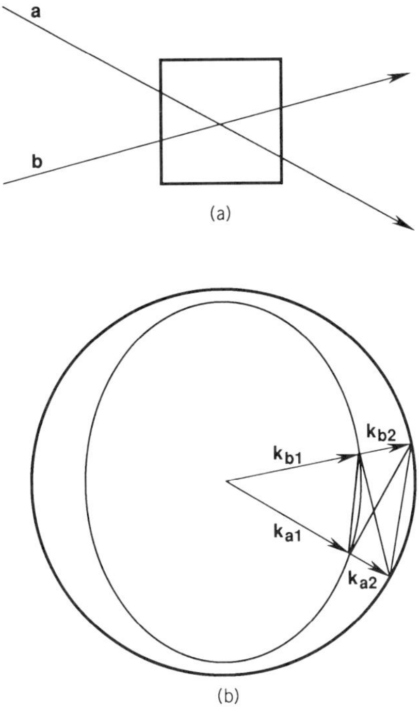

Figure 4.8. (a) Two-beam coupling in anisotropic photorefractive media. (b) The wave vectors of the beams and the wave vectors of the photorefractive gratings. There are, in general, four grating wave vectors.

We consider the case when $\mathbf{k}_{a1}$ and $\mathbf{k}_{a2}$ are parallel and $\mathbf{k}_{b1}$ and $\mathbf{k}_{b2}$ are also parallel. According to the discussion in Chapter 1, we have

$$\mathbf{p}_1 \cdot \mathbf{p}_2 = \mathbf{q}_1 \cdot \mathbf{q}_2 = 0 \tag{4.3-2}$$

In other words, the two independent modes of propagation along each direction are mutually orthogonal. The rest of the pairs are not parallel or orthogonal and their magnitude of $\mathbf{p}_1 \cdot \mathbf{q}_1$, $\mathbf{p}_1 \cdot \mathbf{q}_2$, $\mathbf{p}_2 \cdot \mathbf{q}_1$, and $\mathbf{p}_2 \cdot \mathbf{q}_2$ depend on the direction of propagation of the two beams. In uniaxial crystals, these unit vectors can be easily obtained by using Eqs. (1.3-27) and (1.3-28). For most photorefractive crystals with $n_e \simeq n_o$, the displacement field vector is approximately parallel to the electric field vector.

These two beams intersect inside the photorefractive medium and produce, in general, four interference fringe patterns. The index grating can be written formally as

$$\Delta\epsilon = \sum_{i,j} (\Delta\epsilon)_{ij} A_i B_j^* \mathbf{p}_i \cdot \mathbf{q}_j^* \exp(-i\mathbf{K}_{ij} \cdot \mathbf{r})/I_0 \tag{4.3-3}$$

where the summation is over $i = 1, 2$, $j = 1, 2$, I_0 represents the total intensity, $\mathbf{K}_{ij}$ represents the grating vector formed by waves A_i and B_j and $(\Delta\epsilon)_{ij}$ represents the strength of the gratings. The grating wave vectors can be written

$$\mathbf{K}_{ij} = \mathbf{k}_{ai} - \mathbf{k}_{bj} \tag{4.3-4}$$

and are illustrated in Figure 4.8(b).

There are, in general, four index gratings formed by Beam a and Beam b in the photorefractive medium. Each mode of polarization of one beam is coupled with two modes of the other beam. There are special cases where the number of index gratings reduce to two or only one. In uniaxial crystals such as $BaTiO_3$ or SBN, the ordinary mode of polarization of one beam is orthogonal to the extraordinary mode of the other beam provided that the c-axis is either parallel or perpendicular to the plane of incidence (i.e., the plane formed by the wave vectors of the two beams). Under these incidence conditions, the ordinary wave is coupled with the ordinary wave and the extraordinary wave is coupled with the extraordinary wave. And there are only two index gratings.

Consider the case when the c-axis is parallel to the plane of incidence (plane formed by the two beams). The ordinary modes of polarization of the two beams are the same. They are all perpendicular to the plane of incidence. The extraordinary modes of polarization of the two beams are parallel to the plane of incidence. They are not parallel to each other because of the finite angle between the two beams. As a result of the orthogonality between the modes, that is,

$$\mathbf{p}_1 \cdot \mathbf{p}_2 = \mathbf{q}_1 \cdot \mathbf{q}_2 = \mathbf{p}_1 \cdot \mathbf{q}_2 = \mathbf{p}_2 \cdot \mathbf{q}_1 = 0 \tag{4.3-5}$$

the index grating can be formally written

$$\Delta\epsilon = [(\Delta\epsilon)_1 A_1 B_1^* \exp(-i\mathbf{K}_1 \cdot \mathbf{r}) + (\Delta\epsilon)_2 A_2 B_2^* \cos 2\theta \exp(-i\mathbf{K}_2 \cdot \mathbf{r})] \frac{1}{I_0} \tag{4.3-6}$$

where $(\Delta\epsilon)_1$ and $(\Delta\epsilon)_2$ are tensors representing the amplitude of the photo-induced gratings and 2θ is the angle between the two beams in the medium. In Eq. (4.3-6), Index 1 represents the ordinary mode and Index 2 represents the extraordinary mode. There are only two index gratings with wave vectors $\mathbf{K}_1$ and $\mathbf{K}_2$. They are illustrated in Figure 4.9(a).

In the case when the c-axis is perpendicular to the plane of incidence, the extraordinary modes of the two beams are all parallel to the c-axis. Under such an incidence condition, the same orthogonality conditions given by Eq. (4.3-5) are also true. The index grating can thus be written

$$\Delta\epsilon = [(\Delta\epsilon)_1 A_1 B_1^* \cos 2\theta \exp(-i\mathbf{K}_1 \cdot \mathbf{r}) + (\Delta\epsilon)_2 A_2 B_2^* \exp(-i\mathbf{K}_2 \cdot \mathbf{r})] \frac{1}{I_0} \tag{4.3-7}$$

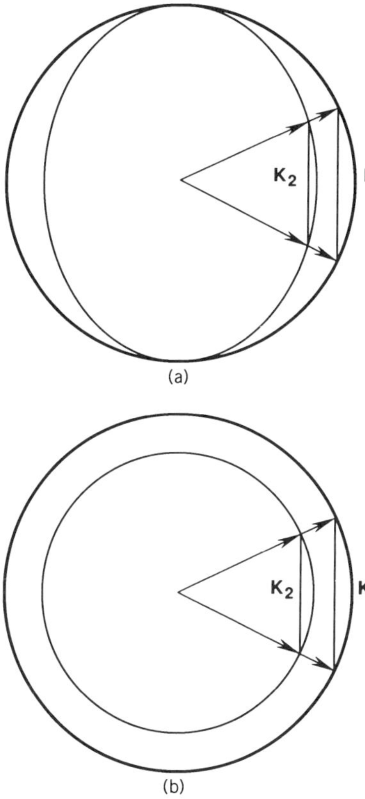

Figure 4.9. Two-beam coupling in uniaxial photorefractive crystals. (a): The c-axis is in the plane of incidence. (b): The c-axis is perpendicular to the plane of incidence.

where again 2θ is the angle between the two beams in the medium. There are again two index gratings with wave vectors $\mathbf{K}_1$ and $\mathbf{K}_2$. They are illustrated in Figure 4.9(b).

The wave vectors $\mathbf{K}_1$ and $\mathbf{K}_2$ in the above two cases are, in general, different from each other. Thus, the ordinary wave is coupled with the ordinary wave and the extraordinary wave is coupled with the extraordinary wave. In other words, there is very little cross-polarization coupling between the two beams because of Bragg mismatch. These two wave vectors become identical (i.e., $\mathbf{K}_1 = \mathbf{K}_2$) when the birefringence disappears (i.e., $n_e = n_o$). Thus in crystals such as GaAs, InP, cross-polarization photorefractive two-wave mixing is possible. We devote the next section to cross-polarization two-beam coupling in cubic crystals.

4.4 TWO-WAVE MIXING IN CUBIC CRYSTALS

Photorefractive two-beam coupling in electro-optic crystals has been studied extensively for its potential in many applications. Much attention has been focused on materials such as barium titanate ($BaTiO_3$), bismuth silicon oxide (BSO), strontium barium niobate (SBN), etc., because of their large coupling constants. Although these oxide materials are very efficient for two-beam coupling, they are relatively slow in response at low operating powers (see Table 3.2). Recently, several experimental investigations have been carried out to study the two-wave mixing in cubic crystals such as GaAs which responds much faster than any of the previously mentioned oxides at the same operating power.

In addition to the faster temporal response, the optical isotropy and the tensor nature of the electro-optic coefficients of GaAs crystals allow the possibility of cross-polarization coupling. Such cross-polarization two-wave mixing is very inefficient in barium titanate and strontium barium niobate because of the optical anisotropy except along certain special directions.

In this section, we describe a nonlinear model of photorefractive two-wave mixing in cubic crystals, especially those with point group symmetry of $\bar{4}3m$. The theory shows that cross-polarization two-wave mixing is possible in crystals such as GaAs. Exact solutions of coupled mode equations are obtained for the case of co-directional coupling.

Referring to Figure 4.10, we consider the intersection of two polarized beams inside a cubic photorefractive crystal. Since the crystal is optically isotropic, the electric field of the two beams can be written as

$$\mathbf{E} = (\mathbf{s}A_s + \mathbf{p}_1 A_p)\exp(-i\mathbf{k}_1 \cdot \mathbf{r}) + (\mathbf{s}B_s + \mathbf{p}_2 B_p)\exp(-i\mathbf{k}_2 \mathbf{r}) \quad (4.4\text{-}1)$$

where $\mathbf{k}_1$ and $\mathbf{k}_2$ are wave vectors of the beams, $\mathbf{s}$ is a unit vector perpendicular to the plane of incidence, $\mathbf{p}_1$, $\mathbf{p}_2$ are unit vectors parallel to the plane of incidence and perpendicular to the wave vectors, respectively.

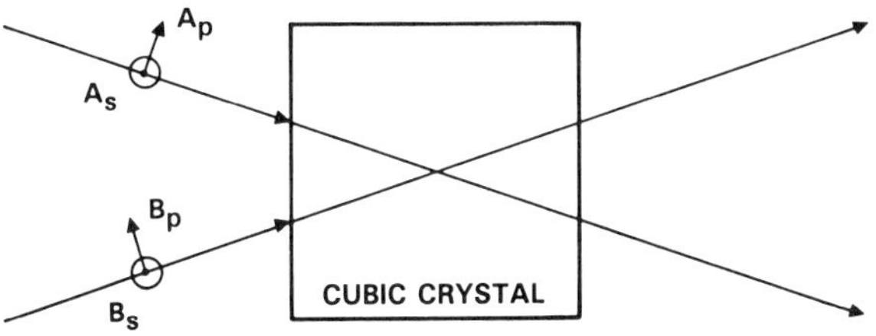

Figure 4.10. Schematic drawing of photorefractive two-beam coupling in cubic crystals.

Since each beam has two polarization components, there are four waves involved and A_s, A_p, B_s, B_p are amplitudes of the waves. All the waves are assumed to have the same frequency. In addition, we assume that the crystal does not exhibit optical rotation.

In the photorefractive crystal (from $z=0$ to $z=L$), these two beams generate an interference pattern,

$$\mathbf{E}^*\mathbf{E} = A_s^* A_s + A_p^* A_p + B_s^* B_s + B_p^* B_p + [(A_s^* B_s + A_p^* B_p \mathbf{p}_1 \cdot \mathbf{p}_2) \exp(-i\mathbf{K}\cdot\mathbf{r}) + \text{c.c.}] \tag{4.4-2}$$

where $\mathbf{K} = \mathbf{k}_2 - \mathbf{k}_1$ is the grating wave vector, and c.c. represents the complex conjugate. We note that there are two contributions to the sinusoidal variation of the intensity pattern. As a result of the photorefractive effect, a space-charge field E^{sc} is formed which induces a volume index grating via the Pockels effect,

$$(\Delta\epsilon)_{ij} = -\epsilon_0 n^4 r_{ijk} E_k \tag{4.4-3}$$

where ϵ_0 is the dielectric permittivity of a vacuum, n is the index of refraction of the crystal, r_{ijk} is the electro-optic coefficient, and E_k is the k-component ($k = x, y, z$) of the space-charge field. The fundamental component of the induced grating can be written

$$\Delta\epsilon = -\epsilon_0\epsilon_1[(A_s^* B_s + A_p^* B_p \cos 2\theta) \exp(-i\mathbf{K}\cdot\mathbf{r} - i\phi) + \text{c.c.}])/I_0 \tag{4.4-4}$$

where ϵ_1 is a 3×3 tensor, ϕ is the spatial phase shift between the index grating and the intensity pattern, and 2θ is the angle between the beams inside the crystal, and I_0 is given by

$$I_0 = A_s^* A_s + A_p^* A_p + B_s^* B_s + B_p^* B_p$$

For cubic crystals with point group symmetry of $\bar{4}3m$, ϵ_1 is given by

$$\epsilon_1 = n^4 r_{41} \begin{bmatrix} 0 & E_z & E_y \\ E_z & 0 & E_x \\ E_y & E_x & 0 \end{bmatrix} \tag{4.4-5}$$

where $r_{41} = r_{231} = r_{312} = r_{123}$, and E_x, E_y, E_z are the three components of the amplitude of the space-charge field.

Substitution of the index grating Eq. (4.4-3) into Maxwell's wave equation leads to the following set of coupled equations,

$$\begin{aligned}
\frac{d}{dz} A_s &= \frac{i}{2\beta_1} e^{i\phi}[\Gamma_{ss} B_s + \Gamma_{sp_2} B_p](A_s B_s^* + A_p B_p^* \cos 2\theta)/I_0 \\
\frac{d}{dz} B_s &= \frac{i}{2\beta_2} e^{-i\phi}[\Gamma_{ss} A_s + \Gamma_{sp_1} A_p](A_s^* B_s + A_p^* B_p \cos 2\theta)/I_0 \\
\frac{d}{dz} A_p &= \frac{i}{2\beta_1} e^{i\phi}[\Gamma_{p_1 s} B_s + \Gamma_{p_1 p_2} B_p](A_s B_s^* + A_p B_p^* \cos 2\theta)/I_0 \\
\frac{d}{dz} B_p &= \frac{i}{2\beta_2} e^{-i\phi}[\Gamma_{p_2 s} A_s + \Gamma_{p_2 p_1} A_p](A_s^* B_s + A_p^* B_p \cos 2\theta)/I_0
\end{aligned} \tag{4.4-6}$$

where

$$\Gamma = \omega^2 \mu \epsilon_0 \epsilon_1 \tag{4.4-7}$$

and

$$\Gamma_{ij} = \langle i|\Gamma|j\rangle, \quad i, j = \mathbf{s}, \mathbf{p}_1, \mathbf{p}_2 \tag{4.4-8}$$

and β_1, β_2 are the z-components of the wave vectors.

As indicated by the subscripts, Γ_{ij} is the coupling constant between the ith and jth polarized waves. Thus Γ_{ss} and $\Gamma_{p_1 p_2}$ are the parallel coupling constants, and Γ_{sp_1}, Γ_{sp_2} are the cross coupling constants. A similar set of equations had been derived by previous workers and exact solutions were obtained for the case of co-directional parallel coupling. Here we focus our attention on the case of cross-polarization coupling. In the cross-polarization coupling, the s-component of beam 1 is coupled with the p-component of beam 2, and the p-component of beam 1 is coupled with the s-component of beam 2.

4.4.1 Co-Directional Cross Coupling

To illustrate the use of coupled Eqs. (4.4-6), in the case of co-directional cross coupling, we consider a special case in which the crystal orientation does not allow the parallel coupling to occur. Such a two-beam coupling configuration is shown in Figure 4.11(a). The two beams enter the crystal in such a way that the grating wave vector is along the (1 1 0) direction of the crystal. In this configuration, the unit vector $\mathbf{s}$ is parallel to (0 0 1) and the unit vectors $\mathbf{p}_1$, $\mathbf{p}_2$ are perpendicular to (0 0 1). The amplitude of the induced index grating ϵ_1 can thus be written, according to Eq. (4.4-5)

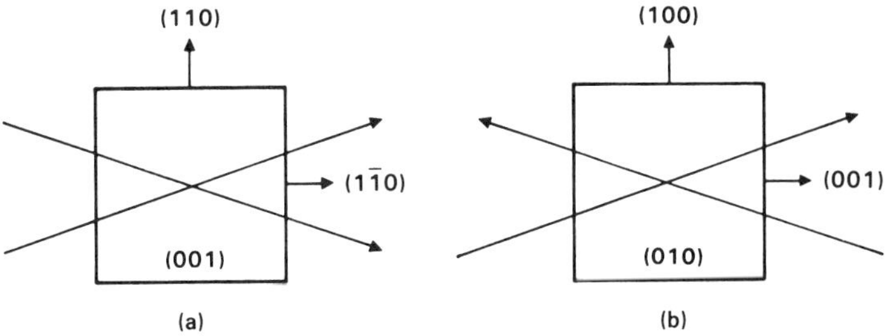

Figure 4.11. (a) A configuration for co-directional cross polarization coupling in a cubic crystal of point group symmetry $\bar{4}3m$; (b) a configuration for contra-directional cross polarization coupling in the same class of crystals.

$$\epsilon_1 = \frac{1}{\sqrt{2}} n^4 r_{41} \begin{bmatrix} 0 & 0 & 1 \\ 0 & 0 & 1 \\ 1 & 1 & 0 \end{bmatrix} E^{sc} \tag{4.4-9}$$

where E^{sc} is the amplitude of the space-charge field.

According to Eqs. (4.4-9) and (4.4-8), and after a few steps of algebra, the coupling constants can be written

$$\begin{aligned} &\Gamma_{ss} = \Gamma_{p_1 p_2} = 0 \\ &\Gamma_{sp_1} = \Gamma_{p_1 s} = \Gamma_{sp_2} = \Gamma_{p_2 s} = (2\pi/\lambda)^2 n^4 r_{41} E^{sc} \cos\theta \end{aligned} \tag{4.4-10}$$

where we assume that the beams enter the crystal symmetrically such that

$$\beta_1 = \beta_2 = (2\pi/\lambda) n \cos\theta \tag{4.4-11}$$

We now substitute Eqs. (4.4-10) and (4.4-11) into the coupled equations (4.4-6). This leads to

$$\begin{aligned} \frac{d}{dz} A_s &= -\gamma B_p (A_s B_s^* + A_p B_p^* \cos 2\theta)/I_0 \\ \frac{d}{dz} B_s &= \gamma A_p (A_s^* B_s + A_p^* B_p \cos 2\theta)/I_0 \\ \frac{d}{dz} A_p &= -\gamma B_s (A_s B_s^* + A_p B_p^* \cos 2\theta)/I_0 \\ \frac{d}{dz} B_p &= \gamma A_s (A_s^* B_s + A_p^* B_p \cos 2\theta)/I_0 \end{aligned} \tag{4.4-12}$$

where we have taken $\phi = \pi/2$, and γ is real and is given by

$$\gamma = \tfrac{1}{2}(2\pi/\lambda) n^3 r_{41} E^{sc} \tag{4.4-13}$$

We notice from the coupled equations (4.4-12) that there are two contributions to the holographic index grating. The relative phase of these two contributions is very important because it determines whether these two parts enhance or destroy the index grating. Such a relative phase is determined by the phases of the four wave amplitudes. Thus the energy exchange among the four waves depends on the input polarization states.

We now derive the solutions of these coupled equations. According to Eqs. (4.4-12), the total intensity I_0 is a constant (i.e., independent of z). Thus it is convenient to normalize the beam amplitudes such that $I_0 = 1$. Here, remember that we neglect the material absorption in the coupled equations. We will obtain closed form solutions of these coupled equations for the case of no absorption. In the case when the material absorption cannot be neglected, the solutions are obtained by simply multiplying an exponential factor accounting for the absorption. This is legitimate provided that all four amplitudes have the same attenuation coefficient.

To obtain the solutions of the coupled equations (4.4-12), it is useful to employ some of the constants of integration which are given by

$$A_s A_s^* + B_p B_p^* = c_1 \tag{4.4-14}$$

$$A_p A_p^* + B_s B_s^* = c_2 \tag{4.4-15}$$

$$A_s A_p^* + B_s^* B_p = c_3 \tag{4.4-16}$$

$$A_s B_s - A_p B_p = c_4 \tag{4.4-17}$$

Using a change of variable, the coupled equations can be written [4],

$$\frac{d}{dz} g = -\gamma(g^2 c_2 \cos 2\theta + g\sigma^* - c_1) \tag{4.4-18}$$

$$\frac{d}{dz} f = \gamma(f^2 c_1 \cos 2\theta + f\sigma - c_2) \tag{4.4-19}$$

where

$$f = A_p / A_s, \qquad g = B_p / B_s \tag{4.4-20}$$

$$\sigma = c_3 - c_3^* \cos 2\theta \tag{4.4-21}$$

Equations (4.4-18) and (4.4-19) can be integrated by using formula

$$\int \frac{dx}{a + bx + cx^2} = \frac{-2}{(b^2 - 4ac)^{1/2}} \tanh^{-1} \frac{b + 2cx}{(b^2 - 4ac)^{1/2}}$$

and the results are

$$\begin{aligned} f = A_p / A_s &= [-\sigma + q \tanh(-q\gamma z/2 + C)] / (2c_1 \cos 2\theta) \\ g = B_p / B_s &= [-\sigma^* + q^* \tanh(q^* \gamma z/2 + C')] / (2c_2 \cos 2\theta) \end{aligned} \tag{4.4-22}$$

with

$$q^2 = 4c_1c_2 + \sigma^2 \tag{4.4-23}$$

where C and C' are constants of integration which are determined by the boundary condition at $z = 0$.

Once f and g are known, the intensity of all four waves can be obtained by using the conservation laws given by Eqs. (4.4-14)–(4.4-17). They are given by

$$\begin{aligned}
|A_p|^2 &= \frac{|f|^2c_1 - |fg|^2c_2}{1 - |fg|^2} \\
|B_p|^2 &= \frac{|g|^2c_2 - |fg|^2c_1}{1 - |fg|^2} \\
|A_s|^2 &= \frac{|A_p|^2}{|f|^2} = \frac{c_1 - |g|^2c_2}{1 - |fg|^2} \\
|B_s|^2 &= \frac{|B_p|^2}{|g|^2} = \frac{c_2 - |f|^2c_1}{1 - |fg|^2}
\end{aligned} \tag{4.4-24}$$

We now consider a special case of particular interest in which the cubic crystal is sandwiched between a pair of cross polarizers. Such a configuration is useful to eliminate unwanted background radiation which often causes noises in the detection of signals. The boundary conditions in this case may be taken as

$$A_p(0) = B_p(0) = 0 \tag{4.4-25}$$

Using Eqs. (4.4-16), (4.4-20)–(4.4-23), we have $f(0) = g(0) = 0$, $c_3 = 0$, and $\sigma = 0$, and the solutions become

$$\begin{aligned}
f &= -q \tanh(q\gamma z/2)/2c_1 \cos 2\theta \\
g &= q \tanh(q\gamma z/2)/2c_2 \cos 2\theta
\end{aligned} \tag{4.4-26}$$

where $q = 2(c_1c_2)^{1/2}$.

Taking $\theta = 0$, and using Eqs. (4.4-24), the intensity of the four waves becomes

$$|A_s|^2 = c_1 \frac{1}{1 + \tanh^2(q\gamma z/2)}$$

$$|A_p|^2 = c_2 \frac{\tanh^2(q\gamma z/2)}{1 + \tanh^2(q\gamma z/2)}$$

$$|B_s|^2 = c_2 \frac{1}{1+\tanh^2(q\gamma z/2)}$$
$$|B_p|^2 = c_1 \frac{\tanh^2(q\gamma z/2)}{1+\tanh^2(q\gamma z/2)} \qquad (4.4\text{-}27)$$

where $c_1 = |A_s(0)|^2$, $c_2 = |B_s(0)|^2$. Figure 4.12(a) shows the variation of these intensities as functions of position for the case of $c_2/c_1 = 0.1$. We note that for strong coupling ($\gamma L \gg 1$), one half of the incident pump energy $A_s(0)$ is coupled to the p-component of the signal beam B_p, and one half of the incident signal energy $B_s(0)$ is coupled to the p-component of the pump beam. As a result of the opposite signs in the wave amplitudes ($f<0$), the two contributions to the index grating tend to cancel each other. Thus when the energy of the p-components reaches one half of the incident energies, the coupling ceases.

It is possible that the p-component of the signal beam B_p receives most of the incident pump energy A_s. Figure 14.12(b) illustrates a case in which the pump beam has both s- and p-components, whereas the signal beam is s-polarized. We note that, for strong coupling ($\gamma L \gg 1$), most of the energy of the s-component of the pump beam is coupled to the p-component of the signal beam. This is very similar to that of conventional two-wave mixing.

The exact solutions (4.4-22) and (4.4-24) are useful when the coupling is strong ($\gamma L \gg 1$) and the energy exchange is significant. For the case of weak coupling ($\gamma L \ll 1$), or very little pump depletion, we may assume that the pump beam amplitudes (A_s, A_p) remain virtually unchanged throughout the interaction. Under these conditions, the coupled equations become

$$\frac{d}{dz} B_s = \gamma a B_p + \gamma b B_s$$
$$\frac{d}{dz} B_p = \gamma c B_p + \gamma d B_s \qquad (4.4\text{-}28)$$

where a, b, c, d are dimensionless constants and are given by

$$a = |A_p|^2 \cos 2\theta / I_0$$
$$b = A_p A_s^* / I_0$$
$$c = A_s A_p^* \cos 2\theta / I_0 \qquad (4.4\text{-}29)$$
$$d = |A_s|^2 / I_0$$

We note that the magnitude of all four of these constants is less than unity.

The coupled equations (4.4-28) can be easily integrated and the results subject to the boundary conditions are

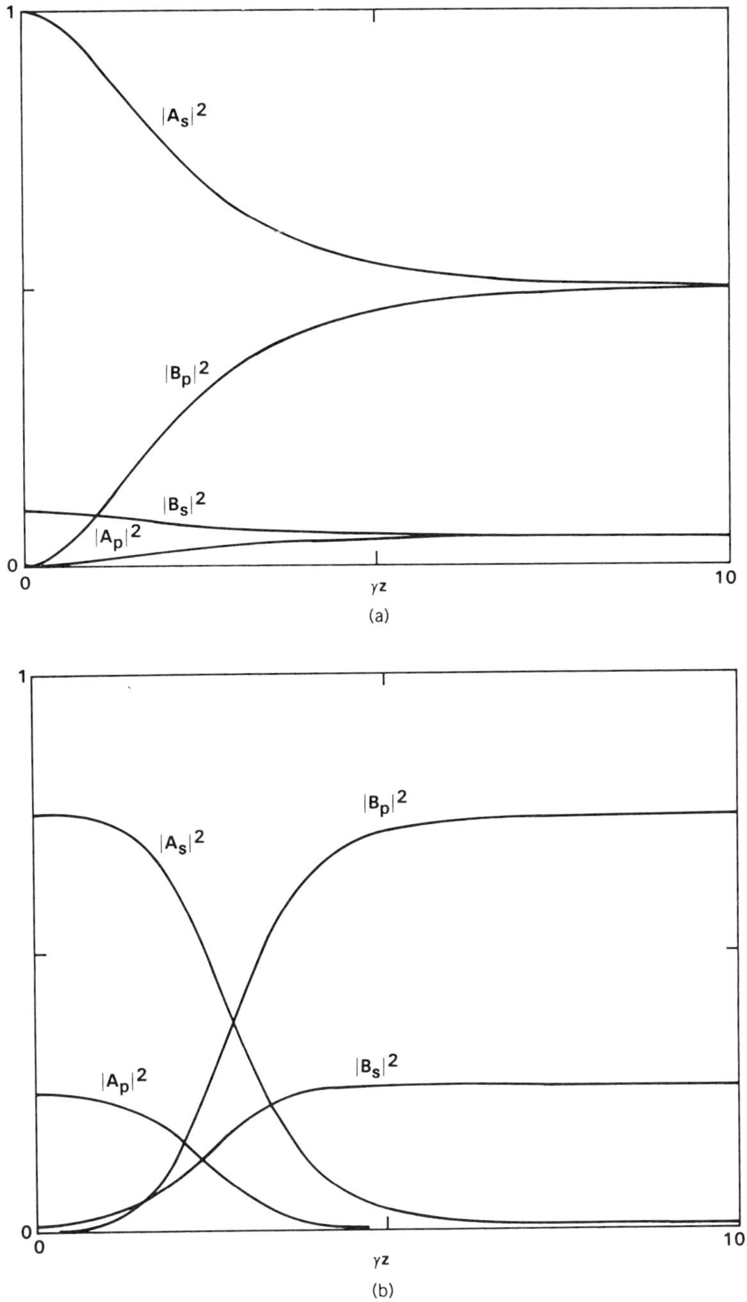

Figure 4.12. Intensity of the four waves are plotted as functions of distance for two interaction situations. (a) Both incident beams are s-polarized (i.e., $A_p(0) = B_p(0) = 0$, $c_2/c_1 = 0.1$). (b) The pump beam is linearly polarized at an azimuth angle of 30° relative to the s-direction, and the signal beam is s-polarized, $c_2/c_1 = 0.01$.

$$\begin{aligned} B_s(z) &= [bB_s(0) + aB_p(0)]\exp[(b+c)\gamma z] \\ &\quad + [cB_s(0) - aB_p(0)]/(b+c) \\ B_p(z) &= c[bB_s(0) + aB_p(0)]\exp[(b+c)\gamma z] \\ &\quad - b[cB_s(0) - aB_p(0)]/[a(b+c)] \end{aligned} \tag{4.4-30}$$

where $B_s(0)$, and $B_p(0)$ are the amplitudes of the signal beams at $z = 0$.

If we set $B_p(0) = 0$ in Eq. (4.4-30), we obtain

$$\begin{aligned} B_s(z) &= B_s(0)\{b\exp[(b+c)\gamma z] + c\}/(b+c) \\ B_p(z) &= B_s(0)bc\{\exp[(b+c)\gamma z] - 1\}/[a(b+c)] \end{aligned} \tag{4.4-31}$$

By examining Eqs. (4.4-30) and (4.4-31), we note that the exponential growth of the signal beam is dependent on the polarization state of the pump beam. If we assume further that $\gamma z \ll 1$, Eq. (4.4-31) can be written approximately as

$$B_s(z) = B_s(0) + B_s(0)\frac{A_pA_s^*}{I_0}\gamma z$$

$$B_p(z) = B_s(0)\frac{|A_s|^2}{I_0}\gamma z$$

We notice that the amplitude of B_s may increase or decrease depending on the polarization state of the pump beam A, whereas the amplitude B_p is an increasing function of γz.

4.4.2 Contra-Directional Cross Coupling

Referring to Figure 4.11(b), we consider a case of contra-directional interaction which does not permit the parallel coupling to occur. The two beams enter the crystal in such a way that the grating wave vector is along the (0 0 1) direction of the crystal. In this configuration, the unit vectors $\mathbf{p}_1$, $\mathbf{p}_2$ are perpendicular to (0 1 0) and the unit vector $\mathbf{s}$ is parallel to (0 1 0). The amplitude of the induced index grating ϵ_1 can thus be written, according to Eq. (4.4-5)

$$\epsilon_1 = n^4 r_{41}\begin{bmatrix} 0 & 1 & 0 \\ 1 & 0 & 0 \\ 0 & 0 & 0 \end{bmatrix} E^{sc} \tag{4.4-32}$$

where E^{sc} is the amplitude of the space-charge field.

According to Eqs. (4.4-8) and (4.4-32), and after a few steps of algebra, the coupling constants can be written

$$\Gamma_{ss} = \Gamma_{p_1 p_2} = 0$$
$$\Gamma_{sp_1} = \Gamma_{p_1 s} = \Gamma_{sp_2} = \Gamma_{p_2 s} = n^4 r_{41} E^{sc} \cos\theta \left(\frac{2\pi}{\lambda}\right)^2 \tag{4.4-33}$$

where we assume that the beams enter the crystal symmetrically (see Figure 4.11(b)) such that

$$\beta_1 = -\beta_2 = -(2\pi/\lambda) n \cos\theta \tag{4.4-34}$$

We now substitute Eqs. (4.4-33) and (4.4-34) into the coupled equations (4.4-6). This leads to

$$\begin{aligned}
\frac{d}{dz} A_s &= \gamma B_p (A_s B_s^* + A_p B_p^* \cos 2\theta)/I_0 \\
\frac{d}{dz} B_s &= \gamma A_p (A_s^* B_s + A_p^* B_p \cos 2\theta)/I_0 \\
\frac{d}{dz} A_p &= \gamma B_s (A_s B_s^* + A_p B_p^* \cos 2\theta)/I_0 \\
\frac{d}{dz} B_p &= \gamma A_s (A_s^* B_s + A_p^* B_p \cos 2\theta)/I_0
\end{aligned} \tag{4.4-35}$$

where we have taken $\phi = \pi/2$, and γ is real and is given by

$$\gamma = \tfrac{1}{2}(2\pi/\lambda) n^3 r_{41} E^{sc} \tag{4.4-36}$$

Notice that the coupled equations (4.4-35) are similar to those of equations (4.4-12), except the signs for A_s and A_p. The difference in signs is due to the direction of propagation of the pump beam (A_s, A_p). As a result of this difference, the total intensity I_0 is no longer a constant. According to Eq. (4.4-35), $(A_s^* A_s + A_p^* A_p - B_s^* B_s - B_p^* B_p)$, which is proportional to the net Poynting power flow along the $+z$ direction, is a constant. There are other constants of integration. These include:

$$A_s^* A_s - B_p^* B_p = c_1 \tag{4.4-37}$$

$$A_p^* A_p - B_s^* B_s = c_2 \tag{4.4-38}$$

$$A_s A_p^* - B_s^* B_p = c_3 \tag{4.4-39}$$

$$A_s B_s - A_p B_p = c_4 \tag{4.4-40}$$

Because of the fact that I_0 is not a constant, the integration of the coupled equations (4.4-35) is not as simple as that of Eq. (4.4-12). As of now, there is no closed form solution available. However, numerical techniques can be used to integrate the coupled equations.

For the case of no pump depletion, we may treat A_s and A_p as constants. In this case the coupled equations for B_s and B_p are identical to those of the co-directional coupling, and the solutions are given by Eqs. (4.4-30) and (4.4-31).

In summary, we have derived a general theory of the coupling of polarized beams in cubic photorefractive crystals. As a result of the optical isotropy of the crystal and the tensor nature of the holographic photorefractive grating, cross-polarization energy coupling occurs. Exact solutions for the case of co-directional coupling are obtained. Such cross-polarization coupling may be useful for the suppression of background noises.

4.5 OPTICAL FOUR-WAVE MIXING

Four-wave mixing (FWM) is a convenient method for the generation of phase conjugated waves. In this method, a nonlinear medium is pumped by a pair of counterpropagating laser beams. When a signal beam is incident into the medium, a fourth beam is generated. This beam is propagating opposite to the signal beam and is a time-reversed replica of the signal beam. In this section, we treat the coupling of these four beams in a photorefractive medium.

Referring to Figure 4.13, we consider the interaction of four beams in a photorefractive medium. We assume that all beams have the same frequency ω. The electric field can be written

$$E = \sum_{j=1}^{4} A_j \exp[i(\omega t - \mathbf{k}_j \cdot \mathbf{r})] \tag{4.5-1}$$

where A_1, A_2, A_3, A_4 are complex amplitudes of the four waves, $\mathbf{k}_1$, $\mathbf{k}_2$, $\mathbf{k}_3$, $\mathbf{k}_4$ are the wave vectors. In Eq. (4.5-1), we assume that all wave vectors are in the same plane and all waves are polarized perpendicular to the plane. In the case when $A_3 = A_4 = 0$, there are only two waves and only one photorefractive grating. The coupled equations (4.1-12) and (4.1-13) can be written

$$\frac{d}{dz} A_1 = -\frac{1}{2} \Gamma(A_1 A_2^*) A_2 / I_0 \tag{4.5-2}$$

$$\frac{d}{dz} A_2 = \frac{1}{2} \Gamma^*(A_1^* A_2) A_1 / I_0 \tag{4.5-3}$$

where we assume $\alpha = 0$ and Γ is given by

$$\Gamma = i \frac{2\pi n_1}{\lambda \cos\theta} e^{-i\phi} = \gamma + i2\beta \tag{4.5-4}$$

with γ and β given by Eq. (4.1-20).

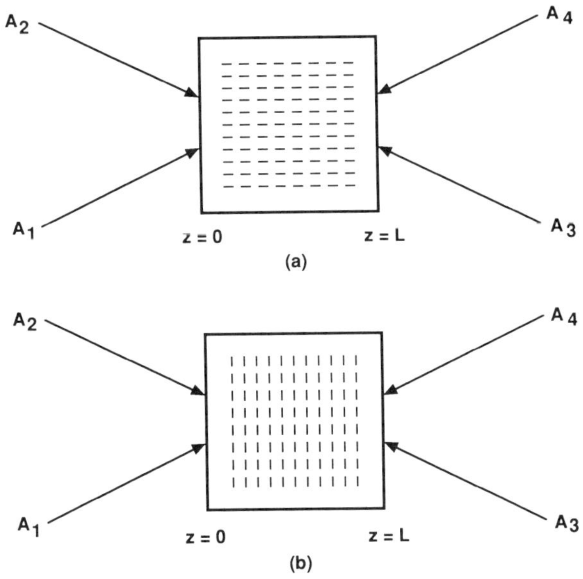

Figure 4.13. Four-wave mixing in photorefractive media. (a) Four-wave mixing via transmission gratings. The gratings are formed by beam pair (A_1, A_2) and/or beam pair (A_3, A_4). (b) Four-wave mixing via reflection gratings. The gratings are formed by beam pair (A_1, A_3) and/or beam pair (A_2, A_4).

In Eqs. (4.5-2) and (4.5-3), the term $A_1A_2^*$ represents the amplitude of the grating formed by A_1 and A_2. Now with the presence of four waves, there are, in general, six gratings in the medium. In the case of four-wave mixing for phase conjugation, the wave vectors come in two oppositely directed pairs, that is,

$$\mathbf{k}_2 = -\mathbf{k}_3 \quad \text{and} \quad \mathbf{k}_4 = -\mathbf{k}_1 \tag{4.5-5}$$

If we further assume that these waves enter the medium symmetrically as shown in Figure 4.13, then there are four different gratings. These gratings are represented by $(A_1A_2^* + A_3A_4^*)$, $(A_2A_4^* + A_1A_3^*)$, $(A_2A_3^*)$, and $(A_1A_4^*)$. The first one is a transmission grating and the second one is a reflection grating. The remaining ones are $2k$ gratings (i.e., gratings with wave vector $2k$). We will assume that only the transmission grating $(A_1A_2^* + A_3A_4^*)$ gives rise to strong interaction among the beams. This is the so-called one grating approximation.

4.5.1 Transmission Gratings

The predominance of one grating is common in practical situations due to the dependence of the space-charge field on the grating periods, grating orientations and coherence between the beams. Under these conditions, the coupled equations for the four complex amplitudes are

$$\frac{d}{dz} A_1 = -\frac{1}{2} \Gamma(A_1 A_2^* + A_3 A_4^*) A_2 / I_0 \tag{4.5-6}$$

$$\frac{d}{dz} A_2 = \frac{1}{2} \Gamma^*(A_1^* A_2 + A_3^* A_4) A_1 / I_0 \tag{4.5-7}$$

$$\frac{d}{dz} A_3 = \frac{1}{2} \Gamma(A_1 A_2^* + A_3 A_4^*) A_4 / I_0 \tag{4.5-8}$$

$$\frac{d}{dz} A_4 = -\frac{1}{2} \Gamma^*(A_1^* A_2 + A_3^* A_4) A_3 / I_0 \tag{4.5-9}$$

where Γ is the complex coupling constant given by Eq. (4.5-4). We note that there are two contributions to the grating, one due to beams 1 and 2, the other due to beams 3 and 4. These two contributions are identical in grating periods and orientations. They have the same coupling constant Γ. I_0 is the total averaged light intensity

$$I_0 = I_1 + I_2 + I_3 + I_4 \tag{4.5-10}$$

In the above set of nonlinear coupled-wave equations, we also assume that the material absorption is negligible. Before we proceed with the solution of these equations, we note that there are conservation laws that govern the variation of these amplitudes. Using Eqs. (4.5-6)–(4.5-9), we obtain

$$|A_1|^2 + |A_2|^2 = c_1 \tag{4.5-11}$$

$$|A_3|^2 + |A_4|^2 = c_2 \tag{4.5-12}$$

$$A_1 A_4 + A_2 A_3 = c \tag{4.5-13}$$

where c_1, c_2, and c are constants of integration. Equations (4.5-11) and (4.5-12) are results of conservation of energy, whereas Eq. (4.5-13) is related to the reciprocity theorem [5]. Using these constants of integration, exact solutions of the coupled equations can be obtained. We will first look at approximate solutions assuming negligible pump depletion. The issue of reciprocity in photorefractive wave mixing is discussed in Chapter 7.

4.5.1.1 Undepleted-Pump Approximation. In optical phase conjugation using four-wave mixing, we can designate beams 2 and 3 as pumps and beam 1 as the signal beam. In the undepleted-pump approximation, we assume that

$$|A_1|^2,\ |A_4|^2 \ll |A_2|^2,\ |A_3|^2 \tag{4.5-14}$$

so that

$$\frac{dA_2}{dz} = \frac{dA_3}{dz} = 0 \tag{4.5-15}$$

becomes a reasonable approximation. Thus, the coupled Eqs. (4.5-6)–(4.5-9) can be written

$$\frac{d}{dz} A_1 = -\frac{1}{2} \Gamma \frac{|A_2|^2}{I_0} A_1 - \frac{1}{2} \Gamma \frac{A_2 A_3}{I_0} A_4^* \tag{4.5-16}$$

$$\frac{d}{dz} A_4^* = -\frac{1}{2} \Gamma \frac{|A_3|^2}{I_0} A_4^* - \frac{1}{2} \Gamma \frac{A_2^* A_3^*}{I_0} A_1 \tag{4.5-17}$$

where we took a complex conjugation of Eq. (4.5-9) to arrive at Eq. (4.5-17).

With A_2 and A_3 being constant, Eqs. (4.5-16) and (4.5-17) can be integrated. The solutions, with the boundary conditions $A_4(L) = 0$ and $A_1(0)$, are

$$A_1(z) = \frac{\exp(-\frac{1}{2}\Gamma z) + q \exp(-\frac{1}{2}\Gamma L)}{1 + q \exp(-\frac{1}{2}\Gamma L)} A_1(0) \tag{4.5-18}$$

$$A_4^*(z) = \left(\frac{A_3^*}{A_2}\right) \frac{\exp(-\frac{1}{2}\Gamma z) - \exp(-\frac{1}{2}\Gamma L)}{1 + q \exp(-\frac{1}{2}\Gamma L)} A_1(0) \tag{4.5-19}$$

where q is the pump intensity ratio

$$q = \frac{|A_3|^2}{|A_2|^2} \tag{4.5-20}$$

For photorefractive media with Γ real and positive, both $A_1(z)$ and $A_4^*(z)$ are decaying exponentially in the medium. Figure 4.14 plots the intensities I_1 and I_4 as functions of z in the medium. We note that A_4 is generated by the four-wave mixing process. It starts with $A_4 = 0$ at $z = L$ and grows exponentially along the $-z$ direction.

The amplitude of the fourth beam A_4 at $z = 0$ is proportional to $A_1^*(0)$ according to Eq. (4.5-19). Thus, beam 4 is the phase conjugate of the signal beam A_1. The phase conjugate reflection coefficient is given by

$$\rho = \frac{A_4(0)}{A_1^*(0)} = \left(\frac{A_3}{A_2^*}\right) \frac{1 - \exp(-\frac{1}{2}\Gamma^* L)}{1 + q \exp(-\frac{1}{2}\Gamma^* L)} \tag{4.5-21}$$

Notice that ρ is a complex number with a phase dependent on $A_2 A_3$ as well as the complex coupling constant Γ. In photorefractive media with strong coupling constants, $\exp(-\frac{1}{2}\Gamma^* L)$ is either very large or very small. Under

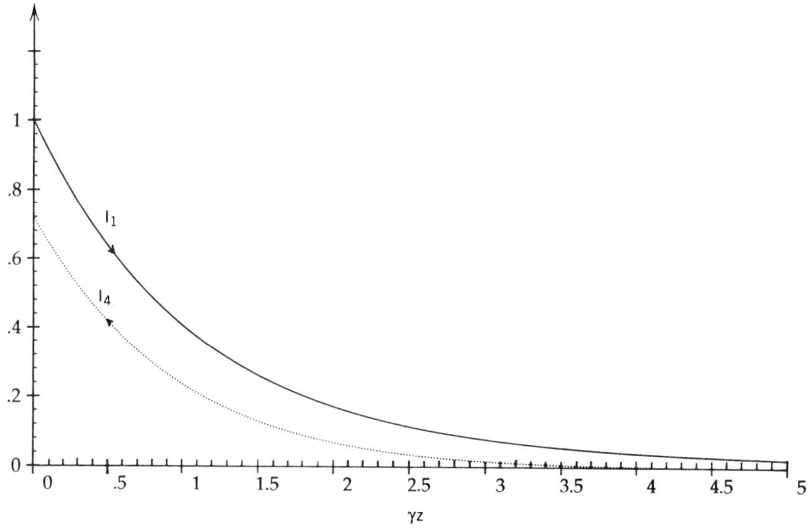

Figure 4.14. Beam intensities I_1 and I_4 versus position z in a non-depleted four-wave mixing with a pump ratio of $q = 1$ and a coupling constant of $\Gamma L = 5$.

these conditions, the phase of ρ equals to that of $A_2 A_3$. Physically, phase information of the pump beams will appear in the phase conjugate beam.

The phase conjugate reflectivity is given by, according to Eq. (4.5-21)

$$R = |\rho|^2 = \left| \frac{\sinh(\frac{1}{4}\Gamma L)}{\cosh(\frac{1}{4}\Gamma L - \ln\sqrt{q})} \right|^2 \tag{4.5-22}$$

To examine the dependence of these parameters, we rewrite the complex coupling constant Γ as

$$\Gamma = \gamma + i2\beta = i\,\frac{2\pi n_1}{\lambda \cos\theta}\, e^{-i\phi} \equiv ige^{-i\phi} \tag{4.5-23}$$

where g is the magnitude of Γ. Figure 4.15 plots the phase conjugate reflectivity R as a function of the pump intensity ratio q for various values of γL. The case in which $\phi = \pi/2$ is of particular interest because it occurs naturally in photorefractive media that operate by diffusion only ($E_0 = 0$).

We note that $\Gamma = \gamma$ for $\phi = \pi/2$. The phase conjugate reflectivity becomes

$$R = \frac{\sinh^2 \frac{1}{4}\gamma L}{\cosh^2(\frac{1}{4}\gamma L - \ln\sqrt{q})} \tag{4.5-24}$$

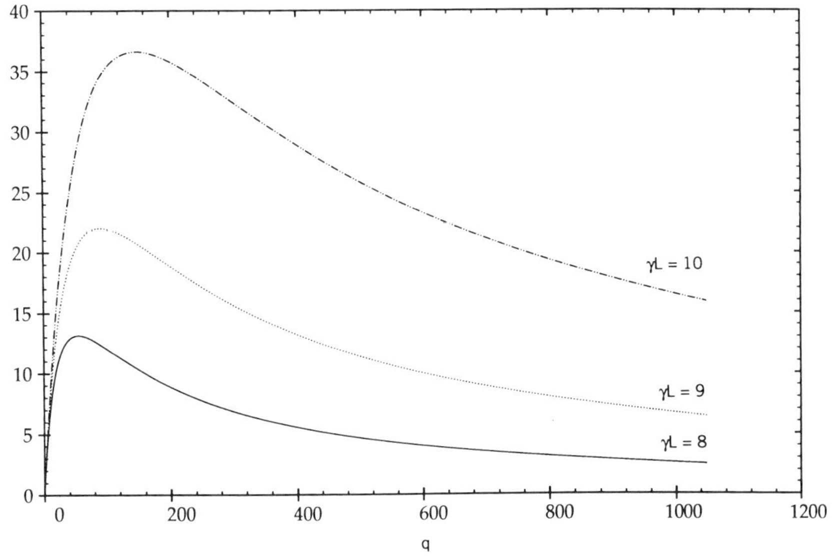

Figure 4.15. Phase conjugate reflectivity versus pump intensity ratio for $\gamma L = 8$, 9, 10 with $\phi = \pi/2$.

The reflectivity is an increasing function of γL. For symmetric pumping ($q = 1$), the reflectivity becomes $R = \tanh^2 \frac{1}{4}\gamma L$, which is always less than unity. Larger reflectivities can be achieved by using asymmetric pumping. In fact, maximum reflectivity of $R = \sinh^2 \frac{1}{4}\gamma L$ is achieved with a pump ratio of $q = \exp(\frac{1}{2}\gamma L)$.

Example: Phase Conjugate Reflectivity Consider an optical phase conjugator which consists of four-wave mixing in a $BaTiO_3$ crystal with $\gamma L = 20$. The maximum reflectivity for phase conjugation is $R = 5.5 \times 10^3$, according to Eq. (4.5-24). This happens at a pump beam ratio of $q = 2.2 \times 10^4$.

In the general case when $\phi \neq \pi/2$, maximum reflectivity occurs again at $q = \exp(\frac{1}{2}\gamma L)$ with $R_{\max}$ given by

$$R_{\max} = \left| \frac{\sinh \frac{1}{4}\Gamma L}{\cos \frac{1}{2}\beta L} \right|^2 \tag{4.5-25}$$

Figure 4.16 plots the phase conjugate reflectivity as a function of q for various values of ϕ. We note that higher reflectivity can be achieved by operating the phase conjugation at $\phi \neq \pm\pi/2$.

According to Eq. (4.5-22), self-oscillation ($R = \infty$) is possible provided

$$\cos \tfrac{1}{2}\beta L = 0 \quad \text{and} \quad q = \exp(\tfrac{1}{2}\gamma L) \tag{4.5-26}$$

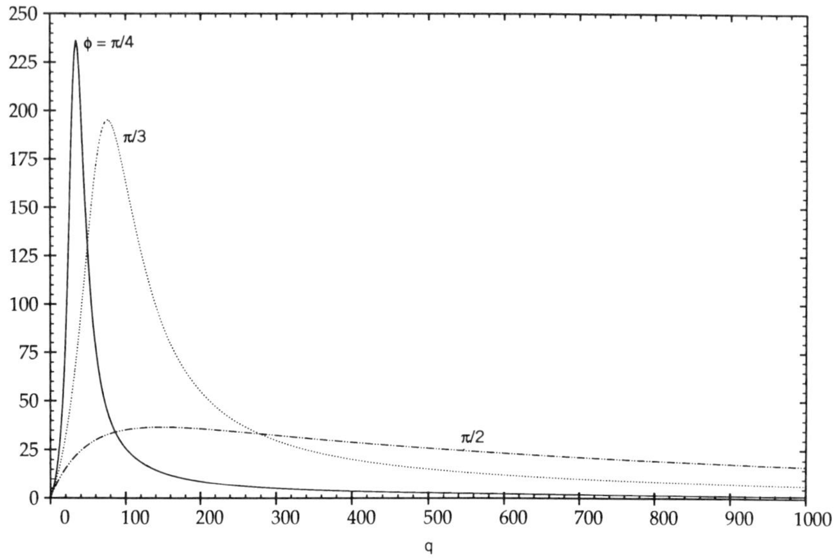

Figure 4.16. Phase conjugate reflectivity versus pump intensity ratio for various ϕ with $|\Gamma L| = 10$.

We note that $\phi \neq \pm\pi/2$ is essential for self-oscillation according to Eqs. (4.5-26) and (4.5-23). At self-oscillation, the boundary conditions become

$$A_1(0) = A_4(L) = 0 \tag{4.5-27}$$

and the solutions of coupled equations are

$$A_1(z) = [1 - \exp(-\tfrac{1}{2}\Gamma z)]B \tag{4.5-28}$$

$$A_4^*(z) = -\left(\frac{A_2^*}{A_3}\right)[1 + q\exp(-\tfrac{1}{2}\Gamma z)]B \tag{4.5-29}$$

where B is an arbitrary constant. Figure 4-17 plots the intensities of these two beams at self-oscillation. We note that both $|A_1|^2$ and $|A_4|^2$ grow exponentially from their zero boundary value (4.5-27). Equation (4.5-29) agrees with the boundary condition $A_4^*(L) = 0$ because $[1 + q\exp \times (-\tfrac{1}{2}\Gamma L)] = 0$ is equivalent to Eq. (4.5-26).

4.5.1.2 Exact Solutions. The undepleted-pump approximation provides simple solutions for the amplitudes A_1 and A_4. There are situations when such an approximation is no longer useful, especially in photorefractive media with large coupling constants γL. For these cases, we need to include

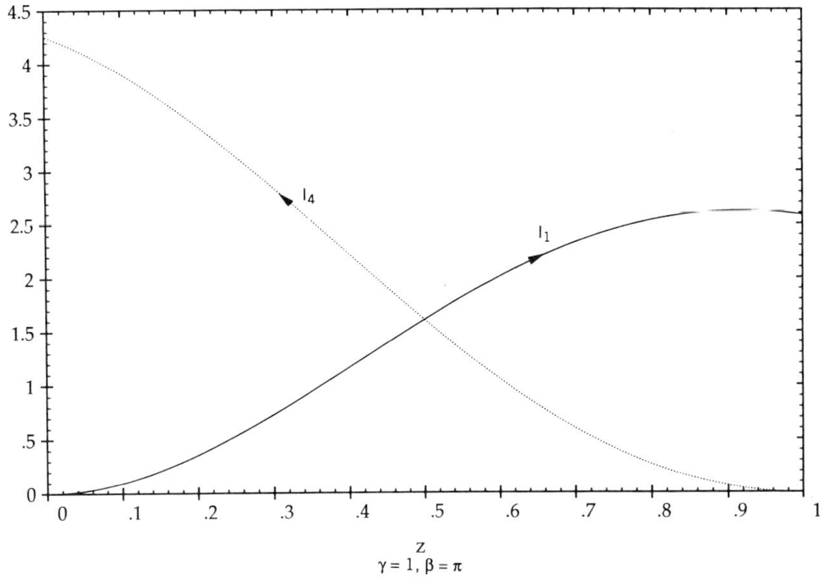

Figure 4.17. Beam intensities I_1 and I_4 as functions of z at self-oscillation in a four-wave mixing with $\gamma L = 1$ and $\beta L = \pi$.

the variation of the pump beams A_2 and A_3. Exact solutions of the coupled equations can be obtained by using a change of the variables, similar to that used in Eq. (4.4-20).

We define two new variables [6]:

$$u = \frac{A_2}{A_3^*} \quad \text{and} \quad v = \frac{A_4}{A_1^*} \tag{4.5-30}$$

By taking differentiation with respect to z, we obtain

$$u' = \frac{A_2'}{A_3^*} - \frac{A_2 A_3^{*\prime}}{A_3^{*2}} \tag{4.5-31}$$

$$v' = \frac{A_4'}{A_1^*} - \frac{A_2 A_1^{*\prime}}{A_1^{*2}} \tag{4.5-32}$$

where the $'$ indicates differentiation with respect to z. We now use the coupled equations and the conservation laws, Eqs. (4.5-6)–(4.5-13) and rewrite the coupled equations in terms of u and v. After a few steps of algebraic manipulation, we obtain

$$\frac{du}{dz} = \frac{\Gamma^*}{2I_0}[c + u(c_1 - c_2) - u^2 c^*] \tag{4.5-33}$$

$$\frac{dv}{dz} = -\frac{\Gamma^*}{2I_0}[c - v(c_2 - c_1) - v^2c^*] \tag{4.5-34}$$

These two equations can be integrated by using the following formula

$$\int \frac{dx}{a + bx + cx^2} = \frac{-2}{(b^2 - 4ac)^{1/2}} \tanh^{-1} \frac{b + 2cx}{(b^2 - 4ac)^{1/2}} \tag{4.5-35}$$

The results are

$$\tanh^{-1} \frac{2c^*u - \sigma}{s} = \frac{s\Gamma^*}{4I_0} z + D \tag{4.5-36}$$

$$\tanh^{-1} \frac{2c^*v + \sigma}{s} = -\frac{s\Gamma^*}{4I_0} z + D' \tag{4.5-37}$$

where D and D' are constants of integration, and s, σ are given by

$$\sigma = c_1 - c_2 \tag{4.5-38}$$

$$s = (\sigma^2 + 4|c|^2)^{1/2} \tag{4.5-39}$$

Physically, the constant σ represents the net power flow along $+z$ direction. The constants are determined by the boundary conditions.

For optical phase conjugation via four-wave mixing, the boundary conditions are $A_4(L) = 0$ and $A_1(0) \neq 0$, which is equivalent to $v = 0$ at $z = L$. Using these conditions and Eq. (4.5-37), we obtain

$$v = \frac{A_4(z)}{A_1^*(z)} = \frac{2c \tanh \kappa(z - L)}{-s + \sigma \tanh \kappa(z - L)} \tag{4.5-40}$$

where the new parameter κ is defined as

$$\kappa = \frac{s\Gamma^*}{4I_0} \tag{4.5-41}$$

The phase conjugate reflection coefficient is given by

$$\rho = \frac{A_4(0)}{A_1^*(0)} = \frac{2c \tanh \kappa L}{s + \sigma \tanh \kappa L} \tag{4.5-42}$$

which reduces to Eq. (4.5-21) in the case when the pump depletion is negligible (see Problem 4.7).

Equation (4.5-42) can now be employed to calculate the reflectivity of phase conjugation. Although the expression seems to be simple, the constants s, σ, c, and κ are not directly related to the input amplitudes $A_1(0)$,

$A_2(0)$, $A_3(L)$, and $A_4(L)$. According to Eqs. (4.5-11)–(4.5-13) and $A_4(L) = 0$, we obtain

$$\sigma = c_1 - c_2 = I_1(0) + I_2(0) - I_3(L) \tag{4.5-43}$$

$$c = A_1(0)A_4(0) + A_2(0)A_3(0) = A_3(L)A_2(L) \tag{4.5-44}$$

where we note that $A_2(L)$, $A_3(0)$, and $A_4(0)$ are not known. Equation (4.5-44) can be rewritten in terms of u and v (or ρ)

$$c = \rho I_1(0) + \frac{1}{u_0^*} I_2(0) = u_L I_3(L) \tag{4.5-45}$$

where u_0 and u_L are values of u at $z = 0$ and $z = L$, respectively. These two values are related by Eq. (4.5-36). The relationship can be written

$$\tanh^{-1}\left(\frac{2c^* u_L - \sigma}{s}\right) - \tanh^{-1}\left(\frac{2c^* u_0 - \sigma}{s}\right) = \kappa L \tag{4.5-46}$$

or equivalently

$$\frac{2c^* s(u_L - u_0)}{s^2 - (\sigma - 2c^* u_0)(\sigma - 2c^* u_L)} = \tanh \kappa L \tag{4.5-47}$$

In arriving at Eq. (4.5-47), we used the following identity.

$$\tanh(x + y) = \frac{\tanh x + \tanh y}{1 + \tanh x \tanh y}$$

Using Eqs. (4.5-45) and (4.5-42), we can solve for u_0 and u_L in terms of σ, c, s, $I_1(0)$, and $I_3(L)$. They are given by

$$u_L = \frac{c}{I_3(L)} \tag{4.5-48}$$

$$u_0^* = \frac{I_2(0)}{c - I_1(0)\dfrac{2c \tanh \kappa L}{s + \sigma \tanh \kappa L}} \tag{4.5-49}$$

These two expressions can now be substituted into Eq. (4.5-47). The result is a transcendental equation for c. The equation can then be solved numerically for a given set of input amplitudes. Once c is obtained, s and κ can all be obtained by using Eqs. (4.5-39) and (4.5-41). These numbers can then be used to evaluate the phase conjugate reflectivity using Eq. (4.5-42). Figure 4.18 plots the phase conjugate reflectivity $|\rho|^2$ as a function of the coupling constant.

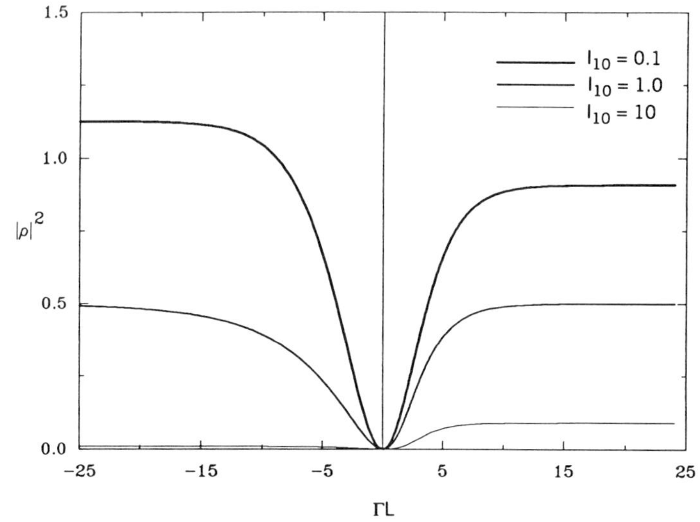

(a)

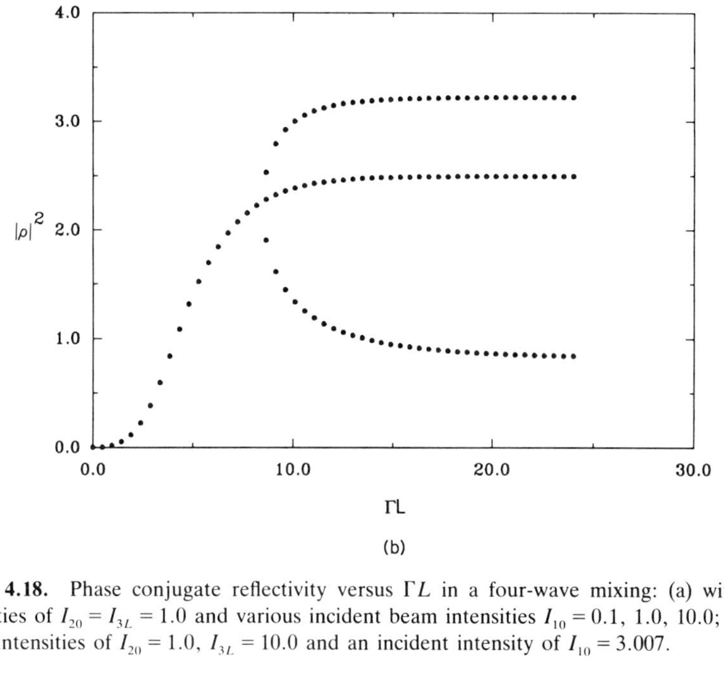

(b)

Figure 4.18. Phase conjugate reflectivity versus ΓL in a four-wave mixing: (a) with pump intensities of $I_{20} = I_{3L} = 1.0$ and various incident beam intensities $I_{10} = 0.1$, 1.0, 10.0; (b) with pump intensities of $I_{20} = 1.0$, $I_{3L} = 10.0$ and an incident intensity of $I_{10} = 3.007$.

We note that $|\rho|^2$ can be a multivalued function of ΓL under the appropriate conditions. In other words, two or more values of reflectivity may exist for a given set of input amplitudes and coupling constant (see Figure 4.18(b)). This may lead to bistability. In addition, phase conjugate reflectivity $|\rho|^2$ is a nonlinear function of the input intensities I_{10}, I_{20}, and I_{3L}. Figure 4.19 plots $|\rho|^2$ versus I_{10} for various incident configurations. We note that the reflectivity decreases as the incident intensity I_{10} becomes greater than either of the pump intensities. Again, multivalued response is possible and this may lead to optical bistability (see Figure 4.19(b)).

We now examine the variation of the beam intensities in the photorefractive medium. The constant D in Eq. (4.5-36) can be eliminated if either u_0 or u_L is known. We obtain from Eq. (4.5-36)

$$u = \frac{A_2(z)}{A_3^*(z)} = \frac{su_0 + (2c + \sigma u_0)\tanh \kappa z}{s - (\sigma - 2c^* u_0)\tanh \kappa z} \tag{4.5-50}$$

Once c is evaluated from the input amplitudes, u and v are obtained as functions of z. The intensities can then be obtained by using the conservation laws Eqs. (4.5-11) and (4.5-12). By squaring Eqs. (4.5-40) and (4.5-50), we obtain

$$I_4 = |v|^2 I_1 \tag{4.5-51}$$

$$I_2 = |u|^2 I_3 \tag{4.5-52}$$

Similarly, Eqs. (4.5-11) and (4.5-12) can be rewritten

$$I_1 + I_2 = c_1 \tag{4.5-53}$$

$$I_3 + I_4 = c_2 \tag{4.5-54}$$

These four equations can be solved for the intensities. The solutions are

$$I_1 = \frac{c_1 - |u|^2 c_2}{1 - |uv|^2} \tag{4.5-55}$$

$$I_2 = \frac{c_2|u|^2 - |uv|^2 c_1}{1 - |uv|^2} \tag{4.5-56}$$

$$I_3 = \frac{c_2 - |v|^2 c_1}{1 - |uv|^2} \tag{4.5-57}$$

$$I_4 = \frac{c_1|v|^2 - |uv|^2 c_2}{1 - |uv|^2} \tag{4.5-58}$$

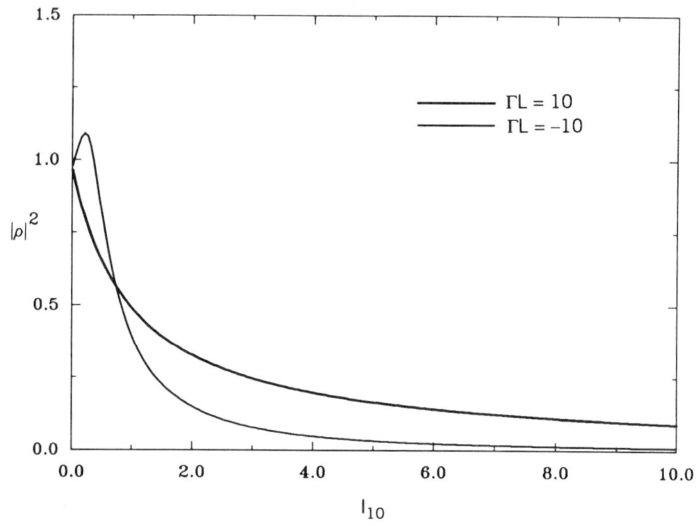

(a)

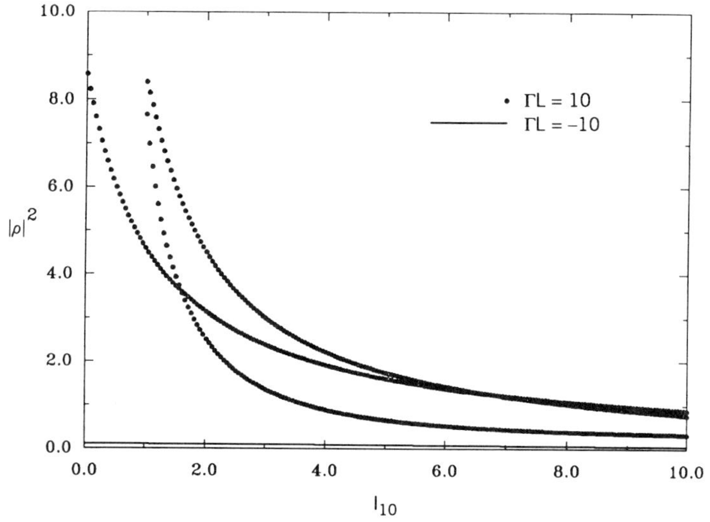

(b)

Figure 4.19. Phase conjugate reflectivity versus incident signal intensity I_{10}: (a) with pump intensities of $I_{20} = I_{3L} = 1.0$ and coupling constants of $\Gamma L = -10.0$, 10.0; (b) with pump intensities of $I_{20} = 1.0$, $I_{3L} = 10.0$ and coupling constants of $\Gamma L = -10.0$, 10.0.

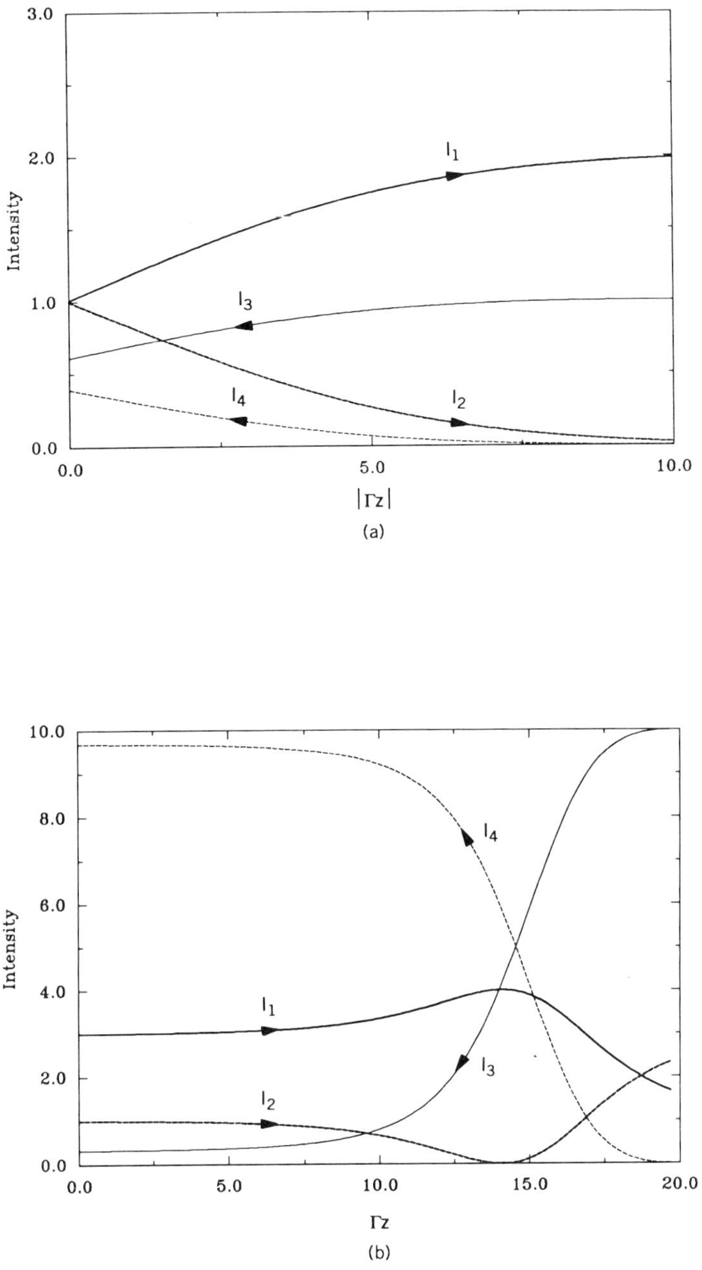

Figure 4.20. Beam intensities $I_1(z)$, $I_2(z)$, $I_3(z)$, $I_4(z)$ for typical four-wave mixing configurations: (a) with boundary conditions of $I_{10} = I_{20} = I_{3L} = 1.0$ and a coupling constant of $\Gamma = -1.0$; (b) with boundary conditions of $I_{10} = 3.007$, $I_{20} = 1.0$, $I_{3L} = 10.0$ and a coupling constant of $\Gamma = 1.0$.

Figure 4.20 plots the spatial variation of these intensities inside the photorefractive medium for various different incident conditions.

4.5.2 Reflection Gratings

In the above discussion, we assume that there is only one grating $(A_1A_2^* + A_3A_4^*)$ which is responsible for the wave mixing. There are circumstances that this grating may be vanishing or negligible due to mutual incoherence, orthogonal polarizations, or vanishing electro-optic coefficients. Under these circumstances, we may have to consider the effect of the reflection grating $(A_1A_3^* + A_2A_4^*)$. This is especially important when beam 1 is only coherent with beam 3 and beam 2 is only coherent with beam 4. The coupled equations are written

$$\begin{aligned}
\frac{d}{dz} A_1 &= -\frac{1}{2} \Gamma(A_1A_3^* + A_2A_4^*)A_3/I_0 \\
\frac{d}{dz} A_2 &= -\frac{1}{2} \Gamma(A_1A_3^* + A_2A_4^*)A_4/I_0 \\
\frac{d}{dz} A_3 &= -\frac{1}{2} \Gamma^*(A_1^*A_3 + A_2^*A_4)A_1/I_0 \\
\frac{d}{dz} A_4 &= -\frac{1}{2} \Gamma^*(A_1^*A_3 + A_2^*A_4)A_2/I_0
\end{aligned} \tag{4.5-59}$$

where Γ is the coupling constants and I_0 is the total intensity (i.e., $I_0 = I_1 + I_2 + I_3 + I_4$). Again, the conservation of energy and the reciprocity theorem lead to the following constants of integration:

$$|A_1|^2 - |A_3|^2 = c_1' \tag{4.5-60}$$

$$|A_2|^2 - |A_4|^2 = c_2' \tag{4.5-61}$$

$$A_1A_4 - A_2A_3 = c' \tag{4.5-62}$$

where c_1', c_2', c' are constants. The integration of these equations is more complicated because I_0 is not a constant. Closed form solutions are available. In the last section of this chapter, we will employ the method of grating integral to obtain closed form solutions for the cases when Γ is real. For general solution with complex Γ, readers are referred to [8].

4.5.3 Self-Oscillation

We consider self-oscillation in four-wave mixing with transmission gratings (see Figure 4.21). In self-oscillation, there is no incident signal beam (i.e., $A_1(0) = 0$). The photorefractive medium is pumped by a pair of counterpropagating beams A_2 and A_3. Such self-oscillation was treated earlier using

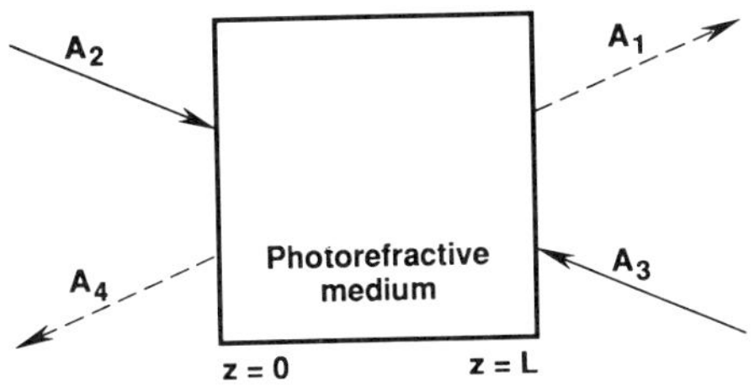

Figure 4.21. Self-oscillation with optical four-wave mixing in a photorefractive medium. The boundary conditions are $A_1(0) = A_4(L) = 0$.

no-pump-depletion approximation. We now investigate the properties of self-oscillation using the results of the exact solutions.

According to Eq. (4.5-42), self-oscillation occurs when

$$s + \sigma \tanh \kappa L = 0 \tag{4.5-63}$$

which corresponds to an infinite phase conjugate reflectivity. By examining Eq. (4.5-63), we find that self-oscillation occurs only when κL is complex. For real κL, $\tanh \kappa L$ is always less than unity and $|\sigma \tanh \kappa L|$ is always less than s, according to Eq. (4.5-39). Using the normalization $I_0 = 1$ and $\Gamma = \gamma + i2\beta$, the oscillation condition can be written

$$\frac{\tanh \dfrac{\gamma L}{4} s + i \tan \dfrac{\beta L}{2} s}{1 + i \tanh \dfrac{\gamma L}{4} s \tan \dfrac{\beta L}{2} s} = -\frac{s}{\sigma} \tag{4.5-64}$$

Since the right side is real and greater than unity, solutions to Eq. (4.5-64) exist only when

$$\tan \frac{\beta L}{2} s = \infty \tag{4.5-65}$$

and

$$\tanh \frac{\gamma L}{4} s = -\frac{\sigma}{s} \tag{4.5-66}$$

These conditions reduce to Eq. (4.5-26) when the pump depletion is negligible. Let I_{20} and I_{3L} be the incident intensities, then σ is given by

$$\sigma = \frac{I_{20} - I_{3L}}{I_{20} + I_{3L}} = \frac{1 - q}{1 + q} \tag{4.5-67}$$

where we used $I_0 = I_{20} + I_{3L}$ and q is the pump intensity ratio.

With σ known and for each given γL, Eq. (4.5-66) can be used to solve for s (or equivalently $|c|$). Once $|c|$ is known, $|u|$ and $|v|$ can be evaluated

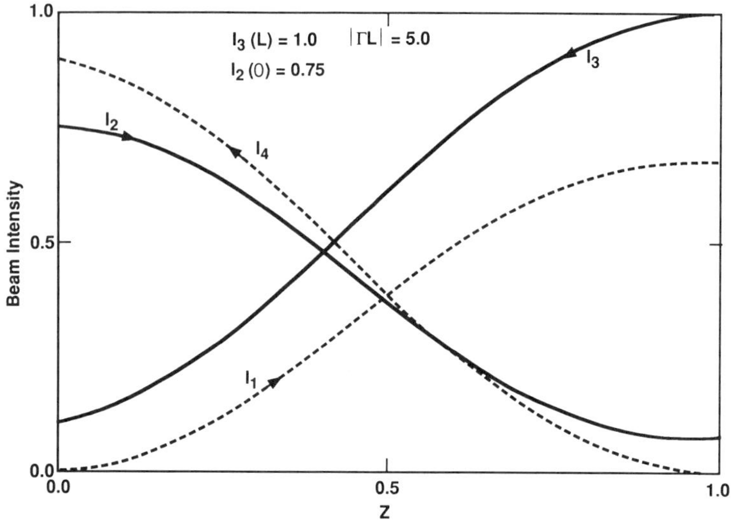

Figure 4.22. Beam intensities I_1, I_2, I_3, and I_4 versus z at self-oscillation. The boundary conditions are $I_1(0) = 0$, $I_2(0) = 0.75$, $I_3(L) = 1.0$, $I_4(L) = 0$ and the coupling constant is $|\Gamma L| = 5.0$. The arrows indicate the direction of propagation of the beams.

for all z and the intensities I_1, I_2, I_3, and I_4 can be obtained by using Eqs. (4.5-55)–(4.5-58). Figure 4.22 plots the intensity variation of these beams inside the photorefractive medium.

By examining the oscillation condition in Eq. (4.5-63) or Eq. (4.5-66), we find that oscillation occurs only when $\sigma \neq 0$, which requires asymmetrical pumping. At oscillation, I_{1L} or I_{40} are functions of γL. The threshold condition can be obtained by putting $I_{1L} = 0$ (or $I_{40} = 0$). This leads to $|c|^2 = I_{20} I_{3L}$ and $s = 1$. Thus, the threshold condition becomes

$$\tanh \frac{\gamma L}{4} = \frac{q-1}{q+1} \tag{4.5-68}$$

which is exactly the same as Eq. (4.5-26). The threshold condition in the exact solution is precisely the oscillation condition in the no-pump-depletion approximation.

By examining the beam intensities in Figure 4.22, we find that beams 1 and 4 grow exponentially from the zero boundary value and eventually deplete the energy of the two pump beams.

4.6 NONDEGENERATE FOUR-WAVE MIXING

In the above discussion of four-wave mixing, we assume that all the waves have exactly the same frequency and the four beams form two pairs of counterpropagating waves. These conditions, along with the one-grating approximation, lead to a set of coupled equations (4.5-6)–(4.5-9) which can then be solved with various boundary conditions. We now consider the case when the frequencies of the waves can be different. Let the electric field be written

$$E = \sum_{j=1}^{4} A_j \exp[i(\omega_j t - \mathbf{k}_j \cdot \mathbf{r})] \tag{4.6-1}$$

where A_1, A_2, A_3, A_4 are the amplitudes of the waves, $\mathbf{k}_1$, $\mathbf{k}_2$, $\mathbf{k}_3$, $\mathbf{k}_4$ are the wave vectors and ω_1, ω_2, ω_3 and ω_4 are the frequencies.

One-grating approximation used before is still valid provided the wave vectors and frequencies satisfy the following conditions:

$$\mathbf{k}_2 - \mathbf{k}_1 = \mathbf{k}_4 - \mathbf{k}_3 = \mathbf{K} \tag{4.6-2}$$

$$\omega_2 - \omega_1 = \omega_4 - \omega_3 = \Omega \tag{4.6-3}$$

where $\mathbf{K}$ is the wave vector of the grating and Ω is the angular frequency of the grating. Under these conditions, the coupled equations for the four amplitudes are formally identical to those given by Eqs. (4.5-6)–(4.5-9). The coupling constant for the nondegenerate case is given by, according to Eq. (3.5-17)

$$\Gamma' = \frac{\Gamma}{1 + i\Omega\tau} \tag{4.6-4}$$

where τ is a characteristic time constant of the photorefractive grating.

The solutions derived earlier in Section 4.5 can thus be used for nondegenerate four-wave mixing, provided we use the appropriate coupling constant as given by Eq. (4.6-4).

4.7 TWM APPROXIMATION

Although exact solutions can be derived for four-wave mixing, the results are difficult to use because of the complicated transcendental equation involved. Equation (4.5-47) must be solved for the constant c using numerical techniques. In Section 4.5, we discuss an approximation which assumes negligible pump depletion. Such approximation is useful when the coupling constant ΓL is small. In many of the photorefractive media such as $BaTiO_3$, BSO, $LiNbO_3$, SBN, the coupling constants are easily in the range of $\Gamma L > 10$. The approximation is not applicable in these media.

In this section, we will introduce an approximation in which two-wave mixing is the dominant process. In the approximation, we assume that the grating is dominated by a pair of waves. By examining the coupled equations (4.5-6)–(4.5-9), we note that the grating consists of two parts, $A_1A_2^*$ and $A_3A_4^*$. If we assume

$$|A_3A_4^*| \ll |A_1A_2^*| \tag{4.7-1}$$

for any z between 0 and L, then the coupled equations can be written

$$\frac{d}{dz} A_1 = -\frac{1}{2} \Gamma(A_2^*A_1)A_2/I_0 \tag{4.7-2}$$

$$\frac{d}{dz} A_2 = \frac{1}{2} \Gamma^*(A_1^*A_2)A_1/I_0 \tag{4.7-3}$$

$$\frac{d}{dz} A_3 = \frac{1}{2} \Gamma(A_2^*A_1)A_4/I_0 \tag{4.7-4}$$

$$\frac{d}{dz} A_4 = -\frac{1}{2} \Gamma^*(A_1^*A_2)A_3/I_0 \tag{4.7-5}$$

The first two equations are identical to Eqs. (4.1-12), except for the absorption term. They can be solved analytically. Once A_1 and A_2 are solved, they can be substituted into Eqs. (4.7-4) and (4.7-5). The amplitudes A_3 and A_4 can then be integrated.

To illustrate the use of this approximation, let us solve $A_1^*A_2$ and then try to integrate the amplitudes A_3 and A_4. Using Eqs. (4.7-2) and (4.7-3), we obtain

$$\frac{d}{dz} (A_1A_2^*) = \frac{1}{2} \Gamma[|A_1|^2 - |A_2|^2]A_1A_2^*/I_0 \tag{4.7-6}$$

which can also be written

$$\frac{d}{dz} \ln(A_1A_2^*) = \frac{1}{2} \Gamma \frac{I_1 - I_2}{I_0} \tag{4.7-7}$$

where $I_1 = |A_1|^2$ and $I_2 = |A_2|^2$. The intensities I_1 and I_2 are given by Eqs. (4.1-21) and (4.1-22). Using these expressions for the intensities and $I_0 = I_1(0) + I_2(0)$, we obtain

$$\frac{d}{dz} \ln(A_1A_2^*) = -\frac{1}{2} \Gamma \frac{\sinh(\gamma z - \ln m)}{1 + \cosh(\gamma z - \ln m)} \tag{4.7-8}$$

where m is the input beam intensity ratio

$$m = \frac{I_1(0)}{I_2(0)} \tag{4.7-9}$$

and γ is the real part of Γ (see Eq. (4.5-4)). An integration of Eq. (4.7-8) yields

$$\ln(A_1A_2^*) = -\frac{1}{2}\frac{\Gamma}{\gamma}\ln[1+\cosh(\gamma z - \ln m)] \qquad (4.7\text{-}10)$$

or equivalently

$$A_1A_2^* = C\left[\frac{1}{1+\cosh(\gamma z - \ln m)}\right]^{\Gamma/2\gamma} \qquad (4.7\text{-}11)$$

where C is a constant. If we define

$$A_1A_2^* = |A_1A_2^*|e^{i\psi} \qquad (4.7\text{-}12)$$

and use $\Gamma = \gamma + 2i\beta$, the magnitude and phase of $A_1A_2^*$ can be written

$$|A_1A_2^*| = \frac{|C|}{[1+\cosh(\gamma z - \ln m)]^{1/2}} \qquad (4.7\text{-}13)$$

and

$$\psi = \psi_c - \frac{\beta}{\gamma}\ln[1+\cosh(\gamma z - \ln m)] \qquad (4.7\text{-}14)$$

where ψ_c is a constant which is also the phase of C.

In the case of pure diffusion with $\beta = 0$, the solutions reduce to

$$\frac{A_1A_2^*}{I_0} = \frac{1}{[2+2\cosh(\gamma z - \ln m)]^{1/2}} \qquad (4.7\text{-}15)$$

or equivalently

$$\frac{A_1A_2^*}{I_0} = \frac{1}{2\cosh\left(\dfrac{\gamma z - \ln m}{2}\right)} \qquad (4.7\text{-}16)$$

Substituting Eq. (4.7-16) for $A_1A_2^*$ in the coupled Eqs. (4.7-4) and (4.7-5) and still assuming $\Gamma = \gamma$, we obtain

$$\frac{d}{dz}A_3 = g(z)A_4 \qquad (4.7\text{-}17)$$

$$\frac{d}{dz}A_4 = -g(z)A_3 \qquad (4.7\text{-}18)$$

with

$$g(z) = \frac{\gamma}{4\cosh\left(\dfrac{\gamma z - \ln m}{2}\right)} \qquad (4.7\text{-}19)$$

The coupled equations represent Bragg diffraction in the periodic medium whose depth of index modulation is governed by Eq. (4.7-19). The equations can be integrated directly or by using the method of grating integral. First, we solve these equations directly. Let

$$A_3 = \sqrt{I_3}\exp(-i\psi_3) \tag{4.7-20}$$

$$A_4 = \sqrt{I_4}\exp(-i\psi_4) \tag{4.7-21}$$

We obtain

$$I_3' = 2g\sqrt{I_3 I_4}\cos\Delta\psi \tag{4.7-22}$$

$$I_4' = -2g\sqrt{I_3 I_4}\cos\Delta\psi \tag{4.7-23}$$

$$\Delta\psi' = g\left[\left(\frac{I_3}{I_4}\right)^{1/2} - \left(\frac{I_4}{I_3}\right)\right]^{1/2} \sin\Delta\psi \tag{4.7-24}$$

where the prime indicates differentiation and

$$\Delta\psi = \psi_4 - \psi_3 \tag{4.7-25}$$

These equations are similar to those used in the lock-in coupling of ring laser gyros [7]. In the case when there is only beam 3 incident at $z = L$ (i.e., $A_4(L) = 0$), a proper solution is

$$\Delta\psi = 0 \tag{4.7-26}$$

and the equation for I_4 becomes

$$I_4' = -2g[I_4(I_{3L} - I_4)]^{1/2} \tag{4.7-27}$$

where I_{3L} is the intensity of beam 3 at $z = L$ (i.e., $I_{3L} = |A_3(L)|^2$). Using Eq. (4.7-19) for g, we obtain

$$\frac{I_4'}{[I_4(I_{3L} - I_4)]^{1/2}} = \frac{-\gamma}{2\cosh\left(\dfrac{\gamma z - \ln m}{2}\right)} \tag{4.7-28}$$

An integration of the above equation yields

$$-\sin^{-1}\left(\frac{I_4}{I_{3L}}\right)^{1/2} = \tan^{-1}\left[\exp\left(\frac{1}{2}\gamma z\right)\frac{1}{\sqrt{m}}\right] + \text{constant} \tag{4.7-29}$$

The boundary condition requires that $I_4 = 0$ at $z = L$. This leads to

$$-\sin^{-1}\left(\frac{I_4}{I_{3L}}\right)^{1/2} = \tan^{-1}\left[\exp\left(\frac{1}{2}\gamma z\right)\frac{1}{\sqrt{m}}\right] - \tan^{-1}\left[\exp\left(\frac{1}{2}\gamma L\right)\frac{1}{\sqrt{m}}\right] \tag{4.7-30}$$

The diffraction efficiency, defined as

$$\eta = \frac{I_4(0)}{I_{3L}} \tag{4.7-31}$$

is then given, according to Eq. (4.7-30), by

$$\eta = \frac{I_4(0)}{I_{3L}} = \sin^2\left\{\tan^{-1}\left[\frac{1}{\sqrt{m}}\exp\left(\frac{1}{2}\gamma L\right)\right] - \tan^{-1}\left(\frac{1}{\sqrt{m}}\right)\right\} \tag{4.7-32}$$

We plot the diffraction efficiency η as a function of γL for various values of m. Notice that η is not a periodic function of γL which appears in fixed gratings. Figure 4.23 shows the diffraction efficiency. The intensities of the beams are given, according to Eq. (4.7-30) and $I_3 = I_{3L} - I_4$, by

$$I_4(z) = I_{3L}\sin^2\left\{\tan^{-1}\left[\frac{1}{\sqrt{m}}\exp\left(\frac{1}{2}\gamma L\right)\right] - \tan^{-1}\left[\frac{1}{\sqrt{m}}\exp\left(\frac{1}{2}\gamma z\right)\right]\right\} \tag{4.7-33}$$

$$I_3(z) = I_{3L}\cos^2\left\{\tan^{-1}\left[\frac{1}{\sqrt{m}}\exp\left(\frac{1}{2}\gamma L\right)\right] - \tan^{-1}\left[\frac{1}{\sqrt{m}}\exp\left(\frac{1}{2}\gamma z\right)\right]\right\} \tag{4.7-34}$$

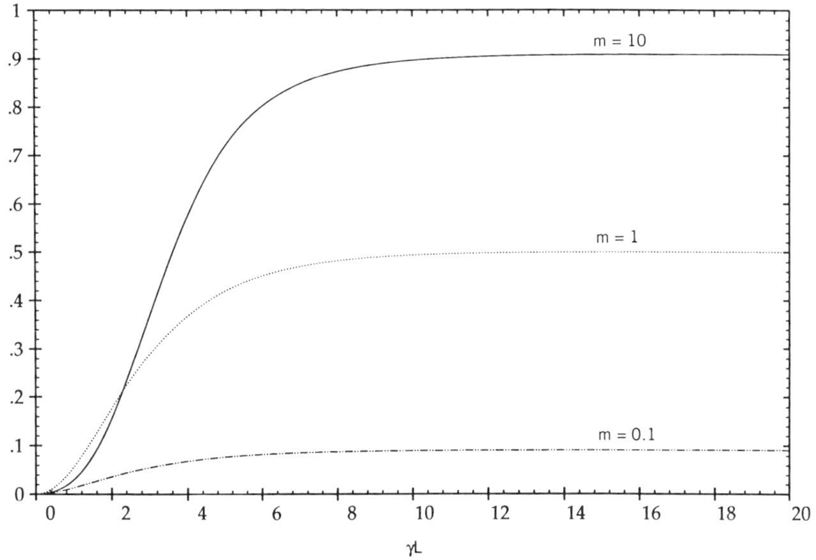

Figure 4.23. Diffraction efficiency η versus coupling constant γL for pump beam ratios $m = 0.1$, 1.0, 10.0. The spatial phase shift ϕ is assumed to be $-\pi/2$.

We now use the method of grating integral (discussed in Chapter 2) to solve the coupled Eqs. (4.7-17) and (4.7-18). Let

$$u = \int_L^z g(z')\, dz' \tag{4.7-35}$$

Using this new variable, the coupled equations become

$$\frac{d}{du} A_3 = A_4 , \qquad \frac{d}{du} A_4 = -A_3 \tag{4.7-36}$$

The solutions subject to the boundary condition $A_4(L) = 0$, are

$$A_3 = A_3(0) \cos u \tag{4.7-37}$$

$$A_4 = A_3(0) \sin u \tag{4.7-38}$$

where $A_3(0)$ is the amplitude of A_3 at $u = 0$ (or equivalently, $z = L$) and u is given by Eq. (4.7-35). Using Eq. (4.7-19) for g, we can carry out the integration and obtain

$$u = \tan^{-1} \frac{1}{\sqrt{m}} \exp\left(\frac{1}{2} \gamma z\right) - \tan^{-1} \frac{1}{\sqrt{m}} \exp\left(\frac{1}{2} \gamma L\right) \tag{4.7-39}$$

By comparing the results, we find that Eqs. (4.7-37)–(4.7-39) are identical to Eqs. (4.7-33) and (4.7-34). This illustrates how useful the method of grating integral can be.

4.8 METHOD OF GRATING INTEGRAL FOR FOUR-WAVE MIXING

4.8.1 Transmission Gratings

The exact solutions discussed in Section 4.5 are valid for any complex coupling constant Γ provided that all the coupling constants are identical. Although the results are useful in solving some of the photorefractive resonators and conjugators, they are difficult to use for general purposes. Many of the constants of integration depend on a two-point boundary condition. In this section, we will employ the technique introduced in Section 2.7, known as the method of grating integral, to obtain closed form solutions of four-wave mixing.

Although the method is applicable to coupled equations with different coupling constants, we will limit ourselves to the case given by Eqs. (4.5-6)–(4.5-9). By defining an interference function

$$G(z) = \frac{A_1 A_2^* + A_3 A_4^*}{I_0} \tag{4.8-1}$$

The coupled equations can be written

$$\frac{d}{dz} A_1 = -\frac{1}{2} \Gamma G A_2 \tag{4.8-2}$$

$$\frac{d}{dz} A_2 = \frac{1}{2} \Gamma^* G^* A_1 \tag{4.8-3}$$

$$\frac{d}{dz} A_3 = \frac{1}{2} \Gamma G A_4 \tag{4.8-4}$$

$$\frac{d}{dz} A_4 = -\frac{1}{2} \Gamma^* G^* A_3 \tag{4.8-5}$$

If the function $G(z)$ is solved and if the phase of ΓG is independent of z, then the above equations can be easily integrated by using the method of grating integral. In what follows, we derive the grating interference function $G(z)$ and show that the phase of ΓG is a constant provided Γ is real. Using the definition and the coupled equations, we obtain

$$\frac{d}{dz} G = \frac{1}{2} \Gamma G \frac{I_1 - I_2 + I_4 - I_3}{I_0} \tag{4.8-6}$$

Let

$$f(z) = \frac{I_1 - I_2 + I_4 - I_3}{I_0} \tag{4.8-7}$$

and carry out few steps of similar algebraic operations, we obtain

$$\frac{d}{dz} |G|^2 = \frac{1}{2} (\Gamma + \Gamma^*)|G|^2 f \tag{4.8-8}$$

$$\frac{d}{dz} f = -(\Gamma + \Gamma^*)|G|^2 \tag{4.8-9}$$

Before we proceed with the integration of these two equations, we notice that

$$4|G|^2 + f^2 = C^2 \tag{4.8-10}$$

where C is a constant. By eliminating $|G|^2$ from the last two equations, we obtain

$$\frac{df}{dz} = -\frac{\Gamma + \Gamma^*}{4} (C^2 - f^2) \tag{4.8-11}$$

An integration of the above equation yields

$$f(z) = -C \tanh\left[\frac{C}{2} \gamma z - D\right] \tag{4.8-12}$$

where γ is the real part of Γ and D is a constant given by

$$D = \tanh^{-1} \frac{f(0)}{C} \tag{4.8-13}$$

The magnitude of the interference function $|G|$ is then obtained by substituting Eq. (4.8-12) for $f(z)$ in Eq. (4.8-9). This leads to

$$|G(z)| = \frac{C}{2 \cosh\left[\dfrac{C}{2} \gamma z - D\right]} \tag{4.8-14}$$

The phase of $G(z)$ is defined as

$$G(z) = |G| e^{-i\phi} \tag{4.8-15}$$

To solve for the phase ϕ, we examine the following equation

$$\frac{d}{dz} \ln G = \frac{1}{G} \frac{d}{dz} G = \frac{d}{dz} \ln|G| - i \frac{d}{dz} \phi \tag{4.8-16}$$

Using Eq. (4.8-6) and the definition of $f(z)$, we obtain

$$\frac{d}{dz} \phi = -\beta f = \beta C \tanh\left(\frac{C}{2} \gamma z - D\right) \tag{4.8-17}$$

where β is 1/2 of the imaginary part of Γ. An integration of the above equation yields

$$\phi(z) = \frac{2\beta}{\gamma} \ln \cosh\left(\frac{C}{2} \gamma z - D\right) + \text{constant} \tag{4.8-18}$$

According to the result, the phase is a constant only when Γ is a real coupling constant (i.e., $\beta = 0$). Thus, the coupled Eqs. (4.8-2)–(4.8-5) can be integrated using the method of grating integral provided Γ is a real coupling constant.

Defining the new variable

$$u = \int_0^z \frac{1}{2} \gamma |G| \, dz' \tag{4.8-19}$$

we can rewrite the coupled equations as

$$\frac{d}{du} A_1 = -e^{-i\phi} A_2 \tag{4.8-20}$$

$$\frac{d}{du} A_2 = e^{i\phi} A_1 \tag{4.8-21}$$

$$\frac{d}{du} A_3 = e^{-i\phi} A_4 \tag{4.8-22}$$

$$\frac{d}{du} A_4 = -e^{i\phi} A_3 \tag{4.8-23}$$

where ϕ is a constant. The solutions are

$$A_1 = A_1(0) \cos u - A_2(0) e^{-i\phi} \sin u \tag{4.8-24}$$

$$A_2 = A_2(0) \cos u + A_1(0) e^{i\phi} \sin u \tag{4.8-25}$$

$$A_3 = A_3(L) \cos(u - U) + A_4(L) e^{-i\phi} \sin(u - U) \tag{4.8-26}$$

$$A_4 = A_4(L) \cos(u - U) - A_3(L) e^{i\phi} \sin(u - U) \tag{4.8-27}$$

with U being the *grating action integral*

$$U = \int_0^L \frac{1}{2} \gamma |G| \, dz \tag{4.8-28}$$

where L is the length of interaction.

Using Eq. (4.8-14) for $|G|$ and carrying out the integration, we obtain

$$u = \tan^{-1}\left[\exp\left(\frac{C}{2} \gamma z - D\right)\right] - \tan^{-1}(e^{-D}) \tag{4.8-29}$$

and the grating action integral

$$U = u(L) = \tan^{-1}\left[\exp\left(\frac{C}{2} \gamma L - D\right)\right] - \tan^{-1}(e^{-D}) \tag{4.8-30}$$

Although u is obtained as a function of z, the constants of integration C and D are not known. If C and D can be easily obtained in terms of the boundary conditions, then Eqs. (4.8-29) and (4.8-30) are useful for the solutions (4.8-24)–(4.8-27). In the event when C and D are not easily available in terms of the boundary condition, we need to obtain u directly from Eq. (4.8-19).

According to Eqs. (4.8-19) and (4.8-15), u satisfies the following differential equation

$$\frac{du}{dz} = \frac{1}{2} \gamma G e^{i\phi} \tag{4.8-31}$$

or equivalently

$$\frac{du}{dz} = \frac{1}{2} \gamma (A_1 A_2^* + A_3 A_4^*) e^{i\phi} / I_0 \tag{4.8-32}$$

where ϕ is a constant and A_1, A_2, A_3, A_4 are functions of u given by Eqs.

(4.8-24)–(4.8-27). Since A_1, A_2, A_3, A_4 are all sinusoidal functions, Eq. (4.8-32) can be integrated in closed form. We rewrite the equation as

$$\frac{I_0\, du}{A_1 A_2^* + A_3 A_4^*} = \frac{1}{2}\gamma e^{i\phi}\, dz \tag{4.8-33}$$

or equivalently

$$\int_0^u \frac{2I_0 e^{-i\phi}}{A_1 A_2^* + A_3 A_4^*}\, du' = \gamma z \tag{4.8-34}$$

where u' is a dummy variable.

The action integral U can be obtained by solving the following equation

$$\int_0^U \frac{2I_0 e^{-i\phi}}{A_1 A_2^* + A_3 A_4^*}\, du = \gamma L \tag{4.8-35}$$

Once U is obtained, the new variable can be integrated in terms of the incident amplitudes and z and Eqs. (4.8-24)–(4.8-27) provide the exact solutions of the wave amplitudes.

4.8.2 Reflection Gratings

For reflection gratings, the interference function is defined as

$$G(z) = \frac{A_1 A_3^* + A_2 A_4^*}{I_0} \tag{4.8-36}$$

and the coupled equations are written

$$\frac{d}{dz} A_1 = -\frac{1}{2}\Gamma G A_3 \tag{4.8-37}$$

$$\frac{d}{dz} A_2 = -\frac{1}{2}\Gamma G A_4 \tag{4.8-38}$$

$$\frac{d}{dz} A_3 = -\frac{1}{2}\Gamma^* G^* A_1 \tag{4.8-39}$$

$$\frac{d}{dz} A_4 = -\frac{1}{2}\Gamma^* G^* A_2 \tag{4.8-40}$$

From these equations, we obtain

$$\frac{d}{dz} G = -\frac{1}{2}\Gamma G \tag{4.8-41}$$

The solution for G is simply

$$G = G(0)\exp(-\tfrac{1}{2}\Gamma z) \tag{4.8-42}$$

where $G(0)$ is a constant. It is clear from the above equation that the phase of $G(z)$ is a constant provided Γ is real. This is often the case in photorefractive crystals such as $BaTiO_3$ and SBN that operate by diffusion only (i.e., $E_0 = 0$). Thus, in this case, the interference function can be written

$$G(z) = |G(z)|e^{-i\phi} \tag{4.8-43}$$

where ϕ is a constant. By introducing the new variable

$$u = \int_0^z \frac{\gamma|G(z')|}{2}\, dz' \tag{4.8-44}$$

where $\gamma = \Gamma$ is real, the coupled equations can be written

$$\frac{dA_1}{du} = -e^{-i\phi}A_3 \tag{4.8-45}$$

$$\frac{dA_2}{du} = -e^{-i\phi}A_4 \tag{4.8-46}$$

$$\frac{dA_3}{du} = -e^{i\phi}A_1 \tag{4.8-47}$$

$$\frac{dA_4}{dz} = -e^{i\phi}A_2 \tag{4.8-48}$$

and the solutions are written

$$A_1 = \frac{A_1(0)\cosh(u-U) - e^{-i\phi}A_3(L)\sinh u}{\cosh U} \tag{4.8-49}$$

$$A_3 = \frac{A_3(L)\cosh u - e^{i\phi}A_1(0)\sinh(u-U)}{\cosh U} \tag{4.8-50}$$

$$A_2 = \frac{A_2(0)\cosh(u-U) - e^{-i\phi}A_4(L)\sinh u}{\cosh U} \tag{4.8-51}$$

$$A_4 = \frac{A_4(L)\cosh u - e^{i\phi}A_2(0)\sinh(u-U)}{\cosh U} \tag{4.8-52}$$

where U is the grating action integral given by

$$U = u(L) = \int_0^L \frac{\gamma|G(z)|}{2}\, dz \tag{4.8-53}$$

The solutions can then be substituted into Eq. (4.8-44) in order to obtain u as a function of z.

PROBLEMS

4.1 Co-directional Two-Wave Mixing

a. Derive Eqs. (4.1-15) and (4.1-16).

b. Show that

$$I_1 + I_2 = [I_1(0) + I_2(0)] \exp(-\alpha z)$$

c. Let $I_1 = u \exp(-\alpha z)$, $I_2 = v \exp(-\alpha z)$. Show that

$$u' = -\gamma uv/(u+v) , \qquad v' = \gamma uv/(u+v)$$

where the prime indicates differentiation with respect to z.

d. Show that $u + v = c$, where c is a constant.

e. Show that

$$\left(\frac{1}{u} + \frac{1}{c-u}\right) du = \gamma \, dz$$

f. Integrate the equation in (**e**) and derive the solutions given by Eqs. (4.1-21) and (4.1-22).

g. Show that $I_1(z) = I_2(z)$ at $z = (1/\gamma) \ln m$.

4.2 Gain Due to Two-Wave Mixing

a. In a crystal with a given absorption of αL, show that net gain (i.e., $g > 1$) requires that

$$m > e^{\alpha L} - 1$$

and

$$e^{\gamma L} > \frac{m}{e^{-\alpha L} + me^{-\alpha L} - 1}$$

b. For very weak signal beams, $I_2(0) \ll I_1(0)$, show that the gain is

$$g \simeq e^{(\gamma - \alpha)L}$$

c. Estimate the small signal gain for $\alpha = 1 \text{ cm}^{-1}$ and $\gamma = 10 \text{ cm}^{-1}$ for a photorefractive crystal of $L = 1$ cm.

4.3 Photorefractive Transmission Gratings

a. Show that the depth of index modulation in photorefractive grating is proportional to the fringe contrast G defined as

$$G \equiv \frac{2A_1A_2^*}{|A_1|^2 + |A_2|^2}$$

[Hint: Use Eq. (4.1-5)].

b. Show that

$$\frac{d}{dz} A_1 = -\frac{1}{4} \Gamma G A_2 - \frac{1}{2} \alpha A_1, \qquad \frac{d}{dz} A_2 = \frac{1}{4} \Gamma^* G^* A_1 - \frac{1}{2} \alpha A_2$$

c. Show that

$$\frac{d}{dz} G = \frac{1}{2} \Gamma G \frac{I_1 - I_2}{I_1 + I_2}$$

d. Show that

$$G(z) = \frac{1}{\cosh\left(\dfrac{\gamma z - \ln m}{2}\right)}$$

provided $\Gamma = \gamma$ is a real number.

e. Derive $G(z)$ for complex Γ.

4.4 Contra-Directional Two-Wave Mixing

a. Derive Eqs. (4.1-30) and (4.1-31)

b. Let $\alpha = 0$, show that

$$I_2 - I_1 = 2c$$

where c is a constant.

c. Show that Eq. (4.1-30) can be written

$$\left(\frac{1}{I_1} + \frac{1}{I_1 + 2c}\right) dI_1 = -\gamma \, dz$$

d. Integrate the equation in **(c)** and derive the solutions in Eqs. (4.1-37) and (4.1-38).

e. Show that $I_1(z) = I_2(z)$ occurs only when $c = 0$, or equivalently

$$I_2(L) = I_1(0) \exp(-\tfrac{1}{2}\gamma L)$$

4.5 Photorefractive Reflection Gratings. The fringe contrast can be written

$$|G| = \frac{2\sqrt{I_1 I_2}}{I_1 + I_2}$$

a. Show that, according to Eqs. (4.1-37) and (4.1-38)

$$|G| = \left(\frac{Be^{-\gamma z}}{C^2 + Be^{-\gamma z}} \right)^{1/2}$$

b. Show that fringe contrast is always less than unity as long as the net flux is finite.

c. Show that $|G| = 1$ for all z, provided $C = 0$.

d. Show that signal gain is

$$g = \exp(\tfrac{1}{2}\gamma L)$$

when the grating contrast is unity.

4.6 **Photorefractive Coupling Constant in $BaTiO_3$.** The coupling constant γ for two-wave mixing in crystals depend on the orientation as well as spacing of the grating.

a. Show that there is no beam coupling for two ordinary waves when their photo-induced grating wave vector is perpendicular to the c-axis.

b. According to Eq. (4.1-71), two-wave mixing for extraordinary waves disappears when the grating is either parallel ($\alpha = 0$) or perpendicular ($\pm\pi/2$) to the c-axis. Assuming E^s is insensitive to the angle α, find the optimum angle for two-wave mixing.

c. The saturated space-charge field E^s can be written

$$E^s = \frac{K \dfrac{k_B T}{q}}{1 + \dfrac{K^2}{k_D^2}}, \quad \text{with } k_D^2 = \frac{q^2 N_A}{\langle \epsilon \rangle k_B T}$$

where $\langle \epsilon \rangle$ is the effective dielectric constant which depends on α, as given by Eq. (4.1-62). Show that E^s is indeed insensitive to the angle α, provided $K \ll k_D$ (i.e., large grating spacing).

d. Show that, for small grating spacing ($K \gg k_D$), E^s is approximately proportional to $\langle \epsilon \rangle$. Using $\epsilon_\parallel = 135\epsilon_0$ and $\epsilon_\perp = 3600\epsilon_0$, find the angle α where E^s is maximized.

e. Find the optimum angle for two-wave mixing of extraordinary waves in $BaTiO_3$.

4.7 Shows that Eq. (4.5-42) reduces to Eq. (4.5-21) in the event when the pump depletion is negligible.

4.8 **Phases of Photorefractive Gratings.** Let $g_{12} = A_1 A_2^*$ and $g_{34} = A_3 A_4^*$ be the components of a transmission grating $G = (A_1 A_2^* + A_3 A_4^*)$ in a photorefraction medium with $\Gamma = \text{real}$.

a. Show that

$$\frac{d}{dz} g_{12} = \frac{1}{2} \Gamma G(I_1 - I_2)$$

b. Show that the solution can be written

$$g_{12}(z) = C_{12} + \int_0^z \frac{1}{2} \Gamma G(I_1 - I_2)\, dz$$

where $C_{12} = g_{12}(0)$ is a complex constant.

c. Show that g_{12} can also be written

$$g_{12}(z) = C_{12} + f_{12}(z) e^{-i\phi}$$

where $f_{12}(z)$ is a real function with $f_{12}(0) = 0$ and ϕ is a constant.

d. Show that the phase of G is exactly the phase of the complex sum $C_{12} + C_{34}$, where $C_{34} = g_{34}(L)$.

4.9 **Nonreciprocal Transmission.** Consider a thin photorefractive crystal plate with its c-axis perpendicular to the plate surface. Assuming that the plate surfaces are antireflection coated and a partial reflecting dielectric mirror is placed behind the photorefractive crystal plate. We assume that the mirror surface is parallel to the plate and both mirror and the photorefractive crystal are lossless.

a. Show that the transmission is independent of the separation between the mirror and the crystal plate.

b. Find the intensity transmittance as a function of γL.

c. Find the intensity transmittance when the light is incident from the mirror side.

4.10 **Slowly Varying Amplitude Approximation.** Show that Eq. (4.1-8) is valid provided the index modulation n_1 is much less than n_0.

4.11 **Conservation of Power Flow**

a. Show that the coupled Eqs. (4.1-15) and (4.1-16) are consistent with the conservation of energy, that is,

$$I_1 + I_2 = \text{constant}$$

provided $\alpha = 0$.

b. Show that in the event of asymmetrical incidence (i.e., $\beta_1 \neq \beta_2$), the conservation of energy becomes

$$\beta_1 I_1 + \beta_2 I_2 = \text{constant}$$

where β_1, β_2 are the z-components of the wave vectors.

REFERENCES

1. See, for example, P. Yeh, "Two-Wave Mixing in Nonlinear Media," IEEE J. Quantum Electron. **25**, 484–519 (1989), and references therein.
2. P. Yeh, "Contradirectional Two-Wave Mixing in Photorefractive Media," Opt. Comm. **45**, 323–326 (1983).
3. P. Yeh, Optical Waves in Layered Media (Wiley, New York, 1988).
4. P. Yeh, "Photorefractive Two-Beam Coupling in Cubic Crystals," J. Opt. Soc. Am. **4**, 1382–1388 (1987).
5. C. Gu and P. Yeh, "Reciprocity in Photorefractive Wave Mixing," Opt. Lett. **16**, 455–457 (1991).
6. M. Cronin-Golomb, J. O. White, B. Fischer, and A. Yariv, "Exact Solution of a Nonlinear Model of Four-Wave Mixing and Phase Conjugation," Opt. Lett. **7**, 313–315 (1982).
7. See, for example, P. Yeh, J. Tracy, and M. Khoshnevisan, "Phase Conjugate Ring Gyroscopes," Proc. SPIE **412**, 240–245 (1983).
8. N. V. Kukhtarev, T. I. Semenets, K. H. Ringhofer, and G. Tomberger, "Phase Conjugation by Reflection Gratings in Electro-Optic Crystals," Appl. Phys. **B41**, 259–263 (1986).

5

Photorefractive Resonators

Wave mixing in photorefractive media provides energy coupling between beams of coherent radiation. Significant beam amplification can be achieved under the appropriate conditions. Gain coefficients of $\Gamma L > 10$ are common in many samples of $BaTiO_3$ or SBN crystals. Such gain coefficients represent enormous intensity amplification (e.g., exp(10)) for small signal beams. If such gains are introduced in resonators, oscillation can be easily achieved. In this chapter, we discuss a new class of resonators in which the oscillations are obtained by using parametric gain from wave mixing in photorefractive media.

5.1 UNIDIRECTIONAL RING RESONATORS

Referring to Figure 5.1, we investigate the oscillation of a ring resonator in which a photorefractive crystal is inserted. When the crystal is pumped by a laser beam, unidirectional oscillation occurs under the appropriate conditions. Such oscillation has been observed by using a $BaTiO_3$ crystal pumped with an argon ion or a HeNe laser [1]. Unlike conventional gain media (e.g., He–Ne), the gain bandwidth of photorefractive two-wave mixing is very narrow (a few Hertz for $BaTiO_3$, see also Figure 4.7). Despite this fact, the ring resonator can still oscillate over a large range of cavity detuning. This phenomenon was not well understood until a theory of photorefractive phase shift was developed [2]. Such a theory is treated in this section. The theory shows that oscillation can occur at almost any cavity length despite the narrowband nature of two-wave mixing gain, provided the coupling is strong.

Let us focus our attention on the region occupied by the photorefractive crystal and examine the gain due to two-wave mixing. The results of nondegenerate two-wave mixing derived in Section 4.2 can be used to explain the ring oscillation.

In a conventional ring resonator, the oscillation occurs at those frequencies

$$f = f_0 + N\frac{c}{S} \tag{5.1-1}$$

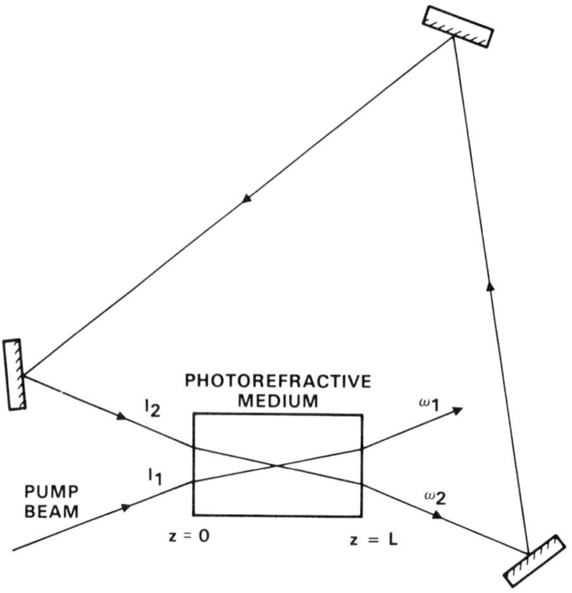

Figure 5.1. Schematic drawing of a unidirectional photorefractive ring resonator. The photorefractive medium is pumped by a laser beam at frequency ω_1. As a result of the signal gain due to two-wave mixing in the photorefractive medium, unidirectional oscillation occurs with a frequency of ω_2 which is determined by the cavity length.

which lie within the gain curve of the laser medium (e.g., He–Ne). Here, S is the effective length of a complete loop, f_0 is a constant, and N is an integer. For $S \leq 30$ cm, these frequencies (Eq. (5.1-1)) are separated by the mode spacing $c/S \geq 1$ GHz. Since the width of the gain curve for the conventional gain medium is typically several GHz due principally to Doppler broadening, oscillation can occur at almost any cavity length S. On the contrary, if the bandwidth of the gain curve is narrower than the mode spacing c/S, then oscillation can sustain, provided the cavity loop is kept at the appropriate length.

Unlike the conventional gain medium, the bandwidth of photorefractive two-wave mixing is very narrow. Using photorefractive crystals that operate by diffusion only, for example, $BaTiO_3$, the coupling constant can be written, according to Eq. (4.2-10)

$$\gamma = \frac{\gamma_0}{1 + (\Omega\tau)^2} \tag{5.1-2}$$

where Ω is the frequency difference between the beams, and γ_0 is the coupling constant for the case of degenerate two-wave mixing (i.e., $\Omega = \omega_2 - \omega_1 = 0$).

The parametric two-wave mixing gain is given, according to Eq. (4.2-12), by

$$g \equiv \frac{I_2(L)}{I_2(0)} = \frac{1+m}{1+me^{-\gamma L}}\, e^{-\alpha L} \tag{5.1-3}$$

where we recall that m is the input beam ratio $m = I_1(0)/I_2(0)$ and L is the length of interaction. Note that amplification ($g > 1$) is possible only when $\gamma > \alpha$ and $m > (1 - e^{-\alpha L})/(e^{-\alpha L} - e^{-\gamma L})$. Also note that g is an increasing function of m (i.e., $\partial g/\partial m > 0$) and g is an increasing function of L, provided $\gamma > \alpha$ and

$$L \le \frac{1}{\gamma} \ln\left[\frac{m(\gamma - \alpha)}{\alpha}\right] \tag{5.1-4}$$

The gain as a function of frequency ω_2 (or equivalently as a function of $\Omega = \omega_2 - \omega_1$), has been plotted in Figure 4.7 for various values of m. Note that gain is significant only when $|\omega_2 - \omega_1|\tau < 1$. For materials such as $BaTiO_3$ and SBN with an operating intensity of $1\ W/cm^2$ τ is between 1 s and 0.1 s. Thus, the gain bandwidth is only a few Hertz. In spite of such an extremely narrow bandwidth, unidirectional oscillation can still be observed easily at "any" cavity length in ring resonators using $BaTiO_3$ crystals as the photorefractive medium. Such a phenomenon can be explained in terms of the additional phase shift (Eqs. (4.1-25) and (4.1-27)) introduced by the photorefractive coupling. This phase shift is a function of the oscillation frequency and is plotted in Figure 5.2 as a function of $\Omega\tau$. For $BaTiO_3$

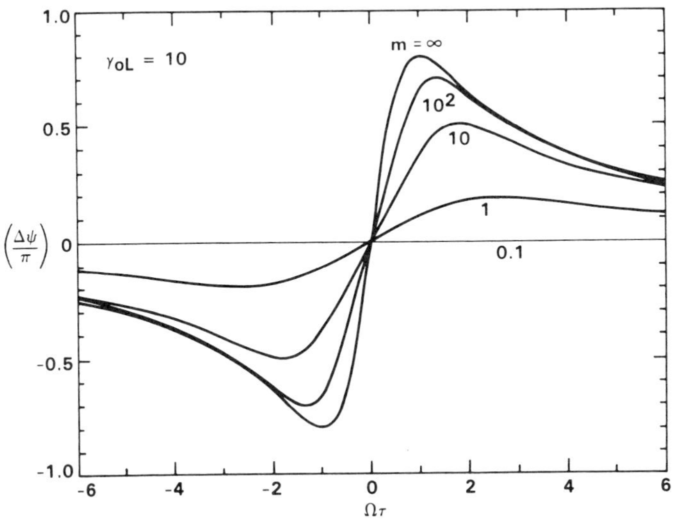

Figure 5.2 Photorefractive phase shift $\Delta\psi$ as a function of $\Omega\tau$ for various values of m.

crystals with $\gamma_0 L > 4\pi$, this phase shift can vary from $-\pi$ to $+\pi$ for a frequency drift of $\Delta\Omega\tau = \pm 1$. Such a phase shift is responsible for the oscillation of the ring resonator which requires a round-trip phase shift of an integer times 2π.

5.1.1 Oscillation Conditions

We now examine the boundary conditions appropriate to a unidirectional ring oscillator. At steady-state oscillation, the electric field must reproduce itself, both in phase and intensity, after each round-trip. In other words, the oscillation conditions can be written

$$\Delta\psi + \int k\, ds = 2N\pi \tag{5.1-5}$$

and

$$gR = 1 \tag{5.1-6}$$

where $\Delta\psi$ is the additional phase shift due to photorefractive coupling, the integration is over a round-trip beam path, the parameter R is the product of the mirror reflectivities, and g is the parametric gain of Eq. (5.1-3).

If we define a cavity detuning parameter $\Delta\Gamma$ as

$$\Delta\Gamma = 2N'\pi - \int k\, ds \tag{5.1-7}$$

where N' is an integer chosen in such a way that $\Delta\Gamma$ lies between $-\pi$ and $+\pi$, then the oscillation condition (5.1-5) can be written

$$\Delta\psi = \Delta\Gamma + 2M\pi \tag{5.1-8}$$

where M is an integer. In other words, oscillation can be achieved only when the cavity detuning can be compensated by the photorefractive phase shift.

Equations (5.1-5) and (5.1-6) may be used to solve for the two unknown quantities $m = I_1(0)/I_2(0)$ and $\Omega = \omega_2 - \omega_1$. If we fix the pump intensity $I_1(0)$ and the pump frequency ω_1, then Eqs. (5.1-5) and (5.1-6) can be solved for the oscillation frequency ω_2 and the oscillation intensity $I_2(0)$. Substituting Eq. (5.1-3) for g in Eq. (5.1-6) and using Eq. (4.1-25), we obtain

$$\Delta\psi = -\frac{\beta}{\gamma} \ln(Re^{-\alpha L}) \tag{5.1-9}$$

This equation can now be used to solve for the oscillation frequency $\Omega\tau$. For the case of pure diffusion, using Eq. (3.5-21) for ϕ and Eq. (4.1-20), we obtain from Eq. (5.1-9)

$$\Omega\tau = \frac{2\,\Delta\psi}{\alpha L - \ln R} = \frac{2(\Delta\Gamma + 2M\pi)}{\alpha L - \ln R} \tag{5.1-10}$$

where $\Delta\Gamma$ is the cavity detuning and is given by Eq. (5.1-7). Substituting Eq. (5.1-3) for g in Eq. (5.1-6), we can solve for m and obtain

$$m = \frac{I_1(0)}{I_2(0)} = \frac{1 - Re^{-\alpha L}}{Re^{\alpha L} - e^{-\gamma L}} \tag{5.1-11}$$

Since m must be positive, we obtain from Eq. (5.1-11) the threshold condition for oscillation,

$$\gamma L > \gamma_t L \equiv \alpha L - \ln R \tag{5.1-12}$$

where γ_t is the threshold parametric gain constant. Since γ is a function of frequency Ω, Eq. (5.1-10) dictates that the parametric gain is above threshold only in a finite spectral regime. Using Eq. (5.1-2) for γ, Eq. (5.1-12) becomes

$$|\Omega\tau| < \left[\frac{\gamma_0 L}{\alpha L - \ln R} - 1\right]^{1/2} \tag{5.1-13}$$

where we recall that γ_0 is the parametric gain at $\Omega = \omega_2 - \omega_1 = 0$. Equation (5.1-13) defines the spectral regime where the parametric gain γ is above threshold (i.e., $\gamma > \gamma_t$).

We have thus far obtained expressions for the oscillation frequency (Eq. (5.1-10)) and the spectral regime where the gain is above threshold. The ring resonator will oscillate only when the oscillation frequency falls within this spectral region. The oscillation frequency $\omega_2 = \omega_1 + \Omega$ is determined by Eq. (5.1-10), with $\Delta\Gamma$ being the cavity detuning (5.1-7).

The same oscillation frequency must also satisfy Eq. (5.1-13). Thus, we obtain the following oscillation condition

$$\frac{2|\Delta\psi|}{\alpha L - \ln R} < \left(\frac{\gamma_0 L}{\alpha L - \ln R} - 1\right)^{1/2} \tag{5.1-14}$$

which can also be written

$$\gamma_0 L > \gamma_t L + \frac{1}{\gamma_t L}(2\,\Delta\psi)^2 \equiv G_t L \tag{5.1-15}$$

where γ_t is the threshold parametric gain of Eq. (5.1-12) for the case when $\Delta\psi = 0$, and G_t may be considered as the threshold gain for the case when $\Delta\psi \neq 0$. According to Eq. (5.1-15), the threshold gain increases as a function of the cavity detuning $\Delta\Gamma$. The cavity detuning $\Delta\Gamma$ not only determines the oscillation frequency (Eq. (5.1-10)), but also the threshold gain G_t.

The $\Delta\Gamma$ in Eq. (5.1-7) is the cavity detuning and is defined between $-\pi$ and π. However, the photorefractive phase shift (Eq. (5.1-9)) can be greater than π. When this happens, the unidirectional ring resonator may oscillate at more than one frequency. These frequencies are given by Eq. (5.1-10), with $M = 0, \pm 1, \pm 2, \ldots$ etc., and with their corresponding threshold gain given by

$$G_t L = \gamma_t L + \frac{1}{\gamma_t L} [2(\Delta\Gamma + 2M\pi)]^2 \tag{5.1-16}$$

In other words, for each cavity detuning $\Delta\Gamma$, the ring resonator can support multimode oscillation, provided the coupling constant γ_0 is large enough. Figure 5.3 shows the oscillation intensity, as well as the oscillation frequency as functions of cavity detuning $\Delta\Gamma$. Note that for larger $\gamma_0 L$, the resonator can oscillate at almost any cavity detuning $\Delta\Gamma$, whereas for small $\gamma_0 L$, oscillation occurs only when the cavity detuning is limited to some small region around $\Delta\Gamma = 0$.

In summary, ring oscillation occurs when the gain due to two-wave mixing overcomes cavity losses and the round-trip optical phase reproduces itself (to within an integer multiple of 2π). The condition on phase is unique because of a significant contribution to the optical phase shift due to

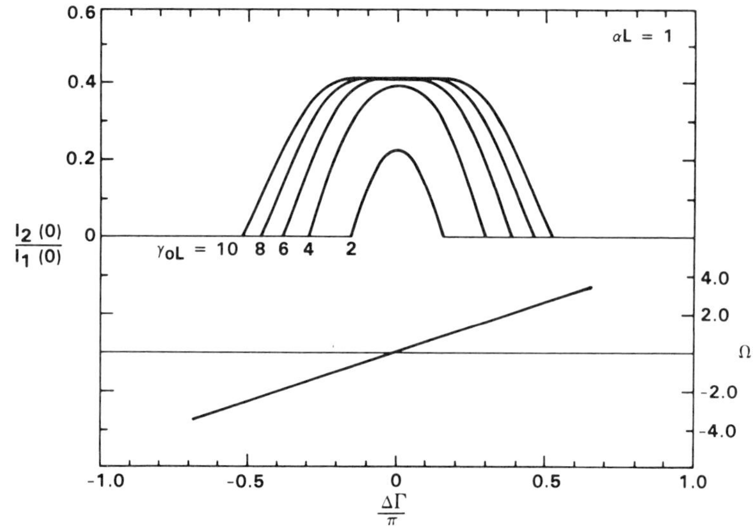

Figure 5.3. Top curves are unidirectional oscillation intensity as a function of cavity detuning $\Delta\Gamma$ for various values of $\gamma_0 L$. The lower curve (a straight line) represents the oscillation frequency shift $\Omega\tau$ as a function of the cavity detuning $\Delta\Gamma$.

nondegenerate photorefractive two-wave mixing. This condition is satisfied at any cavity length if the oscillation frequency is slightly detuned from the pump frequency, since the photorefractive phase shift (Eq. (5.1-9)) depends on the detuning. The frequency difference Ω ($= \omega_2 - \omega_1$) between the pumping and oscillating beams can be written

$$\Omega = [2(\Delta\Gamma + 2M\pi)/\tau A] \qquad (5.1\text{-}17)$$

where $\Delta\Gamma$ is the cavity length detuning defined in Eq. (5.1-7), M is an integer, τ is the photorefractive time response, and A represents the total cavity loss. The detuning ranges are given by (taking M to be zero)

$$|\Omega| \leq (1/\tau)(\gamma_0 L/A - 1)^{1/2} \qquad (5.1\text{-}18)$$

$$|\Delta\Gamma| \leq (A/2)(\gamma_0 L/A - 1)^{1/2} \qquad (5.1\text{-}19)$$

where γ_0 is the degenerate two-wave mixing coupling coefficient, L is the interaction length, and $A = -\ln(RT_sT_p)$ [with R the product of the reflectivities of the cavity mirrors and output coupler; T_s the transmission through the photorefractive crystal accounting for the absorption, Fresnel reflections, and scattering (or beam fanning); and T_p the effective round trip transmission through the empty resonator].

This theory predicts that the unidirectional ring resonator will oscillate at a frequency different from the pump frequency by an amount directly proportional to the cavity length detuning. Furthermore, in a photorefractive material with moderate slow τ, the theory postulates a threshold where oscillation will cease if the cavity detuning (frequency difference) becomes too large. Such a theory has been validated experimentally in a $BaTiO_3$ photorefractive ring resonator. Interested readers are referred to [3].

The interdependence of the optical cavity length and the beat frequency between the oscillating and pumping beams is a general property of photorefractive resonators. These results are not unique to the optical setup shown in Figure 5.1. Similar behavior is observed with other configurations, for example, a linear resonator (Figure 5.4).

The experimental results indicate that the frequency difference between the oscillating and pumping beams in both linear and ring resonator depends on the optical cavity length. This dependence supports the theory that uses a photorefractive phase shift associated with slightly nondegenerate two-wave mixing to satisfy the round-trip phase-oscillation condition for the resonating beam.

Example Consider a ring resonator with total cavity loss of $A = (\alpha L - \ln R) = 1.0$. In other words, the optical intensity drops by a factor of $1/e$ after a round trip. Let the time constant be $\tau = 0.01$ s and $\gamma_0 L = 10$. The frequency detuning range is $|\Omega| \leq 300\ \mathrm{s}^{-1}$ according to Eq. (5.1-18).

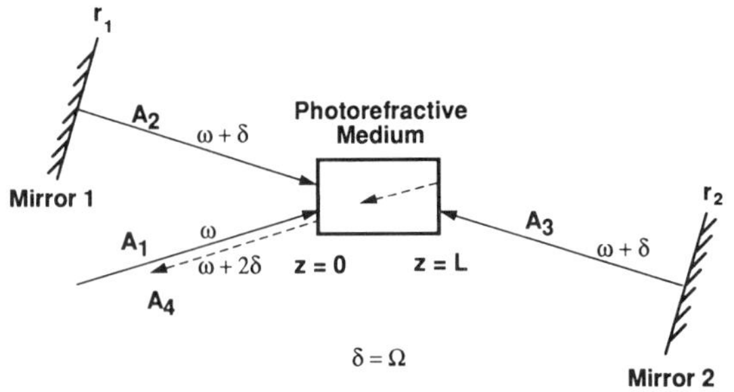

Figure 5.4. A linear photorefractive resonator pumped by a laser beam at frequency ω. As a result of the parametric gain in the photorefractive medium, the oscillation occurs with a frequency of $(\omega + \delta)$ which is determined by the cavity length. A self-generated phase conjugate beam is produced at a frequency of $(\omega + 2\delta)$. $\delta = \Omega$ is the frequency detuning.

5.2 LINEAR RESONATORS

Oscillations can also occur in photorefractive resonators which consist of a photorefractive crystal inside a linear resonator. Referring to Figure 5.4, we consider a linear resonator which consists of two mirrors and a photorefractive medium pumped by a laser beam. The gain due to two-wave mixing allows the oscillation to occur. As soon as oscillation starts, four-wave mixing happens simultaneously. This leads to the return of a phase conjugated beam. Since the counterpropagating beams are generated by self-oscillation, such devices are called self-pumped phase conjugators. In what follows, we investigate the oscillation properties of such photorefractive resonators. Optical phase conjugation and various photorefractive phase conjugators will be treated in Chapter 6.

The coupling in photorefractive linear resonators involves optical four-wave mixing. A vigorous treatment would require the analysis of nondegenerate four-wave mixing discussed in the last chapter. Here, we will first describe the oscillation using a two-wave mixing approximation which is valid provided the intensity of oscillation is much weaker as compared with the pump beam. We will then treat the problem using four-wave mixing. In the two-wave mixing approximation (TWMA), some of the results obtained in the last section can be applied here. The frequency of oscillation is given, according to Eq. (5.1-10), by

$$\omega_2 = \omega_1 + \frac{1}{\tau}\frac{2(\Delta\Gamma + 2M\pi)}{2\alpha L - \ln R} \equiv \omega_1 + \Omega \tag{5.2-1}$$

where $2\alpha L$ is the round trip loss coefficient, ω_1 is the frequency of the pump beam ($\omega_1 = \omega$), and Ω is the frequency difference (i.e., $\Omega = \omega_2 - \omega_1$). We neglect the loss due to diffraction when beam 3 reads the grating and generates beam 4.

The intensity of oscillation is given, according to Eq. (5.1-11), by

$$I_2(0) = I_1(0) \frac{Re^{-2\alpha L} - e^{-\gamma L}}{1 - Re^{-2\alpha L}} \tag{5.2-2}$$

where we recall that R is the product of minor reflectivities and $2\alpha L$ is the round trip absorption coefficient. The intensity of the oscillating beam at $z = L$ is given, according to Eq. (5.1-3) and the condition that $gRe^{-\alpha L} = 1$, by

$$I_2(L) = \frac{I_2(0)}{Re^{-\alpha L}} \tag{5.2-3}$$

The result is consistent with the approximation that the cavity loss due to diffraction into beam 4 is neglected (see Figure 5.4).

As a result of the oscillation and two-wave mixing, a grating formed by beams 1 and 2 is present in the photorefractive medium. When beam 2 hits the mirror (R_2), beam 3 is generated as a retro-reflected beam. This beam will propagate through the photorefractive medium and serve as a readout beam for the grating formed by beams 1 and 2. As a result of holographic reconstruction, beam 4 is generated by Bragg diffraction. This beam is a phase conjugate of beam 1. The intensity of beam 4 can be written

$$I_4(0) = \eta I_3(L) \tag{5.2-4}$$

where η is the diffraction efficiency and $I_3(L)$ is the intensity of beam 3 at $z = L$. Using Eq. (5.2-3) and $I_3(L) = R_2 I_2(L)$, we obtain

$$I_4(0) = \eta \frac{I_2(0)}{R_1 e^{-\alpha L}} \tag{5.2-5}$$

where R_1 and R_2 are mirror reflectivities. The diffraction efficiency can be calculated by using the technique derived in Section 4.7 (see Eqs. (4.7-31), (4.7-32)). The phase conjugate reflectivity is defined as

$$|\rho|^2 = \frac{I_4(0)}{I_1(0)} \tag{5.2-6}$$

which can be written, according to Eqs. (5.2-5) and (5.2-2)

$$|\rho|^2 = \frac{\eta}{R_1 e^{-\alpha L}} \frac{Re^{-2\alpha L} - e^{-\gamma L}}{1 - Re^{-2\alpha L}} \tag{5.2-7}$$

It is important to note that the frequency of beam 4 is shifted relative to the incident beam. Since beam 3 is related to beam 2 by a mirror reflection, $\omega_3 = \omega_2 = \omega_1 + \Omega$. The frequency of beam 4 is given by

$$\omega_4 = \omega_3 + \Omega = \omega_1 + 2\Omega \tag{5.2-8}$$

We note that beam 4 is frequency shifted by 2Ω (or 2δ).

5.2.1 Exact Solution

In the event when the oscillation intensity is not small relative to the pump intensity $I_1(0)$, we need to use the exact solution. The discussion is as follows. Let r_1 and r_2 be the amplitude reflection coefficients of the mirrors. At oscillation, four-wave mixing occurs inside the photorefractive medium. The boundary conditions with $A_1(0)$ known are

$$A_1(0) \neq 0 \tag{5.2-9}$$

$$A_4(L) = 0 \tag{5.2-10}$$

$$A_2(0) = A_3(0) r_1 \exp(-i\phi_1) \tag{5.2-11}$$

$$A_3(L) = A_2(L) r_2 \exp(-i\phi_2) \tag{5.2-12}$$

where ϕ_1, ϕ_2 are phases depending on the distance between the mirrors and the crystal. The constant c in Eq. (4.5-13) can be written

$$c = A_1(0)A_4(0) + A_2(0)A_3(0) = A_2(L)A_3(L) \tag{5.2-13}$$

Using the definition of u, v (see Eq. (4.5-30)), we obtain, according to the boundary conditions (5.2-11) and (5.2-12)

$$|u_0|^2 = \left|\frac{A_2(0)}{A_3^*(0)}\right|^2 = |r_1|^2 \tag{5.2-14}$$

$$|u_L|^2 = \left|\frac{A_2(L)}{A_3^*(L)}\right|^2 = \frac{1}{|r_2|^2} = \frac{|c|^2}{I_{3L}^2}, \qquad u_L = \frac{c}{I_{3L}} = \frac{c}{|cr_2|} \tag{5.2-15}$$

where I_{3L} is the intensity of beam 3 at $z = L$ (i.e., $I_{3L} = |A_3(L)|^2$). The solutions for u and v are given by

$$v = \frac{A_4}{A_1^*} = \frac{2c \tanh \kappa(z-L)}{-s + \sigma \tanh \kappa(z-L)} \tag{5.2-16}$$

$$u = \frac{A_2}{A_3^*} = \frac{u_L[s + \sigma \tanh \kappa(z-L)] + 2c \tanh \kappa(z-L)}{[s - \sigma \tanh \kappa(z-L)] + 2c^* u_L \tanh \kappa(z-L)} \tag{5.2-17}$$

where we recall that $u_L = c/|cr_2|$.

Putting $z = 0$ in Eq. (5.2-17) and using Eqs. (5.2-14) and (5.2-15), we obtain

$$|r_1|^2 = \frac{|[(s - \sigma T) - 2|cr_2|T]|^2}{|r_2|^2 \left| \left[(s + \sigma T) - 2 \left| \dfrac{c}{r_2} \right| T \right] \right|^2} \tag{5.2-18}$$

where

$$T = \tanh \kappa L = \tanh \frac{\Gamma^* s}{4I_0} L \tag{5.2-19}$$

If the intensity is normalized to $I_0 = 1$, so that

$$c_1 + c_2 = 1 \tag{5.2-20}$$

then using $\sigma = c_1 - c_2$, we have

$$1 - \sigma = 2c_2 = 2I_{3L} = 2|cr_2| \tag{5.2-21}$$

Since $s^2 = \sigma^2 + 4|c|^2$ and σ is related to $|c|$ by Eq. (5.2-21), Eq. (5.2-18) can be solved for $|c|$.

The phase conjugate reflection coefficient is given, according to Eq. (5.2-16), by

$$\rho = \frac{A_4(0)}{A_1^*(0)} = \frac{2cT}{s + \sigma T} \tag{5.2-22}$$

Once $|c|$ is solved from Eq. (5.2-18), the magnitude of phase conjugate reflectivity $|\rho|$ is obtained according to Eq. (5.2-22). The phase of ρ remains unresolved.

In addition to $|\rho|$, we also obtain

$$I_{3L} = c_2 = |cr_2| \tag{5.2-23}$$

$$c_1 = 1 - c_2 = 1 - |cr_2| \tag{5.2-24}$$

$$I_{20} = c_1 - I_{10} = 1 - |cr_2| - I_{10} \tag{5.2-25}$$

$$I_{30} = \frac{I_{20}}{|r_1|^2} \tag{5.2-26}$$

$$I_{40} = |\rho|^2 I_{10} \tag{5.2-27}$$

where $I_{i0} = |A_i(0)|^2$. Using these relationships and the normalization

$$I_{10} + I_{20} + I_{30} + I_{40} = I_0 = 1 \tag{5.2-28}$$

we obtain

$$I_{10} = \frac{1 - |cr_2|(1 + R_1)}{1 - |\rho|^2 R_1} \tag{5.2-29}$$

with $R_1 = |r_1|^2$ and $R_2 = |r_2|^2$. The other intensities are related to I_{10} by Eqs. (5.2-25)–(5.2-27).

The threshold for oscillation occurs at $\rho = 0$, which is equivalent to $c = 0$ and $\sigma = 1$. The threshold condition is thus given, according to Eq. (5.2-18), by

$$R_1 R_2 = \left|\frac{1 - T}{1 + T}\right|^2 = \exp\left[-\frac{1}{2}(\Gamma + \Gamma^*)L\right] \tag{5.2-30}$$

This agrees with earlier results of $gR = 1$, assuming $\alpha = 0$.

To study the frequency of oscillation, we need to examine the phase of the amplitude A_2 and A_3. If we write

$$A_2(z) = |A_2(z)| \exp[-i\psi_2(z)] \tag{5.2-31}$$

$$A_3(z) = |A_3(z)| \exp[-i\psi_3(z)] \tag{5.2-32}$$

then the additional phase shift due to photorefractive four-wave mixing can be written

$$\Delta\psi = \psi_2(L) - \psi_2(0) + \psi_3(0) - \psi_3(L) \tag{5.2-33}$$

or

$$\Delta\psi = \int_0^L d(\psi_2 - \psi_3) \tag{5.2-34}$$

Using Eq. (5.2-34) for $\Delta\psi$, the round trip phase condition Eq. (5.1-5) can then be used to determine the frequency of oscillation.

By using the definition for u and v, Eqs. (4.5-7) and (4.5-8) can be rewritten

$$\frac{d}{dz} \ln A_2 = \frac{1}{2} \Gamma^* \frac{u + v}{u} \frac{I_1}{I_0} \tag{5.2-35}$$

$$\frac{d}{dz} \ln A_3 = \frac{1}{2} \Gamma \frac{u^* + v^*}{v^*} \frac{I_4}{I_0} \tag{5.2-36}$$

Substituting Eqs. (5.2-31) and (5.2-32) into the above equations and using Eq. (5.2-34), we obtain

$$\Delta\psi = -\mathrm{Im} \int_0^L \frac{\Gamma^*}{2I_0} \left[I_1(1 + w) + I_4\left(1 + \frac{1}{w}\right)\right] dz \tag{5.2-37}$$

where w is defined as

$$w = \frac{v}{u} = \frac{A_4 A_3^*}{A_1^* A_2} \tag{5.2-38}$$

According to the definition, w is a measure of the relative strength of the contribution to the grating. Using Eqs. (5.2-31) and (5.2-32) and $u_L = c/|cr_2|$, w can be written

$$w = \frac{-2T[|cr_2|(s \quad \sigma T) + 2T|c|^2]}{(s - \sigma T)[(s + \sigma T) + 2T|cr_2|]} \tag{5.2-39}$$

where T is given by

$$T = \tanh \kappa(z - L) = \tanh \frac{\Gamma^* s}{4I_0}(z - L) \tag{5.2-40}$$

Once $|c|$ is solved from Eq. (5.2-18), $|u|^2$ and $|v|^2$ can be used to evaluate all the intensities and w. The phase $\Delta\psi$ can then be obtained by carrying out numerical integration of Eq. (5.2-37). The frequency of oscillation is thus determined by the round trip phase condition

$$\Delta\psi + \int k\, ds = 2N\pi \tag{5.2-41}$$

where the integration is over a round trip inside the resonator. Let $\Delta\Gamma$ be the cavity detuning defined by Eq. (5.1-7), oscillation occurs when

$$\Delta\psi = \Delta\Gamma + 2M\pi \tag{5.2-42}$$

where M is an integer.

Typical results on frequency shift Ω, oscillation intensity and conjugate reflectivity are given in Figure 5.5.

We note that as the cavity is detuned from its origin where degenerate oscillation occurs, the frequency of oscillation deviates from that of the incident laser beam. In general, the frequency shift can be multivalued for a given cavity detuning. This is a bistable (or multistable) phenomenon which may be useful for switching or memory applications. As the cavity is detuned, the phase conjugate reflectivity also varies. For small detuning, the phase conjugate reflectivity remains almost constant until at certain detuning, $|\rho|^2$ drops suddenly. Bistability also occurs here. In Figure 5.5(c), we plot $|\rho|^2$ versus $\Omega\tau$. Generally speaking, $|\rho|^2$ drops significantly when $\Omega\tau > 1$.

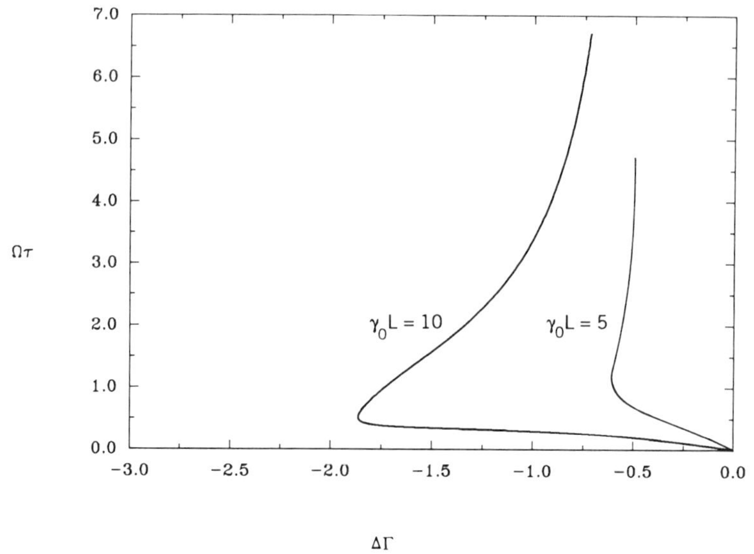

(a)

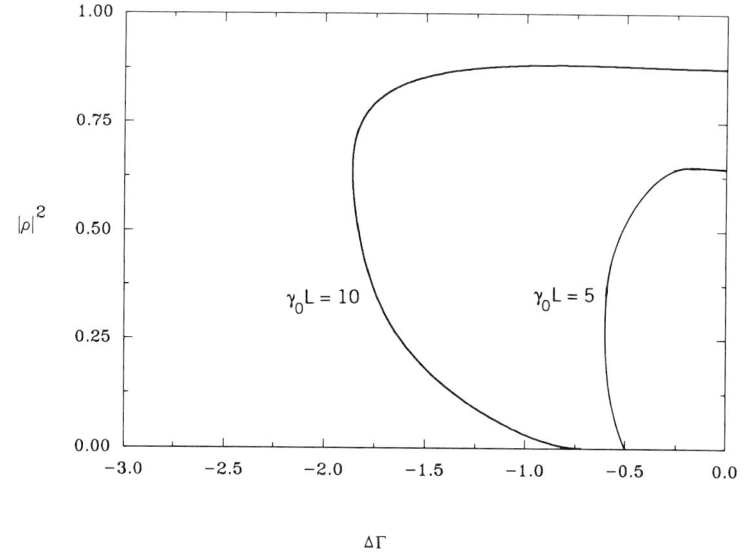

(b)

Figure 5.5. (a) Frequency shift Ω as a function of cavity detuning, (b) phase conjugate reflectivity as functions of cavity detuning, and (c) phase conjugate reflectivity as a function of frequency shift. $R_1 = R_2 = 0.9$.

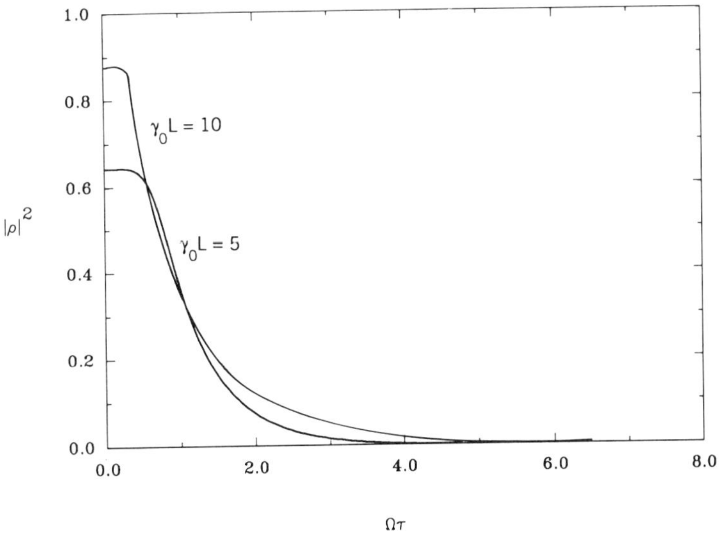

(c)

Figure 5.5. (*Continued*)

5.3 BIDIRECTIONAL OSCILLATION IN RING RESONATORS

Bidirectional oscillation in ring resonators using optical four-wave mixing in nonlinear media was studied previously [4, 5]. Self-oscillation can be achieved when the gain due to parametric four-wave mixing is large enough to overcome the cavity loss. The investigation was aimed at reducing or eliminating the frequency locking in ring laser gyro [5, 6]. Optical four-wave mixing in photorefractive media can also provide the gain for bidirectional oscillation. The oscillation has been observed and some of the special cases has been studied [7–11]. In this section, we treat the problem of four-wave mixing in photorefractive media with the boundary conditions of a ring resonator.

Consider a ring resonator driven by photorefractive four-wave mixing (FWM), as shown in Figure 5.6. Beams 1 and 4 are two external pump waves. Beams 2 and 3 are, respectively, the counterclockwise and clockwise oscillating waves in the resonator. In the case when the transmission gratings are the dominant ones, the coupled mode equations are

$$\frac{dA_1}{dz} = -\frac{1}{2}\Gamma G A_2 \tag{5.3-1}$$

$$\frac{dA_2}{dz} = \frac{1}{2}\Gamma^* G^* A_1 \tag{5.3-2}$$

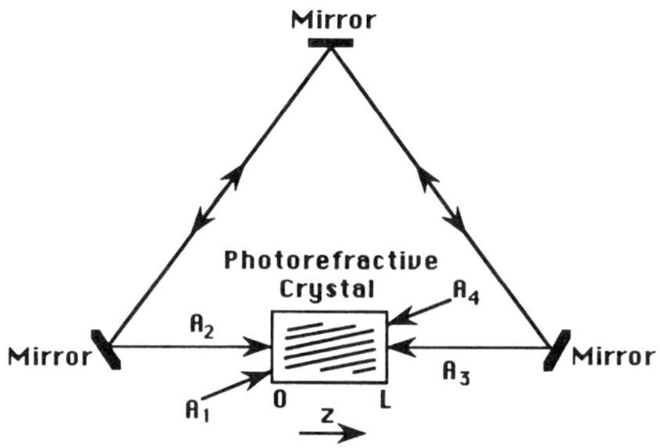

Figure 5.6. Schematic drawing of a bidirectional photorefractive ring resonator. The photorefractive medium is pumped by a pair of counterpropagating laser beams A_1 and A_4. As a result of the parametric gain due to four-wave mixing in the photorefractive medium, bidirectional oscillation occurs.

$$\frac{dA_3}{dz} = \frac{1}{2} \Gamma G A_4 \tag{5.3-3}$$

$$\frac{dA_4}{dz} = -\frac{1}{2} \Gamma^* G^* A_3 \tag{5.3-4}$$

where A_i ($i = 1, 2, 3, 4$) is the amplitude of the ith beam, Γ is the complex coupling constant, and G is the modulation depth of the photorefractive gratings which is given by

$$G = \frac{A_1 A_2^* + A_3 A_4^*}{I_0} \tag{5.3-5}$$

and I_0 is the averaged total intensity, that is,

$$I_0 = |A_1|^2 + |A_2|^2 + |A_3|^2 + |A_4|^2 \tag{5.3-6}$$

In Eqs. (5.3-1)–(5.3-4), we assumed that the absorption is negligible.

To solve Eqs. (5.3-1)–(5.3-4), we first examine the variation of the modulation depth G with respect to z. Using Eqs. (5.3-1)–(5.3-5), we obtain

$$\frac{dG}{dz} = \frac{1}{2} \Gamma G \frac{|A_1|^2 - |A_2|^2 - |A_3|^2 + |A_4|^2}{I_0} \tag{5.3-7}$$

An integration of the equation yields

$$G(z) = G(0) \exp\left[\int_0^z \frac{1}{2} \Gamma \frac{|A_1|^2 - |A_2|^2 - |A_3|^2 + |A_4|^2}{I_0} dz'\right] \tag{5.3-8}$$

Notice that if Γ is real (in other words, the photorefractive medium is in the regime of pure diffusion and the oscillation has no frequency detuning), $G(z)$ has a constant phase, that is,

$$G(z) = |G(z)| \exp(-i\phi_G) \tag{5.3-9}$$

where ϕ_G does not depend on z. We can further examine the modulation depth of the grating formed by beams 1 and 2, and that of the grating formed by beams 3 and 4,

$$G_{12} = \frac{A_1 A_2^*}{I_0} \tag{5.3-10}$$

$$G_{34} = \frac{A_3 A_4^*}{I_0} \tag{5.3-11}$$

According to Eqs. (5.3-1)–(5.3-4), the variation of G_{12} and G_{34} with respect to z are

$$\frac{dG_{12}}{dz} = \frac{1}{2}\Gamma G \frac{|A_1|^2 - |A_2|^2}{I_0} \tag{5.3-12}$$

$$\frac{dG_{34}}{dz} = \frac{1}{2}\Gamma G \frac{-|A_3|^2 + |A_4|^2}{I_0} \tag{5.3-13}$$

Integrating Eqs. (5.3-12) and (5.3-13), we obtain

$$G_{12}(z) = \exp(-i\phi_G) \int_0^z \frac{1}{2} \Gamma |G(z')| \frac{|A_1|^2 - |A_2|^2}{I_0}\, dz' + \frac{A_1(0)A_2^*(0)}{I_0} \tag{5.3-14}$$

$$G_{34}(z) = \exp(-i\phi_G) \int_z^L \frac{1}{2} \Gamma |G(z')| \frac{|A_3|^2 - |A_4|^2}{I_0}\, dz' + \frac{A_3(L)A_4^*(L)}{I_0} \tag{5.3-15}$$

Equations (5.3-14) and (5.3-15) imply that the modulation depth of each grating has a varying part which has a constant phase factor $\exp(-i\phi_G)$ and a constant part which may have a different phase factor depending on the boundary conditions. Since the superposition of the gratings has the phase factor $\exp(-i\phi_G)$, that is,

$$G(z) = G_{12}(z) + G_{34}(z) = |G(z)| \exp(-i\phi_G) \tag{5.3-16}$$

we have

$$A_1(0)A_2^*(0) + A_3(L)A_4^*(L) = \pm|A_1(0)A_2^*(0) + A_3(L)A_4^*(L)| \exp(-i\phi_G) \tag{5.3-17}$$

In other words, the phase of the gratings is determined by their boundary values.

Now we solve the coupled mode equations in the case of real Γ. Using the method of grating integral (see Section 2.7 or [12]), we define a new variable

$$u = \frac{1}{2}\Gamma \int_0^z |G(z')|\, dz' \tag{5.3-18}$$

which is the integral of the grating amplitude. This integral represents the coupling strength as waves propagate through the crystal. Changing variable from z to u, the coupled mode equations can be written as

$$\frac{dA_1}{du} = -\exp(-i\phi_G)A_2 \tag{5.3-19}$$

$$\frac{dA_2}{du} = \exp(i\phi_G)A_1 \tag{5.3-20}$$

$$\frac{dA_3}{du} = \exp(-i\phi_G)A_4 \tag{5.3-21}$$

$$\frac{dA_4}{du} = -\exp(i\phi_G)A_3 \tag{5.3-22}$$

The solutions to the above equations can be easily found, that is,

$$A_1(u) = A_{10}\cos u - \exp(-i\phi_G)A_{20}\sin u \tag{5.3-23}$$

$$A_2(u) = A_{20}\cos u + \exp(i\phi_G)A_{10}\sin u \tag{5.3-24}$$

$$A_3(u) = A_{3L}\cos(u_L - u) - \exp(-i\phi_G)A_{4L}\sin(u_L - u) \tag{5.3-25}$$

$$A_4(u) = A_{4L}\cos(u_L - u) + \exp(i\phi_G)A_{3L}\sin(u_L - u) \tag{5.3-26}$$

where A_{10}, A_{20}, A_{3L}, and A_{4L} are amplitudes at boundaries $z = 0$ and $z = L$.

Now we consider the bidirectional ring resonator in the case of steady state oscillations. Since the oscillating beam must duplicate itself after each round trip, we have

$$A_2(0) = A_2(L)re^{-i\psi} \tag{5.3-27}$$

$$A_3(L) = A_3(0)re^{-i\psi} \tag{5.3-28}$$

where $r = r_1 r_2 r_3$ is the product of reflection coefficients of the three mirrors and ψ is the round trip path integral, that is,

$$\psi = \int k\, ds \tag{5.3-29}$$

Using Eqs. (5.3-23)–(5.3-28), we obtain

$$\frac{A_{20}}{A_{10}} = \frac{r\exp(-i\psi)\exp(i\phi_G)\sin u_L}{1 - r\exp(-i\psi)\cos u_L} \tag{5.3-30}$$

$$\frac{A_{3L}}{A_{4L}} = \frac{-r\exp(-i\psi)\exp(-i\phi_G)\sin u_L}{1 - r\exp(-i\psi)\cos u_L} \tag{5.3-31}$$

Therefore,

$$A_{10}A_{20}^* + A_{3L}A_{4L}^* = \exp(-i\phi_G)\left[I_{10}\left(\frac{r\exp(-i\psi)\sin u_L}{1 - r\exp(-i\psi)\cos u_L}\right)^* - I_{4L}\left(\frac{r\exp(-i\psi)\sin u_L}{1 - r\exp(-i\psi)\cos u_L}\right)\right] \tag{5.3-32}$$

On the other hand, Eq. (5.3-17) requires that $r\exp(-i\psi)\sin u_L/[1 - r \times \exp(-i\psi)\cos u_L]$ must be real. If we write

$$r = |r|\exp(-i\phi_r) \tag{5.3-33}$$

where ϕ_r is a constant phase, the above constrain leads to

$$\psi + \phi_r = M\pi \tag{5.3-34}$$

where M is an integer.

The oscillation intensities (e.g., I_{20}, I_{3L}) can be found by evaluating

$$\frac{A_{20}\exp(-i\phi_G)}{A_{10}} = -\frac{A_{3L}\exp(i\phi_G)}{A_{4L}} = \frac{\pm|r|\sin u_L}{1 \mp |r|\cos u_L} \equiv q \tag{5.3-35}$$

To obtain the grating action u_L, we rewrite Eq. (5.3-18) as

$$\frac{du}{dz} = \frac{\Gamma}{2} G\exp(i\phi_G) \tag{5.3-36}$$

and substitute Eqs. (5.3-23)–(5.3-26) into Eq. (5.3-5) to calculate G. This leads to

$$\frac{du}{dz} = \frac{\Gamma}{2}(b\cos 2u + c\sin 2u) \tag{5.3-37}$$

where the coefficients b and c are given by

$$b = \frac{qI_{10}}{I_0} - \frac{qI_{4L}}{I_0}\cos 2u_L + \frac{I_{4L}(q^2 - 1)}{2I_0}\sin 2u_L \tag{5.3-38}$$

and

$$c = \frac{I_{10}(1-q^2)}{2I_0} - \frac{I_{4L}(q^2-1)}{2I_0}\cos 2u_L - \frac{qI_{4L}}{I_0}\sin 2u_L \tag{5.3-39}$$

Equation (5.3-37) can be integrated exactly. With the boundary condition $u(0) = 0$, the solution can be written

$$\tanh\left[\frac{1}{2}(b^2+c^2)^{1/2}\Gamma z\right] = \frac{(b^2+c^2)^{1/2}\tan u}{b + c\tan u} \tag{5.3-40}$$

When $z = L$, the above solution results in a transcendental equation

$$\Gamma L = \frac{2}{(b^2+c^2)^{1/2}}\tanh^{-1}\frac{(b^2+c^2)^{1/2}\tan u_L}{b + c\tan u_L} \tag{5.3-41}$$

We solve the transcendental equation numerically for various values of ΓL and substitute u_L into Eq. (5.3-35) to obtain the oscillation intensities. Notice that both the + and − signs in Eq. (5.3-35) result in the same intensity for a given ΓL with their solutions to u_L different by π.

Figure 5.7 shows the oscillation intensities I_{20} and I_{3L} as functions of ΓL for various values of the ratio between intensities of the two pump beams I_{10}/I_{4L}. When $|r| < 1$, that is, there is no other gain medium in the cavity, the resonator will oscillate if $\Gamma L > 0$ and $I_{10}/I_{4L} > 1$, or if $\Gamma L < 0$ and $I_{10}/I_{4L} < 1$, as shown in Figure 5.7(a). The oscillation intensity I_{3L} (assuming $I_{4L} = 1$) in the case of $I_{10}/I_{4L} = C < 1$ is symmetric to that in the case of $I_{10}/I_{4L} = 1/C > 1$, with respect to $\Gamma L = 0$. For large pump ratio, there is only one oscillation mode. The oscillation intensities will increase from 0 at certain threshold ΓL and saturate at large ΓL. For small pump ratio, there might be multiple oscillation modes, as shown in Figure 5.7(a) for $I_{10}/I_{4L} =$ 2.0. When $|r| > 1$, that is, there is an additional gain medium in the cavity, the resonator will oscillate if $\Gamma L < 0$ and $I_{10}/I_{4L} > 1$, or if $\Gamma L > 0$ and $I_{10}/I_{4L} < 1$, as shown in Figure 5.7(b). Again, the oscillation intensities are symmetric in the two cases of oscillations.

Figure 5.8 shows the threshold ΓL as a function of the pump beam intensity ratio for various values of $|r|$. As the pump beam ratio increases from 1, the threshold value $|\Gamma L|$ decreases rapidly. For equal pump intensities, that is $I_{10} = I_{4L}$, the factor $(b + c\tan u_L)$ in Eq. (5.3-41) is 0. Thus, there does not exist a solution to Eq. (5.3-41). In this case, there is no steady state oscillation.

The oscillation condition of the bidirectional ring resonator can be used to compare the intensities of the two pump beams. For example, assuming there is no other gain medium in the cavity ($|r| < 1$), if the resonator can establish steady state oscillation for positive ΓL, the pump intensity I_{10} must be greater than I_{4L}. Figure 5.9 shows the oscillation intensity I_{20} as a

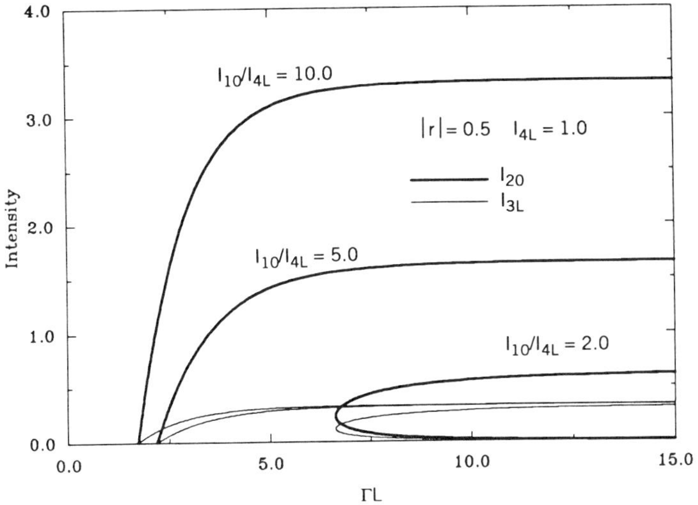

(a)

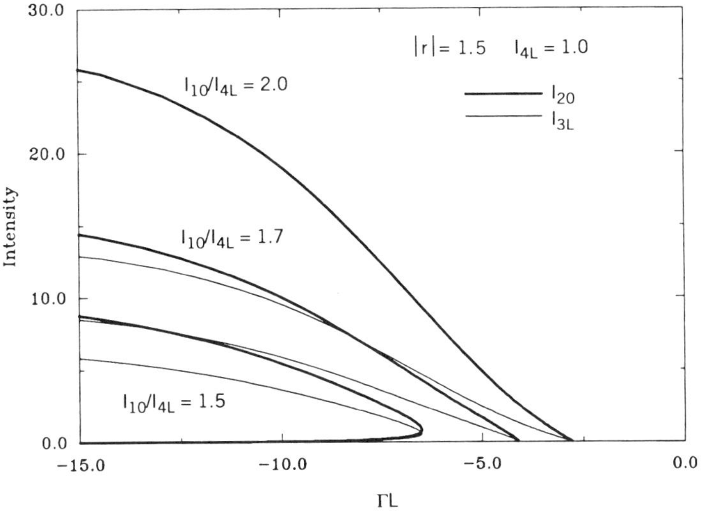

(b)

Figure 5.7. Oscillation intensities I_{20} and I_{3L} as functions of ΓL in a bidirectional ring resonator for various values of the pump ratio I_{10}/I_{4L}. (a) Passive cavity with loss, $|r| < 1$. (b) Active cavity with gain, $|r| > 1$.

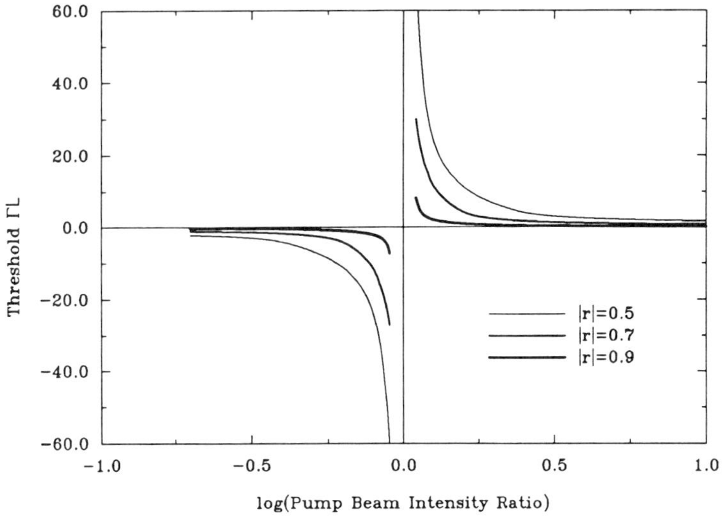

(a)

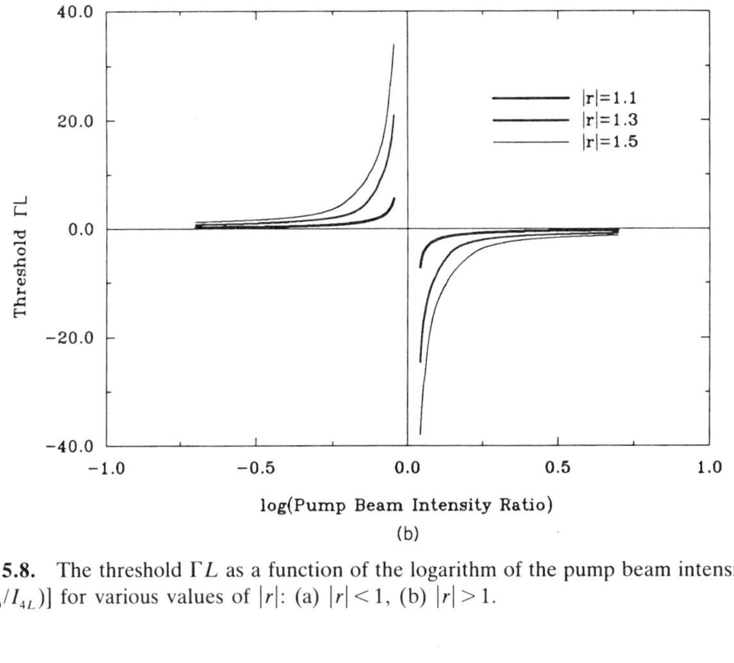

(b)

Figure 5.8. The threshold ΓL as a function of the logarithm of the pump beam intensity ratio $[\log(I_{10}/I_{4L})]$ for various values of $|r|$: (a) $|r|<1$, (b) $|r|>1$.

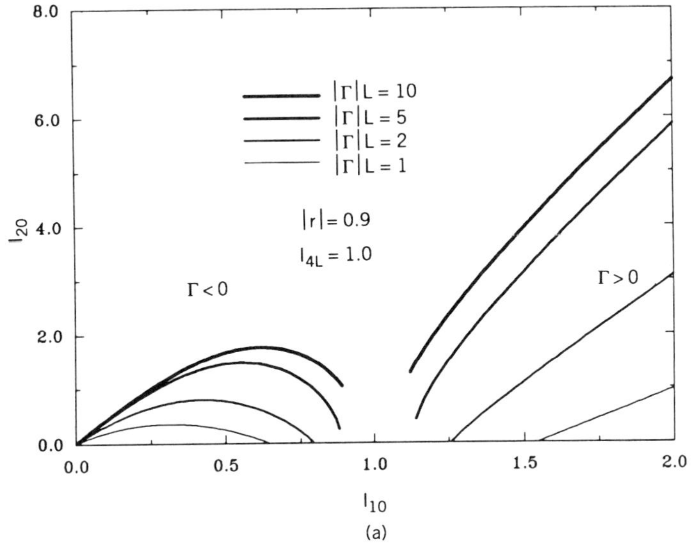

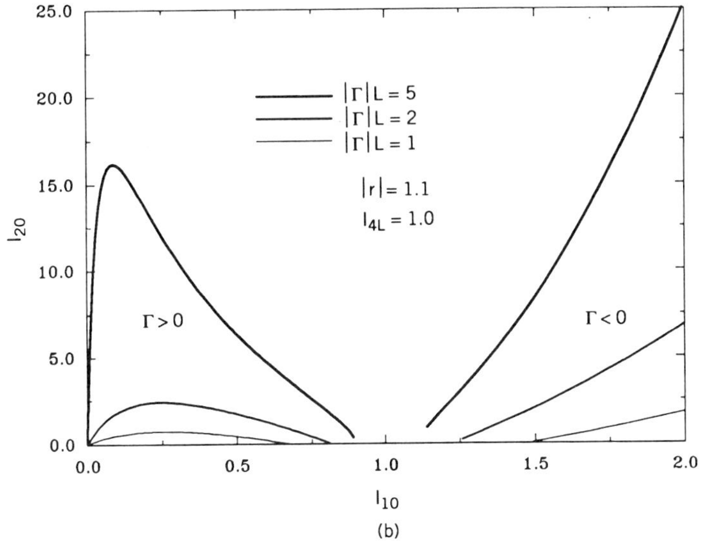

Figure 5.9. Oscillation intensity I_{20} as a function of the pump intensity I_{10} for various values of ΓL in a bidirectional ring resonator. The intensity of the other pump is set at $I_{4L} = 1.0$. (a) Passive cavity with loss, $|r| < 1$. (b) Active cavity with gain, $|r| > 1$. Notice the intensity thresholding. In other words, oscillation occurs only in certain intensity ranges for I_{10}.

function of the pump intensity I_{10} for various values of ΓL. For small ΓL, oscillation occurs only when the intensities of the two pump beams are significantly different; while for large ΓL, oscillation can occur with a small difference between the intensities of the two pump beams. This effect provides a possibility for optical intensity comparison and thresholding.

In summary, we have theoretically solved bidirectional oscillations in phase conjugate resonators in which a photorefractive medium pumped by a pair of counterpropagating beams provides the gain. The results show that steady state oscillations depend strongly on the ratio between the two pump beams. This intensity ratio dependence provides a possible method for optical intensity comparison and thresholding. In this section, we have only considered oscillations without frequency detuning. Further investigations are needed to extend the results to cover cavity detuning, as well as applied field to the photorefractive medium.

PROBLEMS

5.1 Consider the intensity of oscillation $I_2(0)$ in a unidirectional ring resonator.

- **a.** Show that $I_2(0)$ is maximum at $\Omega = 0$.
- **b.** Show that $I_2(0)$ can be larger than the pump intensity $I_1(0)$. Derive an expression for the condition.
- **c.** Show that the output intensity of the resonator is always less than $I_1(0)$.
- **d.** Find the mode spacing in multimode oscillation.

5.2 Derive the threshold condition Eq. (5.2-30).

5.3 Derive Eq. (5.3-41).

5.4 Consider a unidirectional ring resonator filled with two photorefractive crystals. The crystals are each pumped by a laser. Discuss the possibility of power combining of mutually incoherent beams.

5.5 **Coupled Ring Resonators.** Consider two unidirectional ring resonators which are coupled by two-wave mixing in a photorefractive crystal. The first ring resonator is driven by a pump laser beam and a photorefractive crystal. The second resonator is driven by the oscillating beam of the first resonator via two-wave mixing in a photorefractive crystal.

- **a.** Derive the oscillation condition for the case when only the first resonator is oscillating.
- **b.** Derive the oscillation condition for the case when both resonators are oscillating.
- **c.** Discuss the possibility of periodic pulsation.

5.6 Photorefractive Biasing in Ring Laser Gyros. Consider the possibility of using contra-directional two-wave mixing in photorefractive crystals for the biasing of laser gyros [5]. A photorefractive crystal plate is placed inside a ring laser gyro.

a. Discuss the effect of energy coupling due to degenerate two-wave mixing on the oscillation of the oppositely directed beams.

b. If an external electric field is applied, the system may support nondegenerate oscillations even in an inertial frame. Discuss the possibility of biasing. In ring laser gyros, biasing means providing a frequency difference which can be achieved by rotating the gyro at a constant angular velocity.

REFERENCES

1. J. O. White, M. Cronin-Golomb, B. Fischer, and A. Yariv, "Coherent Oscillation by Self-Induced Gratings in Photorefractive Crystals," Appl. Phys. **40**, 450 (1982).
2. P. Yeh, "Theory of Unidirectional Photorefractive Resonators," J. Opt. Soc. Am. **B2**, 1924–1928 (1985).
3. M. D. Ewbank and P. Yeh, "Frequency Shifted Cavity Length in Photorefractive Resonators," Opt. Lett. **10**, 496–498 (1985).
4. P. Yeh, "Theory of Phase-Conjugate Oscillators," J. Opt. Soc. Am. **B2**, 727 (1985).
5. P. Yeh and M. Khoshnevisan, "Phase-Conjugate Ring Gyros and Photorefractive-biased Ring Gyros," SPIE **487**, 102 (1984).
6. J.-C. Diels and I. C. McMichael, "Influence of Wave-Front-Conjugated Coupling on the Operation of a Laser Gyro," Opt. Lett. **6**, 219 (1981).
7. J. O. White, M. Cronin-Golomb, B. Fischer, and A. Yariv, "Coherent Oscillation by Self-Induced Gratings in Photorefractive Crystal $BaTiO_3$," Appl. Phys. Lett. **40**, 450 (1982).
8. P. Yeh, "Photorefractive Coupling in Ring Resonators," Appl. Opt. **23**, 2974 (1984).
9. P. Yeh, "Theory of Unidirectional Photorefractive Ring Oscillators," J. Opt. Soc. Am. **B2**, 1924 (1985).
10. M. D. Ewbank and P. Yeh, "Frequency Shift and Cavity Length in Photorefractive Resonators," Opt. Lett. **10**, 496 (1985).
11. M. D. Ewbank and P. Yeh, "Frequency Shifts of Self-Pumped Phase Conjugators," Proc. SPIE **613**, 59 (1986).
12. A. Bledowski and W. Krolikowski, "Exact Solution of Degenerate Four-Wave Mixing in Photorefractive Media," Opt. Lett. **13**, 146 (1988).
13. P. Yeh and M. Khoshnevisan, "Nonlinear-Optical Bragg Scattering in Kerr Media," J. Opt. Soc. Am. **B4**, 1954 (1987).
14. See, for example, P. Yeh, "Two-Wave Mixing in Nonlinear Media," IEEE J. Quantum Electron. **25**, 484 (1989).

6

Photorefractive Phase Conjugators

Phase conjugators are optical devices that can generate a time-reversed replica of an incident electromagnetic wave [1]. These devices play an important role in many optical systems that require the transmission of optical waves through scattering media such as the atmosphere. Nonlinear optics offer techniques that can be used to generate phase conjugate waves in real time. Photorefractive crystals such as $BaTiO_3$, SBN, BSO, are, by far, the most efficient media for the generation of phase conjugate waves. In this chapter, we start with an introduction on optical phase conjugation. We then discuss various photorefractive phase conjugators.

6.1 INTRODUCTION

Consider the propagation of an electromagnetic wave predominantly along the z-axis. Let the electric field be written

$$\mathbf{E} = \mathbf{A} \cos(\omega t - kz - \phi) \tag{6.1-1}$$

where the amplitude $\mathbf{A}$ and phase ϕ are real functions of position (x, y, z). If the amplitude is a slowly varying function of z compared with $\cos(\omega t - kz - \phi)$, the propagation of the wave can be easily described in terms of the spatial motion of the *wave fronts* (defined as $kz + \phi =$ constants). For any electromagnetic wave given by Eq. (6.1-1), the phase conjugate of it is given by

$$\mathbf{E}_c = \mathbf{A} \cos(\omega t + kz + \phi) \tag{6.1-2}$$

$\mathbf{E}_c$ and $\mathbf{E}$ form a conjugate pair. In other words, one is the phase conjugate of the other.

By examining Eqs. (6.1-1) and (6.1-2), we find that these two waves have exactly the same wave fronts at any point in space. The motion of these two sets of wave fronts are, however, in opposite directions. In fact, if we reverse the sign of t in $\mathbf{E}$, we obtain $\mathbf{E}_c$. Thus, the phase conjugate wave is often referred to as the time-reversed wave. In fact, if $\mathbf{E}$ in Eq. (6.1-1) satisfies the wave equation

$$\left[\nabla^2 - \frac{n^2}{c^2}\frac{\partial^2}{\partial t^2}\right]\mathbf{E} = 0 \tag{6.1-3}$$

it can be shown easily, by virtue of time reversal symmetry, that $\mathbf{E}_c$ given by Eq. (6.1-2) is also a solution to the wave equation provided $n(x, y, z)$ is real and is independent of t.

Using analytic representation, we write Eq. (6.1-1) as

$$\mathbf{E} = \mathbf{A}(\mathbf{r}) \exp\{i[\omega t - kz - \phi(\mathbf{r})]\} \tag{6.1-4}$$

where we remember that only the real part of the right-hand side is the physical value of the electric field. If we further define a complex quantity

$$\mathbf{A}_1(\mathbf{r}) = \mathbf{A}(r)e^{-i\phi(\mathbf{r})} \tag{6.1-5}$$

then the electric field is written

$$\mathbf{E} = \mathbf{A}_1(\mathbf{r})e^{i(\omega t - kz)} \tag{6.1-6}$$

Using these definitions, the phase conjugate wave can be written

$$\mathbf{E}_c = \mathbf{A}_1^*(r)e^{i(\omega t + kz)} \tag{6.1-7}$$

We note that the phase conjugate wave propagates along the $-z$ direction with a complex amplitude $\mathbf{A}_1^*$.

It is important to note that to get $\mathbf{E}_c$ from $\mathbf{E}$, we take the complex conjugate of the spatial part only, leaving the factor $\exp(i\omega t)$ intact. This is equivalent to leaving the spatial part alone but reversing the sign of t.

To appreciate the property of phase conjugation, we consider the propagation of a plane wave through a distorting medium such as atmosphere (Figure 6.1(a)). Let the distorting medium be confined in the region between $z = 0$ and $z = L$. As a result of the index inhomogeneity, the wave fronts of the transmitted wave are no longer planar (see Figure 6.1(a)). Let the electric field of the forward propagating beam $(+z)$ be represented by Eq. (6.1-6). Using the slowly varying amplitude approximation, the complex amplitude must obey the following equation:

$$\nabla_\perp^2 \mathbf{A}_1 + \left[\frac{\omega^2}{c^2} n^2(x, y, z) - k^2\right]\mathbf{A}_1 - 2ik\frac{\partial}{\partial z}\mathbf{A}_1 = 0 \tag{6.1-8}$$

We recall that this approximation is valid provided the fractional variation of $n(x, y, z)$ in one wavelength is small compared to unity. This is the same scalar wave equation discussed in Chapter 1. It is important to note that this is a first order differential equation in z. Thus, if $\mathbf{A}_1$ is known for some $z = z_0$ and for all x, y, then $\mathbf{A}_1$ is determined for all points in space provided $n(x, y, z)$ is known.

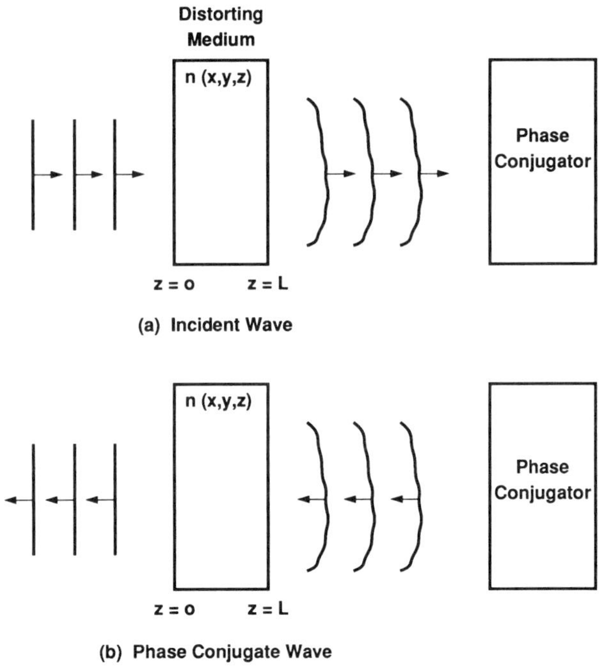

Figure 6.1. (a) An incident beam with planar wave fronts is transmitted through a distorting medium. The wave fronts of the transmitted beam are no longer planar. (b) A time-reversed replica of the transmitted beam is generated by a phase conjugator. The planar wave fronts are restored when the phase conjugate beam retraces the path of the incident beam and transmits through the distorting medium.

We now consider the complex conjugate of Eq. (6.1-8) as a purely mathematical operation:

$$\nabla_\perp^2 \mathbf{A}_1^* + \left[\frac{\omega^2}{c^2} n^2(x, y, z) - k^2 \right] \mathbf{A}_1^* + 2ik \frac{\partial}{\partial z} \mathbf{A}_1^* = 0 \qquad (6.1\text{-}9)$$

This is also a scalar wave equation obeyed by a wave propagating in the $-z$ direction of the form

$$\mathbf{E}_2 = \mathbf{A}_2 e^{i(\omega t + kz)} \qquad (6.1\text{-}10)$$

provided we put

$$\mathbf{A}_2 = \rho \mathbf{A}_1^* \quad \text{for all } x, y, z \qquad (6.1\text{-}11)$$

where ρ is an arbitrary constant. What we have shown is that a wave $\mathbf{A}_2 \exp[i(\omega t + k)]$ whose spatial portion is everywhere complex conjugate of

$\mathbf{A}_1 \exp[i(\omega t - kz)]$, satisfies the same scalar wave equation. Since the scalar wave equation is first order in z, $\mathbf{A}_2(x, y, z)$ for all x, y, z can be determined provided $\mathbf{A}_2$ is known for all x, y at some $z = z_o$. In addition, if

$$\mathbf{A}_2(x, y, z_o) = \rho \mathbf{A}_1^*(x, y, z_o), \quad \text{for all } x, y \tag{6.1-12}$$

then

$$\mathbf{A}_2(x, y, z) = \rho \mathbf{A}_1^*(x, y, z), \quad \text{for all } x, y, z \tag{6.1-13}$$

Physically, this means that if in some plane $z = z_o$ (say $z_o > L$ in Figure 6.1) we generate, by some means, a wave $\mathbf{E}_2$ whose complex amplitude is the complex conjugate of that of $\mathbf{E}$, then $\mathbf{E}_2$ will propagate backwards and remain everywhere the complex conjugate of $\mathbf{E}$. Its wave fronts everywhere thus coincide with those of $\mathbf{E}$. Figure 6.1(b) illustrates the propagation of $\mathbf{E}_2$.

6.2 PHASE CONJUGATION BY FOUR-WAVE MIXING

It is known that wave front reconstruction can be achieved by holographic techniques [2]. Referring to Figure 6.2(a), we consider the interference of two beams inside a holographic recording medium. Beam 1 is called the object beam which contains spatial information and beam 2 is called the reference beam which is simply a plane wave. The electric fields of the beams are written

$$E_1 = A_1 \exp[i(\omega t - \mathbf{k}_1 \cdot \mathbf{r})] \tag{6.2-1}$$

$$E_2 = A_2 \exp[i(\omega t - \mathbf{k}_2 \cdot \mathbf{r})] \tag{6.2-2}$$

where $\mathbf{k}_1, \mathbf{k}_2$ are wave vectors, $A_1(\mathbf{r})$ is the amplitude of the object beam and A_2 is a constant amplitude.

As a result of the beam interference, the intensity can be written as

$$I = |A_1|^2 + |A_2|^2 + A_2 A_1^* e^{-i\mathbf{K}\cdot\mathbf{r}} + A_2^* A_1 e^{i\mathbf{K}\cdot\mathbf{r}} \tag{6.2-3}$$

where

$$\mathbf{K} = \mathbf{k}_2 - \mathbf{k}_1 \tag{6.2-4}$$

In the holographic recording medium, the intensity variation may result in an index variation of the medium. If in addition, the change in the index of refraction is proportional to the intensity, then Δn can be written as

$$\Delta n = n_1|A_1|^2 + n_1|A_2|^2 + \{n_1 A_2 A_1^* e^{-i\mathbf{K}\cdot\mathbf{r}} + \text{c.c.}\} \tag{6.2-5}$$

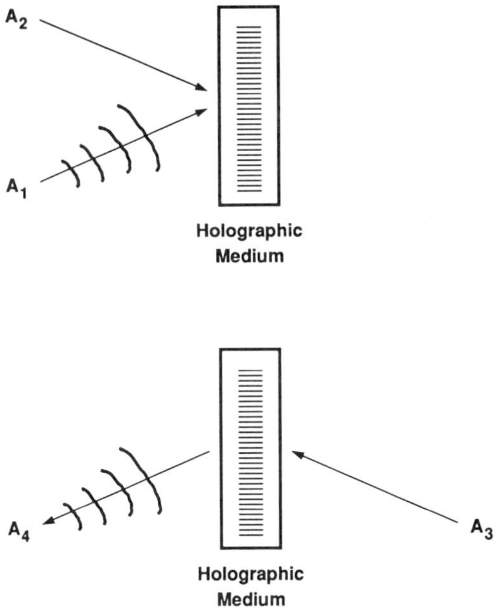

Figure 6.2. *Top*: An object beam A_1 and a reference beam A_2 intersect in a holographic medium. *Bottom*: The recorded hologram is read out by a beam A_3 which is counterpropagating relative to A_2.

where n_1 is a constant. As we can see, the wave front information of the object beam is recorded in the medium in terms of Δn, the change in the index of refraction. We note that with the presence of a reference beam, both the phase and the amplitude of an object beam can be recorded in terms of the intensity of the interference pattern. Aside from a constant term, such a Δn is almost a periodic function of space. It is therefore called a grating. Δn becomes purely periodic only when both A_1 and A_2 are constants. The slight deviation from the perfect periodic variation reflects the spatial variation of the complex amplitude of the object beam. In holography, such a recorded Δn is called a hologram. The hologram records both the amplitude and phase of the object beam.

The complex amplitude (including amplitude and phase) of the object beam can be reconstructed from the hologram by illuminating the medium with a readout beam. Let the electric field of the readout beam be written

$$E_3 = A_3 \exp[i(\omega t - \mathbf{k}_3 \cdot \mathbf{r})] \tag{6.2-6}$$

where $\mathbf{k}_3$ is the wave vector and A_3 is a constant amplitude. In the medium, the polarization will contain the term

$$P = nn_1\epsilon_0[A_2A_1^* e^{-i(\mathbf{K}+\mathbf{k}_3)\cdot\mathbf{r}} + A_1A_2^* e^{i(\mathbf{K}-\mathbf{k}_3)\cdot\mathbf{r}}]A_3 e^{i\omega t} + \text{c.c.} \tag{6.2-7}$$

This polarization can be viewed as a distributed source which radiates optical waves at frequency ω provided either $\mathbf{K}+\mathbf{k}_3$ or $\mathbf{K}-\mathbf{k}_3$ equals the wave vector of the radiated waves. If the readout beam is counterpropagating relative to beam 2 (see Figure 6.2), that is,

$$\mathbf{k}_3 = -\mathbf{k}_2 \tag{6.2-8}$$

then $\mathbf{K}+\mathbf{k}_3 = -\mathbf{k}_1$ and the first term in Eq. (6.2-7) exactly satisfies the wave equation at ω for any $\mathbf{k}_1$. Thus the first term in the distributed polarization will radiate efficiently. Let the electric field of the radiated beam be written

$$E_4 = A_4 \exp[i(\omega t - \mathbf{k}_4 \cdot \mathbf{r})] \tag{6.2-9}$$

where A_4 is the amplitude and $\mathbf{k}_4$ is the wave vector. The electric field E_4 is linearly proportional to the first term of the polarization which exactly satisfies the wave equation at ω, provided the medium is infinite in its transverse dimension and finite in thickness. The amplitude and the wave vector are written

$$A_4 = cn_1(A_2A_3)A_1^* \tag{6.2-10}$$

$$\mathbf{k}_4 = -\mathbf{k}_1 \tag{6.2-11}$$

where c is a constant. Using Eqs. (6.2-10) and (6.2-11), we note that E_4 is exactly the phase conjugate of the object beam E_1. We also note that incidence condition Eq. (6.2-8) automatically fulfills the Bragg condition

$$\mathbf{k}_4 - \mathbf{k}_3 = \mathbf{K} \tag{6.2-12}$$

In summary, phase conjugate waves can be generated by using holographic techniques. The complex amplitude of an object beam is first recorded by using holographic interference inside a nonlinear medium. When the recorded hologram is read by a beam which is counterpropagating relative to the reference beam, the phase conjugate of the object beam is generated.

Photorefractive crystals can be used as the recording media in holography. According to Eq. (4.1-5), the change in the index refraction after the recording is written

$$\Delta n = \frac{1}{2} n_1 e^{i\phi} \frac{A_1^* A_2}{|A_1|^2 + |A_2|^2} e^{-i\mathbf{K}\cdot\mathbf{r}} + \text{c.c.} \tag{6.2-13}$$

which is slightly different from the one given by Eq. (6.2-5) because of the intensity denominator $|A_1|^2 + |A_2|^2$. If the intensity variation of beam 1 is significant and $|A_1|^2$ is not small compared to $|A_2|^2$, then the recorded hologram and the reconstructed phase conjugate wave are not linearly proportional to A_1^*. This may lead to distortion and noise. A high fidelity generation of the phase conjugate wave would require that the change of $|A_1|^2$ over the transverse dimension of the object beam is small relative to $|A_1|^2 + |A_2|^2$.

In four-wave mixing, the recording and the readout happen simultaneously. Phase conjugation by optical four-wave mixing is thus often referred to as real-time holography. Photorefractive four-wave mixing has been treated in Chapter 4. According to the results obtained there, the complex amplitude of the fourth wave at $z = 0$ is given by

$$A_4(0) = \rho A_1^*(0) \tag{6.2-14}$$

with ρ given by Eq. (4.5-42). The results are obtained by assuming all the waves are plane waves.

For the application as a phase conjugator, it is important that A_2 and A_3 are kept constant throughout the medium. The incident wave A_1 is not, in general, a plane wave. It may consist of many spatial components (or Fourier components). If each of these components are conjugated equally, the result is the generation of the phase conjugate wave. Let us write

$$A_1(\mathbf{r}) = \int A(\mathbf{k}) e^{-i\mathbf{k}\cdot\mathbf{r}}\, d\mathbf{k} \tag{6.2-15}$$

where $A(\mathbf{k})$ is the Fourier transform of the incident wave. If each of these components are conjugated, the electric field of the fourth wave can be written

$$A_4(\mathbf{r}) = \int \rho A^*(\mathbf{k}) e^{i\mathbf{k}\cdot\mathbf{r}}\, d\mathbf{k} \tag{6.2-16}$$

In addition, if the phase conjugate reflection coefficient ρ is independent of the wave vector, then it follows directly from Eqs. (6.2-16) and (6.2-15) that

$$A_4(\mathbf{r}) = \rho A_1^*(\mathbf{r}), \quad \text{for all } \mathbf{r} \tag{6.2-17}$$

In the above derivation, we assume that each of the Fourier components undergoes four-wave mixing individually. In other words, the process that leads to Eq. (6.2-14) is linear. Although there is an intensity denominator I_0 which is the sum of all the intensities, it appears as a common factor in the phase conjugation process.

In the above discussion, we assume that all the waves involved have exactly the same frequency. In reality, the frequencies can sometimes be slightly different due to external conditions such as motion, frequency drift in lasers, etc. If the frequency differences are very small, that is,

$$|\omega_i - \omega_j| \ll \omega_k, \quad \text{for all } i, j, k = 1, \ldots, 4 \tag{6.2-18}$$

then efficient wave mixing can still occur and phase conjugate waves can still be generated. Under these circumstances, conservation of energy and momentum require that

$$\omega_4 = (\omega_2 + \omega_3) - \omega_1 \tag{6.2-19}$$

$$\mathbf{k}_4 = (\mathbf{k}_2 + \mathbf{k}_3) - \mathbf{k}_1 \tag{6.2-20}$$

In addition, $\mathbf{k}_4$ is related to ω_4 by

$$|\mathbf{k}_4| = n \frac{\omega_4}{c} \tag{6.2-21}$$

These equations are useful in determining the frequency of the phase conjugate wave. Let the frequency of the incident beam be slightly different from that of the pump beams, that is,

$$\omega_1 = \omega + \delta \tag{6.2-22}$$

where $\omega = \omega_2 = \omega_3$ is the frequency of beams 2 and 3, and δ is a small frequency shift. The two pump beams are counterpropagating so that $\mathbf{k}_2 + \mathbf{k}_3 = 0$. According to Eq. (6.2-19), the frequency of the phase conjugate beam is

$$\omega_4 = \omega - \delta \tag{6.2-23}$$

We note that ω_4 is slightly different from that of the incident beam by 2δ. In addition to the frequency shift, there will be a momentum mismatch of

$$\Delta k = \frac{2\delta}{c} n \tag{6.2-24}$$

according to Eqs. (6.2-20) and (6.2-21). The momentum mismatch will lead to a decrease in the coupling.

Phase conjugation via four-wave mixing is a straightforward method for the generation of time-reversed waves. It requires a pair of counter-propagating beams (A_2, A_3). The alignment of these two beams to achieve $\mathbf{k}_2 + \mathbf{k}_3 = 0$ is essential to obtain high fidelity phase conjugation. In the next section, we will treat a new class of phase conjugators that does not need external pumping beams.

6.3 SELF-PUMPED PHASE CONJUGATORS (SPPC)

In a self-pumped phase conjugation, an incident wave will generate its own phase conjugate wave. There are several physical mechanisms that can yield self-pumped phase conjugation. We will describe these different phase conjugations separately.

6.3.1 SPPC with Resonators

In the last chapter, we studied various photorefractive resonators in which the oscillation is sustained by the parametric gain due to two-wave mixing. In the case of a linear resonator (see Figure 6.3(a)), the oscillation between the two mirrors provides the counterpropagating beams which are needed in phase conjugation via four-wave mixing. To achieve high fidelity in such a phase conjugator, it is important that a single mode (spatial mode) oscillation is maintained in the resonator. As mentioned earlier in the last chapter, there is a frequency shift associated with such a phase conjugator. Let ω be the frequency of the incident beam. The frequency of oscillation in the resonator is determined by the cavity length. Let $\omega + \delta$ be the frequency of the oscillation. The phase conjugate wave will be shifted in frequency by 2δ. For photorefractive crystals at an operating power of the order of 1 W/cm^2, δ is in the range of 1 Hz.

Under the appropriate conditions, oscillation can also occur between two parallel faces of a photorefractive crystal, provided the surfaces are polished. Photorefractive crystals such as $BaTiO_3$ and SBN have an index of refraction of the order of 2.5. Fresnel reflectivity at the crystal–air interface is approximately 18%. Oscillation in such a lossy resonator is still possible as a result of the high gain due to two-wave mixing. Small signal gains in these

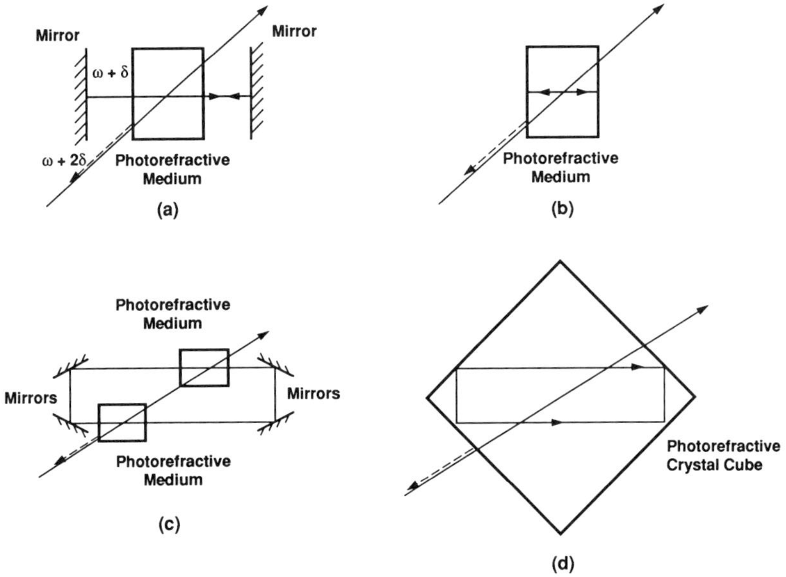

Figure 6.3. Self-pumped phase conjugators (SPPC) with resonators. (a) Linear resonators formed by mirrors. (b) Linear resonators formed by crystal surfaces. (c) Ring resonators formed by mirrors. (d) Ring resonators formed by crystal corners in cubes or rectangular prisms via total internal reflections.

materials in a 1 cm length can be as high as e^{10}. Such oscillation and self-pumped phase conjugation have been observed in photorefractive SBN crystals. Figure 6.3(b) shows such a self-pumped conjugator.

SPPC can also be achieved by using bidirectional oscillation in ring resonators. Referring to Figure 6.3(c), we consider a ring resonator in which gains for bidirectional oscillation are provided. The oppositely directed oscillating beams in the ring resonator provide the pumping needed in four-wave mixing. The process thus leads to self-pumped phase conjugation. Such self-pumped conjugation can occur in a single cube of photorefractive crystal under appropriate conditions [3]. As indicated in Figure 6.3(d), crystal surfaces provide the mirror reflection needed in bidirectional ring oscillation [4]. In fact, total internal reflections occur in many of the oscillations. With an index of refraction around $n = 2.5$, the incident angle for total internal reflection is $\theta > 21.8°$. Thus, ring oscillations in photorefractive crystal cubes involve total internal reflections. These internal ring oscillations only suffer loss due to material absorption and scattering. The frequency of oscillation of these oscillations is determined by the path length of oscillation and can be shifted relative to that of the pump beam. This leads to a small frequency shift in the phase conjugate wave.

6.3.2 SPPC with Stimulated Backward Scattering

One of the most interesting phenomena in photorefractive media is the strong energy transfer in two-wave mixing. For beams with low intensities, the small signal gain can be as big as

$$g = e^{\gamma L} \tag{6.3-1}$$

where γ is the intensity coupling constant and L is the interaction length. For $\gamma L = 20$, this gain is of the order of 10^9. In such a high gain medium, scattered light from an incident laser beam can get amplified. This is known as "fanning" [5]. When a laser beam of finite width enters a photorefractive crystal, scattering occurs due to surface pits, imperfections and defects in the crystal. The scattered light overlaps with the incident beam and they undergo two-wave mixing. Because of the dependence of γ on angles (θ, ϕ), amplification only occurs along those directions for which $\gamma(\theta, \phi) > \alpha$, where α is the absorption coefficient.

Under the appropriate conditions, the backward scattered light can get amplified. It is known for some time that the backward scattered light in stimulated Brillouin scattering (SBS) exhibits phase conjugation. In other words, the wave front of the stimulated backward scattering is identical to that of the incident wave. Such phase conjugation also occurs in photorefractive media [6]. In what follows, we will show that stimulated backward photorefractive scattering indeed exhibits phase conjugation.

To prove that the stimulated backward scattering is a phase conjugate wave, we have to show that the spatial gain coefficient for the phase conjugate wave is much higher compared with other scattered light. Let A_1 be the intensity of the incident beam, the differential equation for the growth of backward scattering is given by

$$\frac{d}{dz} A_2 = -\frac{\gamma}{2} \frac{A_1^* A_1}{I_1 + I_2} A_2 \tag{6.3-2}$$

where γ is the coupling constant, A_2 is the amplitude of the backscattered wave and I_1, I_2 are intensities (i.e., $I_1 = |A_1|^2$, $I_2 = |A_2|^2$). This equation was derived earlier using plane wave solutions. To discuss phase conjugation, we have to assume A_1 and A_2 are both functions of x, y, z. The electric field is thus written

$$E = A_1(x, y, z)e^{i(\omega t - kz)} + A_2(x, y, z)e^{i(\omega t + kz)} \tag{6.3-3}$$

where we assume that both A_1 and A_2 propagate predominantly along the z-axis. In the absence of photorefractive coupling, these two amplitudes satisfy the following equations:

$$\nabla_\perp^2 A_1 - 2ik \frac{\partial}{\partial z} A_1 = 0 \tag{6.3-4}$$

$$\nabla_\perp^2 A_2 + 2ik \frac{\partial}{\partial z} A_2 = 0 \tag{6.3-5}$$

where we recall that $\nabla_\perp^2$ is the transverse Laplacian operator. To examine the growth of the backscattered wave, we combine Eqs. (6.3-2) and (6.3-5) and obtain

$$\frac{\partial}{\partial z} A_2 - \frac{i}{2k} \nabla_\perp^2 A_2 = -\frac{\gamma}{2} \frac{I_1}{I_1 + I_2} A_2 \tag{6.3-6}$$

We now consider a complete set of wave functions $\psi_i(x, y, z)$, $i = 0, 1, 2, \ldots$ which satisfy the orthonormalization relation at a plane $z = z_o$ and the scalar-wave equation describing the complex conjugate of A_1:

$$\int \psi_i^* \psi_j \, dx \, dy = \delta_{ij} \quad \text{at } z = z_o \tag{6.3-7}$$

$$\frac{\partial}{\partial z} \psi_i - \frac{i}{2k} \nabla_\perp^2 \psi_i = 0 \tag{6.3-8}$$

Again, as a result of the linear nature of Eq. (6.3-8), the orthonormalization condition (Eq. (6.3-7)) will hold at any plane z = constant. We now choose a function $\psi_0^*(x, y, z)$ which coincides with the input wave, that is,

$$\beta \psi_0^*(x, y, z) = A_1(x, y, z) \tag{6.3-9}$$

where β is a normalization constant. The remaining wave functions ψ_i^*, $i = 1, 2, 3, \ldots$, are chosen arbitrarily starting from the orthonormalization condition (Eq. (6.3-7)). Using these wave functions as the basis, we can write

$$A_2(x, y, z) = \sum_{i=0}^{\infty} C_i(z)\psi_i(x, y, z) \tag{6.3-10}$$

where C_i are expansion coefficients which are functions of z. We now substitute Eq. (6.3-10) for A_2 into Eq. (6.3-6). This leads to

$$\frac{d}{dz} C_i = -\frac{1}{2} \sum_{j=0}^{\infty} g_{ij} C_j(z) \tag{6.3-11}$$

where

$$g_{ij} = \gamma\beta^2 \int \frac{|\psi_0|^2}{I_1 + I_2} \psi_i^* \psi_j \, dx \, dy \tag{6.3-12}$$

In the above derivation, we assume that $I_2 \ll I_1$, so that $(I_1 + I_2)$ may be treated as a constant.

Equation (6.3-12) represents the growth of the mode amplitudes C_i. Instead of solving the equations of C_i, we will only examine them and compare the growth coefficients which are given by Eq. (6.3-12). As a result of the orthogonality condition of Eq. (6.3-7), we expect that the diagonal coefficient g_{00} exceeds by 2–3 times the gains g_{ii} of the remaining wave functions and the values of the off-diagonal coefficients $|g_{0i}|$, $|g_{ik}|$, $i, k \neq 0$. Therefore, the mode amplitude $C_0(z)$ will grow most rapidly. This proves that the backscattered light is dominated by phase conjugate waves [7].

It is important to note that if the incident wave has an amplitude profile which is constant over the cross section (i.e., $|A_1(x, y, z)|^2 = \text{constant}$), then there should be no preferred growth for the phase conjugate wave. Phase conjugation via stimulated photorefractive backscattering is sometimes referred to as phase conjugation via $2k$ gratings. The photorefractive gratings in such phase conjugation have a grating wave number of $K = 2k$. Figure 6.4 depicts a self-pumped conjugation with $2k$ gratings. We also note

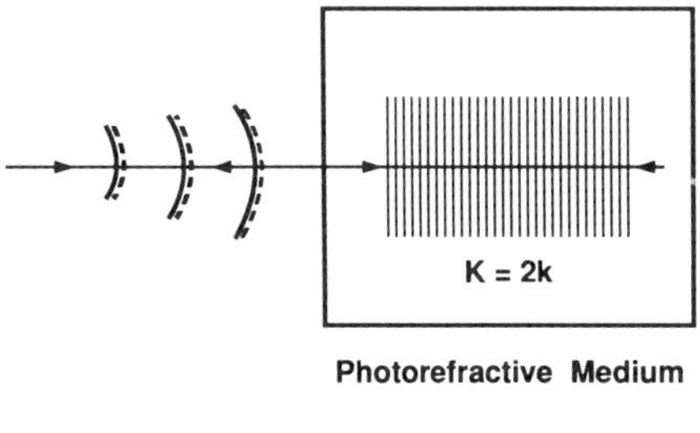

Figure 6.4. SPPC with $2k$ gratings. The solid wave fronts are for the incident beam. The dashed wave fronts are those of the phase conjugate wave.

that in practice when a laser beam of finite width is used, backscattering with $2k$ gratings has the longest interaction length with the incident beam.

6.3.3 Ring Self-Pumped Phase Conjugator

Referring to Figure 6.5, we consider a unique type of self-pumped phase conjugator which involves a ring configuration [8]. A beam A_3 is incident into a photorefractive crystal at $z = L$. The transmitted beam is redirected into the crystal from the opposite face at $z = 0$ by using two mirrors. Under the appropriate conditions, self-pumped phase conjugation occurs. A phase conjugate beam is generated which has wave fronts identical to that of the incident beam and propagate along the opposite direction. The process can be explained in terms of self-oscillation in four-wave mixing or in terms of hologram sharing. The latter is discussed in Section 6.4.

When phase conjugation occurs, all four waves are present inside the crystal. The results obtained in Chapter 4 can be used to study such a phase conjugator. The boundary conditions in this case are

$$A_1(0) = rA_3(0)e^{i\phi} \tag{6.3-13}$$

$$A_2(0) = rA_4(0)e^{i\phi} \tag{6.3-14}$$

$$A_4(L) = 0 \tag{6.3-15}$$

where r is the product of the mirror reflection coefficients, and ϕ is the phase shift due to the propagation in air.

We will normalize the intensity so that

$$I_0 = I_1 + I_2 + I_3 + I_4 = 1 \tag{6.3-16}$$

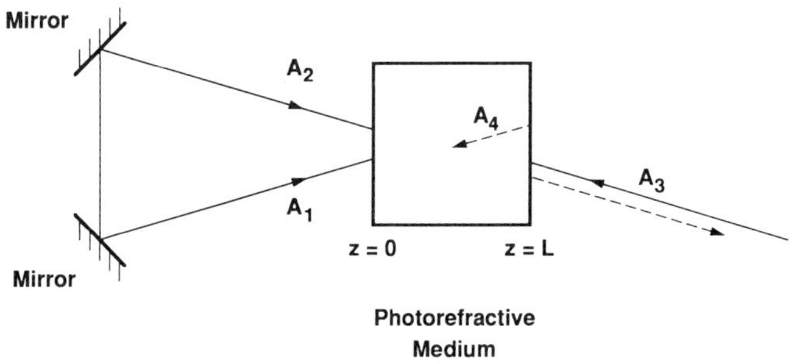

Figure 6.5. A ring self-pumped phase conjugator formed by a photorefractive medium and two mirrors. The probe beam A_3 is incident into the photorefractive medium at $z = L$. The transmitted beam is redirected into the medium at $z = 0$ by the mirrors.

Other constants of integration are

$$I_1 + I_2 = c_1 \tag{6.3-17}$$

$$I_3 + I_4 = c_2 \tag{6.3-18}$$

$$\sigma = c_1 - c_2 \tag{6.3-19}$$

$$c = A_1A_4 + A_2A_3 \tag{6.3-20}$$

Using the boundary conditions Eq. (6.3-13)–(6.3-15), the new variables $u = A_2/A_3^*$, $v = A_4/A_1^*$ are related by

$$u_0 = v_0|r|^2 \tag{6.3-21}$$

As a result of the boundary condition $A_4(L) = 0$, v_0 is given, according to Eq. (4.5-40), by

$$v_0 = \frac{2c \tanh \kappa L}{s + \sigma \tanh \kappa L} \tag{6.3-22}$$

where we recall that

$$s = (\sigma^2 + 4|c|^2)^{1/2} \tag{6.3-23}$$

$$\kappa = \frac{s\Gamma^*}{4I_0} \tag{6.3-24}$$

Using Eq. (4.5-47), we obtain a relationship between u_0 and u_L:

$$u_0 = \frac{u_L(s - \sigma \tanh \kappa L) - 2c \tanh \kappa L}{s + \sigma \tanh \kappa L - 2c^* u_L \tanh \kappa L} \tag{6.3-25}$$

In addition, u_L can also be expressed in terms of the constants of integration. Using Eqs. (6.3-15) and (6.3-20), we have

$$c = A_2(L)A_3(L) = I_{3L}u_L \tag{6.3-26}$$

where $I_{3L} = |A_3(L)|^2$. Similarly, we obtain, according to Eqs. (6.3-16)–(6.3-19)

$$c_2 = I_{3L} = \frac{1 - \sigma}{2} \tag{6.3-27}$$

Thus, u_L is written

$$u_L = \frac{2c}{1 - \sigma} \tag{6.3-28}$$

Now we substitute Eq. (6.3-28) for u_L in Eq. (6.3-25) and use Eqs. (6.3-2) and (6.3-22). This yields

$$\frac{(s-\sigma T)\dfrac{2c}{1-\sigma}-2cT}{(s+\sigma T)-T\dfrac{4|c|^2}{1-\sigma}}=\frac{2cT}{s+\sigma T}|r|^2 \tag{6.3-29}$$

with

$$T \equiv \tanh \kappa L \tag{6.3-30}$$

By further simplifying Eq. (6.3-29), we obtain

$$\frac{s-T}{s(1-sT)-\sigma(s-T)}=\frac{T}{s+\sigma T}|r|^2 \tag{6.3-31}$$

This equation can now be used to solve for $|c|$ numerically, provided σ is known. We recall that s and T are functions of $|c|$. The parameter σ, representing the net power flow along z-direction, can be expressed in terms of $|r|$. According to Eqs. (6.3-16)–(6.3-19), we can write

$$\frac{1-\sigma}{1+\sigma}=\frac{2c_2}{2c_1}=\frac{I_3+I_4}{I_1+I_2} \tag{6.3-32}$$

Using the boundary conditions of Eqs. (6.3-13) and (6.3-14), we obtain

$$\sigma=\frac{|r|^2-1}{|r|^2+1} \tag{6.3-33}$$

Notice that σ is negative, indicating that the net power flow is along the z-direction. This is due to the loss at mirrors on the left (see Figure 6.5). In the event when the mirror reflectivities are unity, $|r|^2=1$, there is no net power flow.

When $|c|$ is solved from Eq. (6.3-31), phase conjugate reflectivity is given, according to Eq. (6.3-28), by

$$|\rho|^2=\left|\frac{A_2(L)}{A_3^*(L)}\right|^2=\frac{4|c|^2}{(1-\sigma)^2} \tag{6.3-34}$$

It is important to note that phase conjugation does not occur all the time. It depends on the coupling constant and the boundary conditions. In other words, there is a threshold condition. At the threshold, $A_2(z)$ is zero, which is equivalent to $|\rho|=0$ or $|c|=0$, according to Eq. (6.3-34). Putting $|c|=0$ in Eq. (6.3-31), which is equivalent to $s=-\sigma$, we obtain the following threshold condition

$$\tanh \kappa L = \frac{1 - |r|^2}{1 + 3|r|^2} \tag{6.3-35}$$

or equivalently

$$\exp\left(\frac{\sigma\Gamma^*}{2} L\right) = \frac{2R}{1+R} \tag{6.3-36}$$

where R is the product of mirror reflectivities,

$$R = |r|^2 \tag{6.3-37}$$

The threshold condition can be further simplified. Using Eq. (6.3-33), we obtain

$$\Gamma^* L = 2\,\frac{(1+R)}{(1-R)} \ln\left(\frac{1+R}{2R}\right) \tag{6.3-38}$$

A plot of the threshold coupling constant versus mirror reflectivity R indicates that a minimum of

$$\Gamma^* L = 2 \tag{6.3-39}$$

is needed for the operation of such conjugators with $R = 1$. Figure 6.6 plots the threshold coupling constant as a function of R. We note that the threshold decreases as the mirror reflectivity R increases.

At a given mirror reflectivity R, the phase conjugate reflectivity $|\rho|^2$ is a function of ΓL. Figure 6.7 plots $|\rho|^2$ as a function of ΓL. The phase conjugate reflectivity $|\rho|^2$ is an increasing function of ΓL. It is also an increasing function of R at a given ΓL.

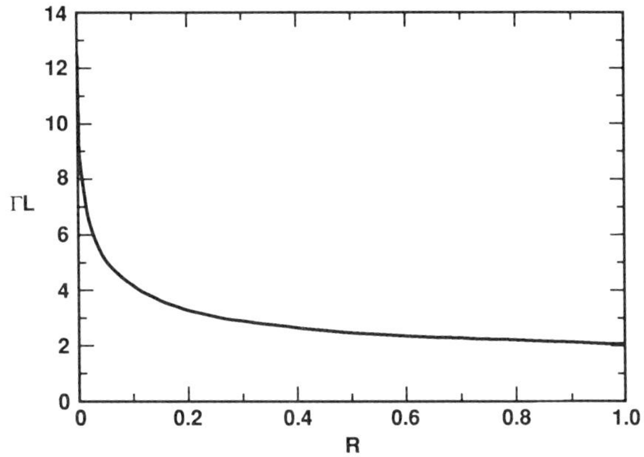

Figure 6.6. Threshold coupling constant for ring SPPC as a function of mirror reflectivity R.

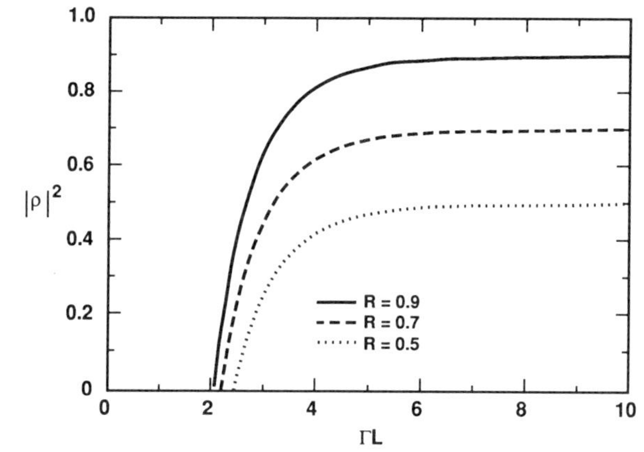

Figure 6.7. Phase conjugate reflectivity $|\rho|^2$ of ring SPPC as a function of ΓL.

6.3.4 SPPC with Semi-Resonators

Referring to Figure 6.8, we consider a unique type of phase conjugators which involve a resonator formed by a single mirror and a photorefractive medium [8]. This may be considered as a special case of SPPC with a linear resonator in which one of the mirrors is removed. Similar to the previous case, a laser beam is incident into a photorefractive crystal at $z = L$. Under the appropriate conditions, an oscillation occurs between the mirror and the crystal and a phase conjugate wave is generated. The process can be explained in terms of self-oscillation in four-wave mixing or in terms of hologram sharing. The latter is discussed in the next section.

We now treat the phase conjugator using optical four-wave mixing. When phase conjugation occurs, all four beams are present in the crystal and the boundary conditions are

$$A_2(0) = 0 \tag{6.3-40}$$

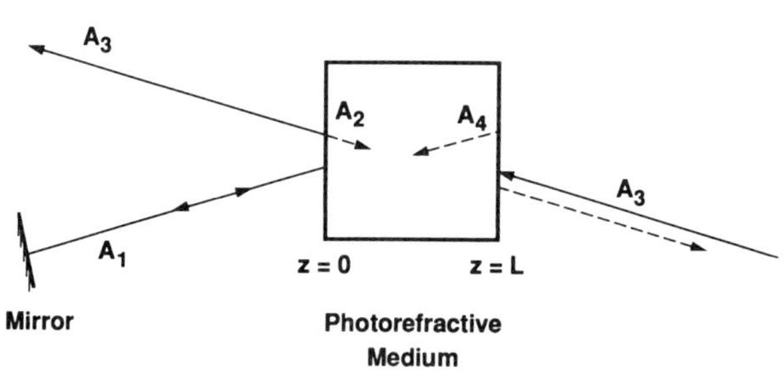

Figure 6.8. A self-pumped phase conjugator with semi-resonators. Oscillation occurs between the mirror and the photorefractive medium when a probe beam A_3 is incident into the medium at $z = L$ under the appropriate conditions.

$$A_4(L) = 0 \tag{6.3-41}$$

$$A_1(0) = A_4(0) r e^{-i\phi} \tag{6.3-42}$$

where r is the reflection coefficient of mirror and ϕ is a phase. Using the intensity normalization of $I_0 = 1$, the constants of integration can be written

$$c_1 = I_{10} = \frac{1+\sigma}{2} \tag{6.3-43}$$

$$c_2 = I_{3L} = \frac{1-\sigma}{2} \tag{6.3-44}$$

$$c = u_L I_{3L} = v_0 I_{10} \tag{6.3-45}$$

In addition, the boundary conditions of Eqs. (6.3-40)–(6.3-42) can be used to simplify expressions for u and v. This leads to

$$u_L = \frac{A_2(L)}{A_3^*(L)} = \frac{2cT}{s - \sigma T} \tag{6.3-46}$$

$$v_0 = \frac{A_4(0)}{A_1^*(0)} = \frac{2cT}{s + \sigma T} \tag{6.3-47}$$

and

$$|v_0|^2 = \frac{1}{|r|^2} = \frac{1}{R} \tag{6.3-48}$$

where we recall that $T = \tanh \kappa L$ and R is the mirror reflectivity. We need to solve for $|c|$ and σ to find phase conjugation reflectivity. Using Eqs. (6.3-44)–(6.3-46), we obtain

$$s = T = \tanh \frac{s\Gamma^*}{4} L \tag{6.3-49}$$

This equation can be used to solve for s. Substituting $s = T$ into Eq. (6.3-47), we obtain from Eq. (6.3-48)

$$\frac{1}{R} = \left| \frac{2c}{1+\sigma} \right|^2 \tag{6.3-50}$$

Once s is obtained by solving Eq. (6.3-49), σ is given by

$$\sigma^2 = s^2 - 4|c|^2 \tag{6.3-51}$$

By eliminating $4|c|^2$ from Eqs. (6.3-50) and (6.3-51), we obtain

$$R\sigma^2 + (1+\sigma)^2 = Rs^2 \tag{6.3-52}$$

which is a quadratic equation in σ. The solutions are

$$\sigma = \frac{-1 \pm (R^2 s^2 + Rs^2 - R)^{1/2}}{1 + R} \tag{6.3-53}$$

Phase conjugation reflectivity, according to Eq. (6.3-46), is given by

$$|\rho|^2 = \frac{4|c|^2}{(1-\sigma)^2} = \frac{s^2 - \sigma^2}{(1-\sigma)^2} = \left(\frac{1+\sigma}{1-\sigma}\right)^2 \frac{1}{R} \tag{6.3-54}$$

with s given by the solution of Eq. (6.3-49) and σ given by Eq. (6.3-53).

To find the parameter s, we must solve the transcendental equation (6.3-49). The solution for real ΓL is relatively simple. We plot both sides of Eq. (6.3-49) as functions of s. The left side is a straight line of slope 1. The right side is a hyperbolic tangent function. Nontrivial solutions exist only when

$$\Gamma L > 4 \tag{6.3-55}$$

The solutions depend only on ΓL and are independent of mirror reflectivity. We note that $|\tanh x|$ is always less than 1. Thus, the solutions are always between $(-1, 1)$. Once s is solved, σ is given by Eq. (6.3-53). Being a net power flow, σ has to be real. This puts a constraint on s^2, that is,

$$s^2 > \frac{1}{1+R} \tag{6.3-56}$$

In other words, if s^2 is too small, there will be no phase conjugation. Figure 6.9 plots phase conjugate reflectivity $|\rho|^2$ as a function of ΓL for various values of R. We note that multiple solutions may lead to bistable operations.

When ΓL is complex, the solutions can be obtained by using the technique discussed in Section 4.5.3. We take $I_0 = 1$, $\Gamma = \gamma + i2\beta$, the oscillation condition becomes

$$\frac{\tanh \dfrac{\gamma L}{4} s - i \tan \dfrac{\beta L}{2} s}{1 - i \tanh \dfrac{\gamma L}{4} s \tan \dfrac{\beta L}{2} s} = s \tag{6.3-57}$$

Since the right side is real, solutions exist when

$$\tan \frac{\beta L}{2} s = \infty \tag{6.3-58}$$

and

$$\tanh \frac{\gamma L}{4} s = \frac{1}{s} \tag{6.3-59}$$

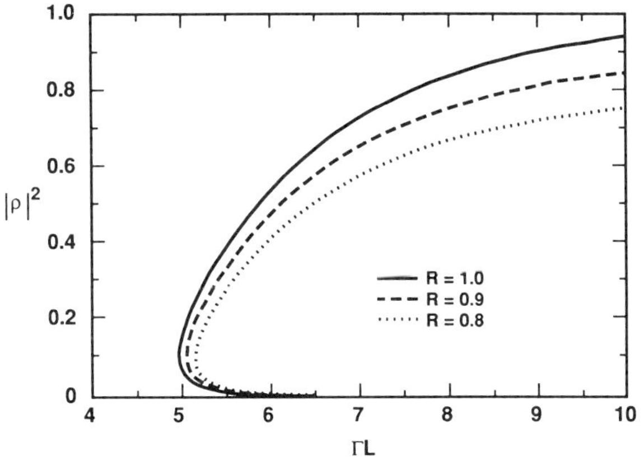

Figure 6.9. Reflectivity $|\rho|^2$ of a self-pumped phase conjugator with semi-resonators as a function of the coupling constant ΓL for mirror reflectivity $R = 0.8, 0.9, 1.0$.

The last equation can be solved using the graphic method discussed earlier. Solutions exist for arbitrary γL. Complex coupling constant ΓL corresponds oscillations with frequency detuning or situations when external field is applied. Solutions to Eq. (6.3-57) also exist when

$$\tan \frac{\beta L}{2} s = 0 \tag{6.3-60}$$

and

$$\tanh \frac{\gamma L}{4} s = s \tag{6.3-61}$$

The last equation can be solved by using graphic techniques. Again, solutions for real s exist only when $\gamma L > 4$.

6.4 MUTUALLY PUMPED PHASE CONJUGATORS (MPPC)

In self-pumped phase conjugators, a single incident laser beam is responsible for the oscillation and the subsequent generation of the phase conjugate wave. In this section, we consider another class of phase conjugators in which two incident laser beams mutually pump the photorefractive medium and produce phase conjugation [9]. Referring to Figure 6.10, we consider the incidence of two beams, one at each face of the photorefractive medium. As a result of mutual phase conjugation, two phase conjugate waves are generated. The process can be explained in terms of self-oscillation in four-wave mixing, resonator model or hologram sharing. We will describe each of the theories, starting from four-wave mixing.

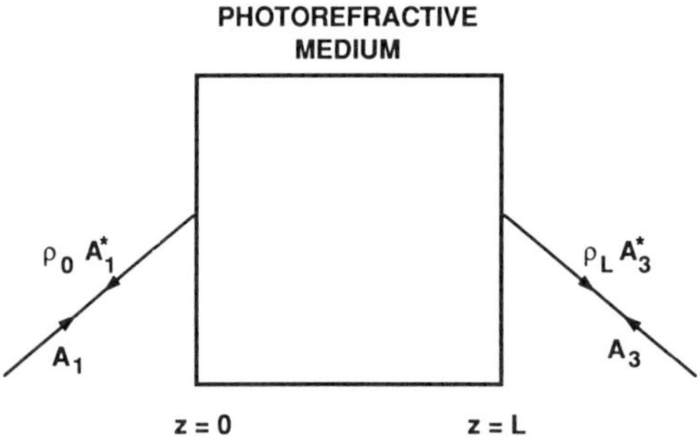

Figure 6.10. Schematic drawing of mutually pumped phase conjugation (MPPC).

6.4.1 MPPC Via Self-Oscillation in Four-Wave Mixing

When mutual phase conjugation occurs, there are four waves in the medium. The results obtained in Chapter 4 are now applied to the problem. The boundary conditions are

$$A_2(0) = A_4(L) = 0 \tag{6.4-1}$$

reflecting the fact that there are no incidence of beam 2 and 4 at $z = 0$ and $z = L$, respectively. These boundary conditions lead, immediately, to

$$u_0 = \frac{A_2(0)}{A_3^*(0)} = 0 \tag{6.4-2}$$

$$v_L = \frac{A_4(L)}{A_1^*(L)} = 0 \tag{6.4-3}$$

The constants of integration are

$$c_1 = I_1 + I_2 = |A_1(0)|^2 = I_{10} \tag{6.4-4}$$

$$c_2 = I_3 + I_4 = |A_3(L)|^2 = I_{3L} \tag{6.4-5}$$

$$\sigma = c_1 - c_2 = |A_1(0)|^2 - |A_3(L)|^2 = I_{10} - I_{3L} \tag{6.4-6}$$

$$c = A_1A_4 + A_2A_3 = A_1(0)A_4(0) = A_2(L)A_3(L) \tag{6.4-7}$$

Using the definition of u and v, c can also be written

$$c = I_{10}v_0 = I_{3L}u_L \tag{6.4-8}$$

where $v_0 = v(0)$ and $u_L = u(L)$.

According to Eqs. (4.5-42) and (4.5-50), phase conjugate reflection coefficients are given by

$$\rho_0 = \frac{A_4(0)}{A_1^*(0)} = v_0 = \frac{2c \tanh \kappa L}{s + \sigma \tanh \kappa L} \tag{6.4-9}$$

$$\rho_L = \frac{A_2(L)}{A_3^*(L)} = u_L = \frac{2c \tanh \kappa L}{s - \sigma \tanh \kappa L} \tag{6.4-10}$$

where we recall that $\kappa = s\Gamma^*/4I_0$. Using Eqs. (6.4-8) and (6.4-9), we obtain

$$s = \tanh \frac{s\Gamma^*}{4} L \tag{6.4-11}$$

where we used intensity normalization $I_0 = I_{10} + I_{3L} = 1$ and $2I_{10} = 1 + \sigma$. We note that this equation is formally identical to the one discussed in the previous section. This equation can be used to solve for s or $|c|$ in terms of ΓL. Once s or $|c|$ is obtained, phase conjugate reflectivities $|\rho_0|^2$ and $|\rho_L|^2$ can be found from Eqs. (6.4-9) and (6.4-10), or directly from

$$\rho_0 = \frac{c}{I_{10}} = \frac{2c}{1+\sigma} \tag{6.4-12}$$

and

$$\rho_L = \frac{c}{I_{3L}} = \frac{2c}{1-\sigma} \tag{6.4-13}$$

It is interesting to note that ρ_0 and ρ_L are related by

$$\frac{\rho_0}{\rho_L} = \frac{I_{3L}}{I_{10}} \equiv q \tag{6.4-14}$$

where q is defined as the input intensity ratio. In other words, the ratio of phase conjugate reflection coefficients is exactly the ratio of incident intensities. In addition, they are related by

$$\rho_0 \rho_L^* = \frac{4|c|^2}{1-\sigma^2} = \frac{s^2 - \sigma^2}{1-\sigma^2} \tag{6.4-15}$$

The definition q and the normalization $I_0 = I_{10} + I_{3L} = 1$ can be used to rewrite σ as

$$\sigma = \frac{1-q}{1+q} \tag{6.4-16}$$

Substituting Eq. (6.4-16) for σ. Eq. (6.4-15) can be rewritten

$$\rho_0 \rho_L^* = \frac{(1+q)^2 s^2 - (1-q)^2}{4q} \tag{6.4-17}$$

We now examine the threshold condition for the phase conjugation. According to Eq. (6.4-11), real solutions for s exist only when

$$\tan \tfrac{1}{2} s\beta L = 0 \quad \text{or} \quad \tan \tfrac{1}{2} s\beta L = \infty \tag{6.4-18}$$

where 2β is the imaginary part of Γ. Using the same technique discussed in the previous section, the solutions for s can be obtained. For a real coupling constant (i.e., pure diffusion case without frequency detuning), $\beta L = 0$, Eq. 6.4-11 becomes

$$s = \tanh \frac{\gamma L}{4} s \tag{6.4-19}$$

which support nontrivial real solutions for s provided

$$\gamma L > 4 \tag{6.4-20}$$

When the coupling constant ΓL is complex, solutions exist when $\tan \frac{1}{2} s\beta L = \infty$. In this case, Eq. (6.4-11) becomes

$$\frac{1}{s} = \tanh \frac{s\gamma}{4} L \tag{6.4-21}$$

which has real solutions for s for any γL. In addition to the conditions Eqs. (6.4-18) and (6.4-20), the solution s must also satisfy the following condition:

$$s^2 > \sigma^2 \tag{6.4-22}$$

which ensures a real $|c|$ and real phase conjugate reflectivities $|\rho_0|^2$ and $|\rho_L|^2$. The threshold condition is thus, according to Eqs. (6.4-16) and (6.4-22)

$$|s| = \left| \frac{1-q}{1+q} \right| \tag{6.4-23}$$

which corresponds to zero reflectivities. Figure 6.11 plots the threshold coupling constant γL as a function of beam intensity ratio q for the case when Γ is real.

We note that the threshold coupling constant for the case of photorefractive medium that operates by diffusion only (e.g., $BaTiO_3$) is $\gamma L = 4$ at $q = 1$. The threshold coupling constant for MPPC increases as q deviates from unity.

Figure 6.12 plots the phase conjugate reflectivity $|\rho_0|^2$ as a function of ΓL at various incident beam ratios for the case when Γ is real.

Since only transmission gratings are considered in the four-wave mixing, phase conjugations in Figure 6.10 are actually cross readings of holograms.

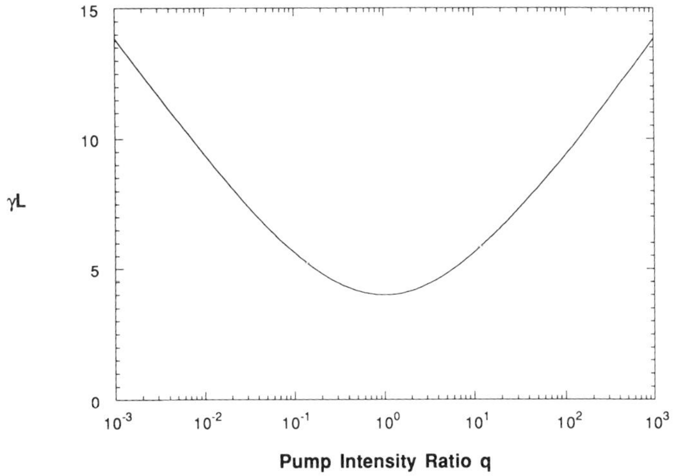

Figure 6.11. Threshold coupling constant γL as a function of incident beam intensity ratio q in a mutually pumped phase conjugator via four-wave mixing. The photorefractive coupling constant ΓL is assumed to be real.

The holograms record amplitude and wave front information of the incident beams A_1 and A_3. When the holograms diffract these two beams, two scattered beams are obtained, A_4 and A_2. We define the diffraction efficiencies as

$$\eta_{12} = \left| \frac{A_2(L)}{A_1(0)} \right|^2, \qquad \eta_{34} = \left| \frac{A_4(0)}{A_3(L)} \right|^2 \tag{6.4-24}$$

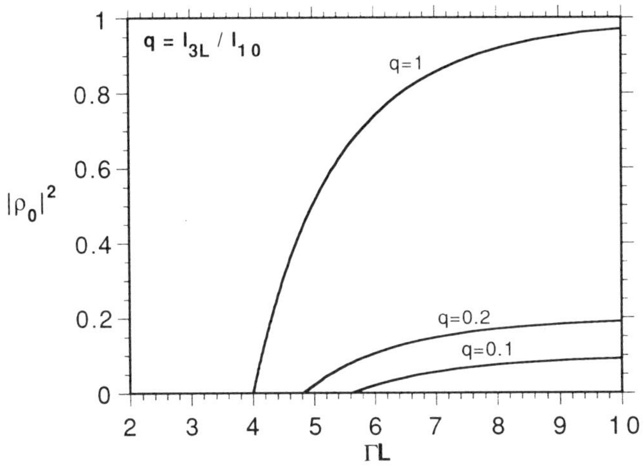

Figure 6.12. Reflectivity $|\rho_0|^2$ of a mutually pumped phase conjugator via four-wave mixing as a function of the coupling constant ΓL for various values of the beam ratio q.

According to Eq. (6.4-7), these two diffraction efficiencies are identical. This is a direct consequence of the reciprocity theorem. In fact, they are always less than unity. Although the diffraction efficiencies are less than unity, phase conjugate reflectivities can be greater than unity.

Figure 6.13 plots the variation of intensities inside the photorefractive medium. We note that as the read beam I_1 enters the medium at $z = 0$, the intensity starts to decrease reflecting the fact that power is transferred to the diffracted beam I_2. The diffracted beam I_2 starts with a zero intensity at $z = 0$. Similarly, beam I_3 diffracts off the grating and produces beam I_4. Mutually pumped phase conjugation can operate with incoherent beams with $\omega_1 \neq \omega_3$. In this situation, the frequency of beam 2 is identical to that to beam 1 (i.e., $\omega_2 = \omega_1$) and the frequency of beam 4 is identical to that of beam 3 (i.e., $\omega_4 = \omega_3$).

In the above analysis of MPPC with four-wave mixing, we assume that only one transmission grating is involved. Although there are two contributions to the grating, there is only one wave vector. In practice, MPPC can occur in many different configurations involving more than one grating. Figure 6.14 shows some of the MPPCs that are observed in photorefractive media [10–16].

With reflectivities greater than unity, MPPC can be combined with a regular mirror and form a resonator. In fact, the SPPC with the semi-resonator described in Figure 6.8 can be viewed as an oscillation between a regular mirror and an MPPC. Oscillation may occur provided the roundtrip reflectivity product is unity. Thus, if the mirror is put on the left side ($z = 0$), oscillation requires that

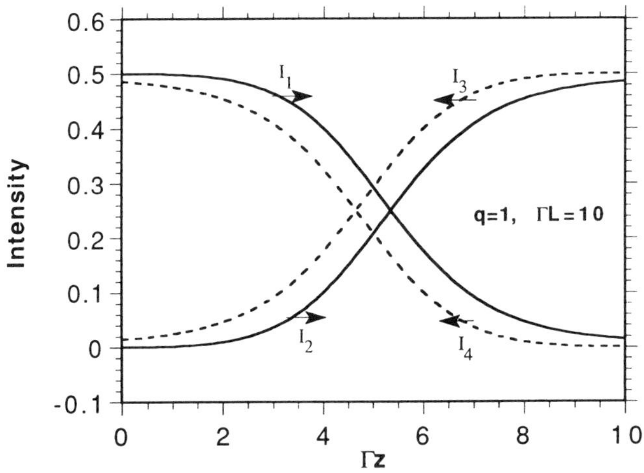

Figure 6.13. Beam intensities I_1, I_2, I_3, and I_4 versus z in a mutually pumped phase conjugator via four-wave mixing with $\Gamma L = 10.0$ and $q = 1$. The arrows indicate the direction of propagation of the beams.

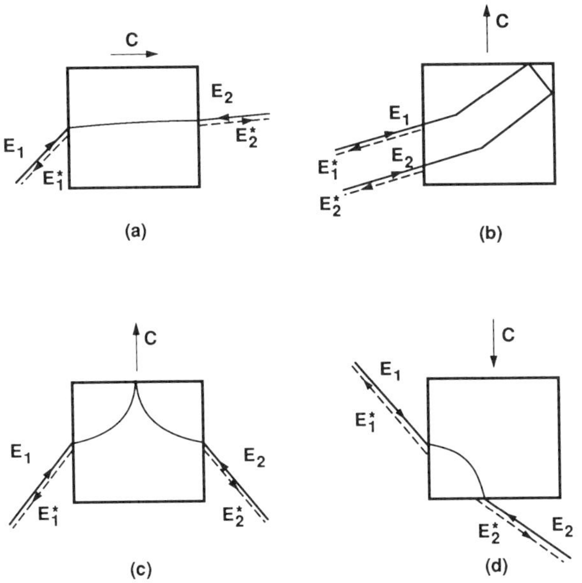

Figure 6.14. Drawings of some of the previously demonstrated MPPCs. In MPPCs, two mutually incoherent laser beams, E_1 and E_2, pump each other to generate a pair of phase-conjugate reflections, E_1^* and E_2^*, by cross readout of photorefractive holograms. Here, E_1, E_2, E_1^*, and E_2^* are short notations for $E_1(\mathbf{r})\exp(i\mathbf{k}_1\cdot\mathbf{r}-i\omega_1 t)$, $E_2(\mathbf{r})\exp(i\mathbf{k}_2\cdot\mathbf{r}-i\omega_2 t)$, $E_1^*(\mathbf{r})\exp(-i\mathbf{k}_1\cdot\mathbf{r}-i\omega_2 t)$, and $E_2^*(\mathbf{r})\exp(-i\mathbf{k}_2\cdot\mathbf{r}-i\omega_1 t)$, respectively; note the temporal-frequency cross talk. The MPPCs demonstrated are (a) the double phase-conjugate mirror [10], (b) the mutually incoherent beam coupling [11], (c) the bird-wing phase conjugator [12], and (d) the bridge phase conjugator [13, 14].

$$|\rho_0|^2 R = 1 \tag{6.4-25}$$

where R is the mirror reflectivity. This condition agrees with Eq. (6.3-48) (Problem 6.7).

6.4.2 MPPC with Stimulated Scattering

In the above discussion, we treat MPPC as a result of self-oscillation in optical four-wave mixing. The plane wave approach, however, does not seem to explain the phase conjugate nature of the process. To illustrate this, we may want to notice that the self-oscillation model using plane waves does not explain why the conical scattered lights collapse as soon as the wave fronts of the incident beams deviate from the planar ones. In addition, experimental observation in various photorefractive crystals indicates that fanning plays an important role and that MPPC could be due to a stimulated scattering process which occurs naturally in high gain media. Experimental studies also indicate that most of the MPPCs require a gain-length product of much higher than that predicted by the self-oscillation model. It was proposed that MPPC in some photorefractive media could be a result of hologram sharing in stimulated scattering [15–17]. The model is based on the fact that cross erasure of index gratings generally occurs except when two sets of beams share a common hologram. When two beams of laser enter a photorefractive medium, fanning (or stimulated scattering) of light occurs for both beams. Holograms are formed by each of the incident laser beam and its scattered beam. Thus, there are two sets of holograms responsible for the two sets of fanning. These two holograms tend to kill each other via the erasure process. When MPPC occurs, these two sets of holograms are identical and the cross erasure disappears. In what follows, we present a coupled-mode analysis for the hologram sharing in MPPC.

Referring to Figure 6.15, let us consider the interaction of two sets of beams in a photorefractive medium. Beam 1 and beam 2 are mutually coherent; beam 3 and beam 4 are also mutually coherent. But beam 2 and beam 3 (or 2 and 4) are mutually incoherent. The frequencies of the beam satisfy the following condition:

$$\omega_1 = \omega_2 \neq \omega_3 = \omega_4 \tag{6.4-26}$$

This condition indicates that beams 1 and 2 are mutually coherent, beams 3 and 4 are also mutually coherent. In addition, these two sets of beams are mutually incoherent. Inside the photorefractive medium, these four beams intersect and form volume holograms. We assume that beams 1 and 2 enter the medium at the face $z = 0$, and beams 3 and 4 enter the medium from the back face $z = L$. If we assume that the frequencies are very different such that $(\omega_3 - \omega_1)\tau \gg 1$ (where τ is the photorefractive grating decay time), the fundamental component of the index grating can be written

$$\Delta n = \frac{1}{2} n_1 e^{i\phi} \frac{A_1^* A_2 \exp(-i\mathbf{K}_{21} \cdot \mathbf{r}) + A_3^* A_4 \exp(-i\mathbf{K}_{43} \cdot \mathbf{r})}{I_0} + \text{c.c.} \tag{6.4-27}$$

where I_0 is proportional to the total averaged intensity and $\mathbf{K}_{21}$ and $\mathbf{K}_{43}$ are the grating wave vectors

$$\mathbf{K}_{21} = \mathbf{k}_2 - \mathbf{k}_1 , \qquad \mathbf{K}_{43} = \mathbf{k}_4 - \mathbf{k}_3 \tag{6.4-28}$$

and c.c. denotes the complex conjugate term. We note that there are only two contributions to the index grating since beams 1 and 2 are incoherent with respect to beams 3 and 4. If $\mathbf{K}_{21}$ is distinct from $\mathbf{K}_{43}$, then the coupled mode equations can be written

$$\begin{aligned} \frac{d}{dz} A_1 &= -\frac{1}{2} \gamma_{21} |A_2|^2 A_1 / I_0 \\ \frac{d}{dz} A_2 &= \frac{1}{2} \gamma_{21}^* |A_1|^2 A_2 / I_0 \end{aligned} \tag{6.4-29}$$

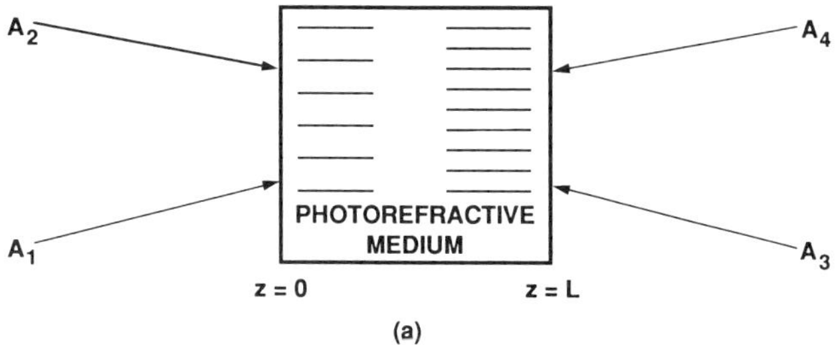

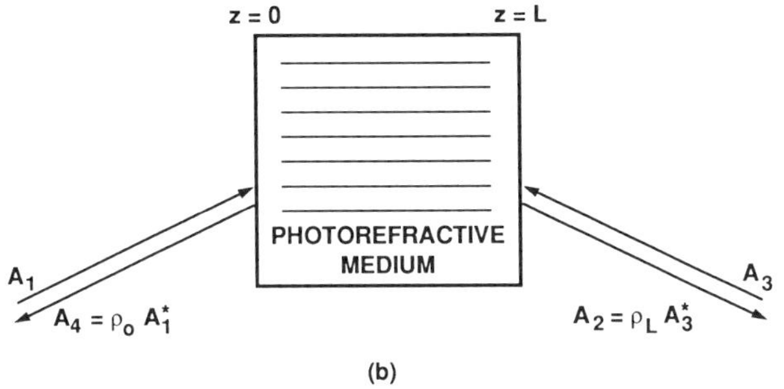

Figure 6.15. Hologram sharing in photorefractive phase conjugators. (a) Two sets of photorefractive holograms with different wave vectors. (b) A common photorefractive hologram in a mutually pumped phase conjugator.

$$\begin{aligned} \frac{d}{dz} A_3 &= \frac{1}{2} \gamma_{43} |A_4|^2 A_3 / I_0 \\ \frac{d}{dz} A_4 &= -\frac{1}{2} \gamma_{43}^* |A_3|^2 A_4 / I_0 \end{aligned} \tag{6.4-30}$$

where γ_{21} and γ_{43} are coupling constants. Here the γ are real intensity coupling constants. In deriving the coupled-mode equations, we assume that the photorefractive medium operates by diffusion only so that $\phi = \pi/2$. In addition, we neglect the material absorption. We notice that there are two sets of two-beam coupled-mode equations. In other words, the two sets of beams undergo two-beam coupling independent of each other, except the denominator I_0 which accounts for the cross erasure of the gratings due to the presence of the other sets of beams. In the usual two-wave mixing (TWM), I_0 is either $|A_1|^2 + |A_2|^2$ or $|A_3|^2 + |A_4|^2$. The I_0 in Eqs. (6.4-27)–(6.4-30) is, however, the sum of four intensities which can be as big as a factor of 2 larger for beams with similar intensities. The cross erasure is a result of two gratings sharing the same volume of photorefractive medium.

When $\mathbf{k}_4 = -\mathbf{k}_1$ and $\mathbf{k}_2 = -\mathbf{k}_3$, wave vectors of the two gratings are identical, that is,

$$\mathbf{K}_{21} = \mathbf{K}_{43} \tag{6.4-31}$$

The coupled-mode equations can be written

$$\frac{d}{dz} A_1 = -\frac{1}{2} \gamma[A_2^* A_1 + A_4^* A_3] A_2 / I_0 \tag{6.4-32}$$

$$\frac{d}{dz} A_2 = \frac{1}{2} \gamma[A_1^* A_2 + A_3^* A_4] A_1 / I_0 \tag{6.4-33}$$

$$\frac{d}{dz} A_3 = \frac{1}{2} \gamma[A_4^* A_3 + A_2^* A_1] A_4 / I_0 \tag{6.4-34}$$

$$\frac{d}{dz} A_4 = -\frac{1}{2} \gamma[A_3^* A_4 + A_1^* A_2] A_3 / I_0 \tag{6.4-35}$$

where $\gamma = \gamma_{21} = \gamma_{43}$ is the coupling constant. We notice that there are two terms on the right side of each equation in Eqs. (6.4-32)–(6.4-35). Each term represents Bragg scattering from one grating. Since these two gratings have the same wave vector, they contribute equally to the Bragg scattering. The magnitude of the grating is, however, determined by the algebraic sum of the two complex numbers $A_1^* A_2$ and $A_3^* A_4$. If in addition,

$$A_4 = \rho A_1^* \tag{6.4-36}$$

$$A_2 = \rho A_3^* \tag{6.4-37}$$

where ρ is an arbitrary constant, then the two terms in the square brackets in Eqs. (6.4-22)–(6.4-35) are identical, and the coupled equations become

$$\frac{d}{dz} A_1 = -\gamma |A_2|^2 A_1 / I_0 \tag{6.4-38}$$

$$\frac{d}{dz} A_2 = \gamma |A_1|^2 A_2 / I_0 \tag{6.4-39}$$

$$\frac{d}{dz} A_3 = \gamma |A_4|^2 A_3 / I_0 \tag{6.4-40}$$

$$\frac{d}{dz} A_4 = -\gamma |A_3|^2 A_4 / I_0 \tag{6.4-41}$$

Let us assume that beams 2 and 4 are the ones that get amplified due to two-beam coupling (i.e., $\gamma > 0$). We now examine the spatial growth of beams 2 and 4. The exponential growth constant for these two beams in Eqs. (6.4-38)–(6.4-41) is twice as large as that of the same beams in Eqs. (6.4-27)–(6.4-30) as a result of the hologram sharing. This factor of 2 is similar to the theoretical proof of the phase-conjugated stimulated backward scattering discussed earlier. It was shown that the gain coefficient for the conjugated wave in SBS is two or three times larger than that of any other scattered wave. In fact the spatial gain constant shown in Eqs. (6.4-38)–

(6.4-41) is exactly the same as that of the usual TWM. The cross erasure, which lowers the gain coefficient by a factor of 2 for other scattered light, disappears as a result of the hologram sharing. In photorefractive media such as $BaTiO_3$ or SBN, the exponential gain coefficient can be as large as $40\,cm^{-1}$ [14]. Thus in a sample of 1 cm, the ratio between the gain of mutually phase conjugated beams and that of the randomly scattered beams can be as large as a factor of exp(20). This enormous factor indicates that the mutually phase conjugate beams are the dominant ones in terms of spatial growth and that the conjugated beams can be generated from the noise.

We now consider the effect of transverse spatial variation (i.e., A_i are functions of x, y) on the process of phase conjugation. For plane waves with $A_i = \text{constant}$, Eqs. (6.4-31), (6.4-36) and (6.4-37) alone do not uniquely yield phase conjugation. In fact, given a set of incident wave vectors, Eq. (6.4-31) yields an infinite set of solutions for $\mathbf{k}_2$ and $\mathbf{k}_4$. These wave vectors are cylindrically symmetric with respect to an axis along the direction of $\mathbf{k}_1 - \mathbf{k}_3$. Thus for plane waves (with A_i independent of x, y), conical scattering may have exactly the same gain coefficient as the conjugated wave provided that their amplitudes satisfy Eqs. (6.4-36) and (6.4-37). This explains the arcs that are often observed in attempts to build mutually pumped phase conjugators. Let $\mathbf{k}_2'$ and $\mathbf{k}_4'$ be an arbitrary set of these solutions. For plane waves, $A_2 \exp[i(\omega_2 t - \mathbf{k}_2' \cdot \mathbf{r})]$ and $A_4 \exp[i(\omega_4 t - \mathbf{k}_4' \cdot \mathbf{r})]$ are also solutions of the wave equation because the A_i are independent of $\mathbf{k}_i$. This, however, is no longer true for waves with transverse spatial variation. For these waves, the complex amplitudes and the wave vectors are related by

$$\nabla^2 A - 2i(\mathbf{k} \cdot \nabla) A = 0 \qquad (6.4\text{-}42)$$

For incident waves with nonplanar wave fronts and spatial amplitude variations, A_1 and A_3 are complex functions of x and y. If Eqs. (6.4-38)–(6.4-41) are satisfied for all x and y, then they uniquely determine the mutually phase-conjugated waves. As a result of the constraint by Eq. (6.4-42),

$$A_2(x, y) \exp[i(\omega_2 t - \mathbf{k}_2' \cdot \mathbf{r})] \quad \text{and} \quad A_4(x, y) \exp[i(\omega_4 t - \mathbf{k}_4' \cdot \mathbf{r})]$$

are, in general, not acceptable solutions of the wave equation (see Problem 6.2). According to this argument, spatial wave front variation plays an important role in the fidelity of mutual phase conjugation. Experimentally, it was found that the fidelity of mutual phase conjugation increases drastically as soon as the wave fronts of the incident waves deviate from the planar one.

In photorefractive crystals such as $BaTiO_3$ or SBN, mutually pumped phase conjugation starts from scattering of the two incident beams (beams 1 and 3) (see Figure 6.15(b)). The amplitudes of these scattered waves are represented by A_2 and A_4. As a result of photorefractive coupling, these scattered waves will be amplified in the medium. Initially these scattered waves A_2 and A_4 may have arbitrary wave fronts. These scattered waves may be decomposed into two parts

$$A_2(0) = c_2 A_3^*(0) + A_2'(0) , \qquad A_4(L) = c_4 A_1^*(L) + A_4'(L) \quad (6.4\text{-}43)$$

where c_2 and c_4 are constants, $A_2'(0)$ and $A_4'(L)$ represent the portions which are orthogonal to the conjugate waves. Here, we recall that these amplitudes are also functions of x and y. As indicated earlier, these nonconjugated parts have smaller gain coefficients compared to the conjugated parts. Although these nonconjugated parts may be significant in the beginning, the conjugated parts dominate at the end due to the exponential growth. Thus in high gain media such as $BaTiO_3$ and SBN, the mutually phase conjugated beams are often many orders of magnitude higher than other scattered beams.

Although the present analysis assumes only one interaction region, the result can be easily extended to include two or multiple interaction regions. In some of the mutually pumped phase conjugation which involves two or more spatial regions of interaction, specular reflection and total internal reflection play an important role in maximizing the grating strength and matching the Bragg angles. It is important to note that, in some cases, the specular reflection from the surfaces may interfere with the hologram and disrupt the process of phase conjugation. The analysis of this effect is beyond the scope of the book.

6.4.3 MPPC with Ring Resonator

Mutually pumped phase conjugation can also be produced by photorefractive oscillation in ring resonators. Referring to Figure 6.16, we consider a ring resonator containing two-beam coupling gain media, for example, photorefractive $BaTiO_3$ crystals, inserted into the cavity. Each crystal is pumped by a laser beam to produce a unidirectional ring oscillation from the two-beam-coupling gain, resulting in two counterpropagating ring oscillations. This arrangement closely follows the geometry for a unidirectional oscillator (a ring resonator with a single photorefractive crystal pumped by one laser beam), which was previously studied in Chapter 5. In Figure 6.16, the two crystals are oriented in such a way that photorefractive crystal no. 1 will provide two-wave mixing gain for a counterclockwise unidirectional oscillation when it is pumped by laser beam 1 (P_1) and photorefractive crystal no. 2 will provide two-wave mixing gain for a clockwise unidirectional oscillation when it is pumped by laser beam 2 (P_2). If the gains are large enough to overcome the cavity losses, as is the case for most photorefractive crystals with proper crystal orientation and beam polarization, then oscillations will occur.

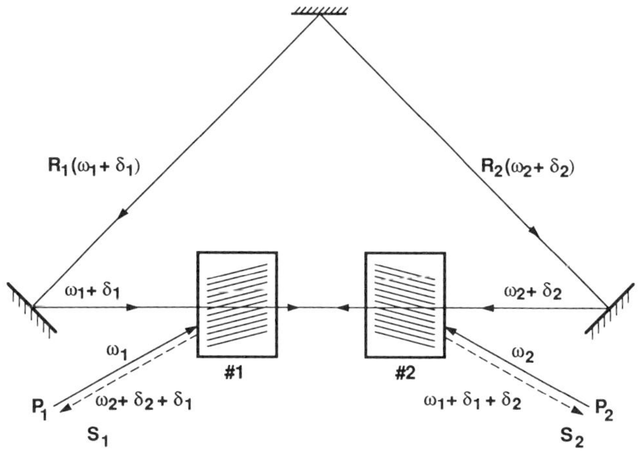

Figure 6.16. MPPC with ring resonators.

Let ω_1 and ω_2 be the frequencies of the two pumping laser beams with the amplitudes P_1 and P_2, respectively. When the cavity oscillations occur, the frequencies of the oscillations can be slightly detuned from those of the pumping beams and are designated $\omega_1 + \delta_1$ and $\omega_2 + \delta_2$.

We now examine the cross readout of the holograms inside the photorefractive crystals. Within each crystal, a volume hologram is formed by the pumping beam and the corresponding unidirectional oscillation beam. For example, a hologram (moving with a velocity proportional to δ_1) is formed by the pump beam P_1 (ω_1) and the counterclockwise oscillation beam, with the amplitude $R_1(\omega_1 + \delta_1)$, inside photorefractive crystal no. 1. The amplitude of the index grating is proportional to $R_1 P_1^*$. When the clockwise oscillation, with the amplitude $R_2(\omega_2 + \delta_2)$, intersects the hologram inside photorefractive crystal no. 1, it is Bragg diffracted in the direction counterpropagating to P_1 with the frequency $\omega_2 + \delta_1 + \delta_2$, provided that ω_1 and ω_2 are close enough to maintain the Bragg condition. We denote the amplitude of the signal generated from crystal no. 1 as $S_1(\omega_2 + \delta_1 + \delta_2)$. (Note that $\omega_1 - \omega_2$ can be tens of gigaHertz, and the Bragg phase-matching condition will still be well satisfied.) A similar situation occurs in photorefractive crystal no. 2, with $R_1(\omega_1 + \delta_1)$ reading out the hologram in photorefractive crystal no. 2 to produce a second signal beam $S_2(\omega_1 + \delta_1 + \delta_2)$ counterpropagating with respect to P_2. According to the holographic interpretation of four-wave mixing, the relationships between the pumps and the oscillations are given by

$$\begin{aligned} S_1(\omega_2 + \delta_1 + \delta_2) &\propto R_1(\omega_1 + \delta_1) P_1^*(\omega_1) R_2(\omega_2 + \delta_2) \\ S_2(\omega_1 + \delta_1 + \delta_2) &\propto R_2(\omega_2 + \delta_2) P_2^*(\omega_2) R_1(\omega_1 + \delta_1) \end{aligned} \tag{6.4-44}$$

Here, it is important to note that the signal beams S_1 and S_2 can be spatial phase conjugates of the incident beams P_1 and P_2, respectively, if the two ring oscillations are phase conjugates of each other. That is, if $R_1 = R_2^*$, then $S_2 \propto P_2^*$. Because the oscillation beam path is defined by the resonator cavity, the two modes of oscillations are exactly counterpropagating. In addition, the two counterpropagating oscillations in a stable cavity are exactly a phase-conjugate pair, provided that they oscillate at the same transverse mode. The cavity can easily be forced to oscillate at the same transverse mode. In addition, the cavity can also be prepared to oscillate in the fundamental transverse mode by placing a pinhole in the beam path, as was done in most experiments. The important consequence of the two ring oscillations being a phase-conjugate pair in our model is that it results in the absence of image cross talk. On the other hand, if the ring oscillations did not form a phase-conjugate pair, but instead consisted of orthogonal transverse modes, then an image cross talk could occur by transfer of the dark regions in the nonuniform intensity distributions on the ring oscillations to the signal beams during the readout processes.

The temporal characteristics of MPPCs are different from that of phase conjugation using a single pump beam, such as self-pumped phase conjugation or stimulated photorefractive backscattering. Although the spatial wave fronts are reconstructed, the frequencies of the phase-conjugated beams from MPPCs are exchanged and modified. The phase conjugate of pump beam 1 has a frequency of $\omega_2 + \Delta$, and the phase conjugate of pump beam 2 has a frequency of $\omega_1 + \Delta$, where $\Delta = \delta_1 + \delta_2$. In other words, the phase conjugate of beam 1 originates from beam 2, and vice versa; the temporal characteristic of one pump beam is transferred to the phase conjugate of the other pump beam. The situation is similar to that of four-wave mixing when the grating written by two mutually coherent beams is read out by a third (reading) beam that is mutually incoherent with respect to the writing beams.

The counterpropagating ring oscillations illustrated in Figure 6.16 and described above can occur within a single $BaTiO_3$ crystal. By virtue of total internal reflections at the surfaces, the crystals can act as an optical cavity that is capable of supporting many modes. These types of internal oscillation were suggested to explain the frequency shift in a self-pumped phase conjugator. For the MPPC, one may consider that each incident beam initiates a unidirectional ring oscillation within the crystal, in accordance with the model given above. When both beams are present, counterpropagating oscillations occur, which lead to mutually pumped phase conjugation.

6.5 POLARIZATION-PRESERVING PHASE CONJUGATORS (PPPC)

Up to now, we have been discussing various schemes of optical phase conjugation concerning the reconstruction of the wave front. We assumed that the polarization states are linear and the phase conjugators are sensitive to light with certain polarization states. For phase conjugators made of photorefractive $BaTiO_3$ crystals, only the extraordinary waves are conjugated efficiently.

Most demonstrations of the correction of wave front distortions by phase conjugation use linearly polarized light and optically isotropic wave front-distorting media. In some cases, however, one may also wish to correct for the changes in polarization induced by optically anisotropic media. For example, one may wish to correct for the polarization scrambling due to environmentally dependent birefringence that is a source of noise and signal fading in fiber optic gyros [19]. Self-pumped phase conjugation is very efficient in barium titanate [20], but only for extraordinary light since it utilizes the large electro-optic coefficients r_{42} and r_{51}. In this section, we describe a scheme for using a single self-pumped crystal of barium titanate to produce the phase conjugate of a wave having arbitrary polarization.

In general, the reflectivity of a phase-conjugate mirror is described by a 2×2 tensor that determines the amplitude, phase, and polarization state of the reflected light. Only when this tensor reduces to a scalar can a phase-conjugate mirror correct for both the wave front distortions and the changes in polarization induced by optically anisotropic wave front-distorting media [21]. We call a phase-conjugate mirror for which the reflectivity tensor reduces to a scalar, a polarization-preserving phase conjugator (PPPC). To show that a PPPC can correct for the change in polarization state introduced by optically anisotropic media, consider the situation of a plane wave with polarization state described by a vector amplitude A,

$$A = \begin{bmatrix} A_x \\ A_y \end{bmatrix} \tag{6.5-1}$$

that is incident on an optically anisotropic medium, followed by a phase-conjugate mirror. After passing through the anisotropic medium, the amplitude becomes

$$A_1 = TA , \quad \text{with } T = \begin{bmatrix} t_{xx} & t_{xy} \\ t_{yx} & t_{yy} \end{bmatrix} \tag{6.5-2}$$

where T is a Jones matrix [22] that describes transmission through the anisotropic medium, and where t_{xx}, t_{xy}, t_{yx}, and t_{yy} are the tensor elements. After reflection from the phase-conjugate mirror, the amplitude becomes

$$A_2 = R(TA)^* , \qquad R = \begin{bmatrix} r_{xx} & r_{xy} \\ r_{yx} & r_{yy} \end{bmatrix} \tag{6.5-3}$$

where R is a tensor that describes reflection from the phase conjugate mirror, and where r_{xx}, r_{xy}, r_{yx}, and r_{yy} are the tensor elements. After passing back through the anisotropic medium, the amplitude of the returning wave is given by

$$A_3 = TRT^* A^* \tag{6.5-4}$$

Since in general this is not proportional to the phase conjugate of the amplitude of the incident wave, the returning wave does not have the same polarization as the incident wave. However, if the reflectivity tensor reduces to a scalar r, then the amplitude of the reflected wave becomes

$$A_3 = rTT^* A^* , \quad \text{for } R = \begin{bmatrix} r & 0 \\ 0 & r \end{bmatrix} = r \tag{6.5-5}$$

For optically anisotropic media, such as birefringent or optically active media, the Jones matrix has the property, $TT^* = 1$. In this case, the amplitude of the reflected wave becomes

$$A_3 = rA^* , \quad \text{for } R = r, \ TT^* = 1 \tag{6.5-6}$$

Since this is proportional to the phase conjugate of the incident wave, the polarization of the reflected wave is identical to that of the incident wave. Thus, we have shown that a PPPC can correct for the change in polarization state introduced by optically anisotropic media. The PPPC cannot correct for the changes in polarization introduced by nonreciprocal phenomena such as the Faraday effect, where $TT^* \neq 1$.

Several methods have been proposed for polarization-preserving phase conjugation [21, 23, 24]. Polarization-preserving phase conjugators can be made using media that have a tensor reflectivity, by decomposing the incident wave into two components having orthogonal polarizations, phase-conjugating these components, and recombining the conjugates [23]. The result is a polarization-preserving phase conjugator only if the components recombine with the correct amplitude and phase. This is insured if the losses and the amplitudes and phases of the complex phase-conjugate reflectivities are the same for both components. A schematic of the polarization-preserving phase conjugator that uses this method is shown in Figure 6.17. To understand how it works, consider an incident wave with a polarization described by a complex vector amplitude $A = (a_1, a_2)$. This wave is decomposed into two orthogonally polarized components, $(a_1, 0)$ and $(0, a_2)$, by the polarizing beam splitter (PBS). The component with polarization orthogonal to the plane of the figure $(0, a_2)$ has its polarization rotated by the half-wave retarder $\lambda/2$ such that its polarization state becomes $(a_2, 0)$. Both components are incident on a barium titanate crystal at an angle such that self-pumped phase conjugation occurs [25]. Since both components have the same polarization and are reflected from the same phase-conjugate mirror, they experience the same complex phase-conjugate reflectivity r, becoming $r(a_1^*, 0)$ and $r(a_2^*, 0)$. The half-wave retarder restores the second of these components to its original polarization, $r(0, a_2^*)$. When the two components recombine at the polarizing beam splitter, they form the phase conjugate

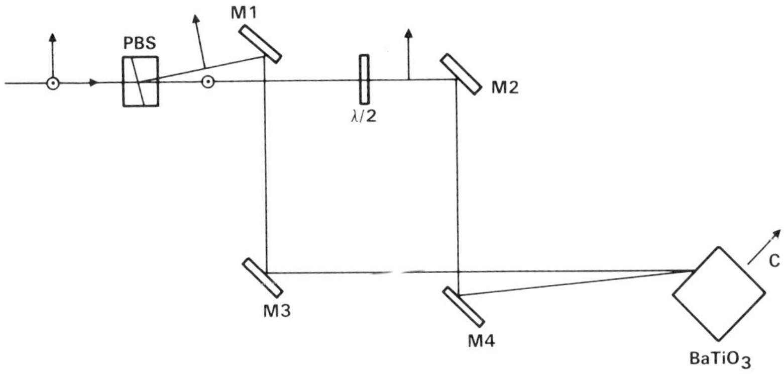

Figure 6.17. Polarization-preserving phase conjugator (PPPC). Light incident on the polarizing beam splitter PBS is decomposed into two orthogonally polarized components. One component is rotated 90° by the half-wave retarder $\lambda/2$. Both components are incident on a barium titanate crystal such that self-pumped phase conjugation occurs. When the two conjugates recombine at PBS, they form a conjugate wave that reproduces the polarization of the incident light.

with the same polarization as the incident wave, $r(a_1^*, a_2^*) = rA^*$. Thus, the configuration shown in Figure 6.17 acts like a polarization-preserving phase conjugator, and henceforth we refer to it as "the PPPC".

Using the self-pumped configuration in the PPPC results in a frequency shift between the components incident on the crystal and their conjugates [26–28]. However, since both components are incident on the same region of the crystal, they see the same pumping waves in the crystal, and therefore they experience the same frequency shift and phase-conjugate reflectivity. In fact, the two beams can be thought of as two components of a single probe beam. Since the two beams entering the crystal see the same pump waves, they experience the same phase conjugate reflectivity. It should be noted that the two beams entering the crystal must be coherent to within the response time of the crystal. In the experiment, the path lengths from the polarizing beam splitter to the crystal were made equal to within the coherence length of the laser. The excellent frequency locking [29] of the conjugates when both beams are present is indicated by the fact that no beating is detected in the return from the PPPC.

Polarization-preserving phase conjugators employing such a scheme have been demonstrated by using a photorefractive $BaTiO_3$ crystal as the phase conjugator [30]. In order to quantify how well the technique works when there is an aberration in the beam, a multimode fiber is placed before the PPPC. The multimode fiber scrambles the polarization of the input wave. With a normal mirror at the end of the fiber, approximately 50% of the power returns in each of the two linear polarization states (parallel and orthogonal to the linear input polarization). With the polarization preserving phase conjugator at the end of the fiber, less than 1% of the light returns with the orthogonal polarization.

We consider next the restoration of image information in multimode fibers by using polarization-preserving phase conjugators. An interesting question would be: can the image information be restored by using a nonpolarization-preserving conjugator?

6.5.1 Image Restoration in Multimode Fibers by Phase Conjugation

When polarized light is incident into a long multimode fiber, the exiting light is randomly distributed among the spatial and polarization modes due to modal scrambling. If a phase conjugator is placed at the end of the fiber, a phase conjugate wave of the scrambled image is generated. The phase conjugate wave propagates backward through the fiber and the original input image is restored by virtue of the nature of phase conjugation. The recovery of the original spatial and polarization modes of the image only takes place when the reflection from the phase conjugator preserves the polarization of the incident light. Here we present a simple theoretical analysis on the correction of modal scrambling in multimode fibers for the cases when both polarization components are conjugated at the fiber output (polarization-preserving phase conjugator [30], PPPC), and when only one polarization component is conjugated (nonpolarization-preserving phase conjugator, NPPPC).

A PPPC produces a time-reversed wave that can correct for modal scrambling as well as polarization scrambling. In the following paragraph, we consider the result of conjugating only one polarization component. Consider a multimode fiber that supports N confined modes with wave functions

$$\mathbf{E}_n(x, y)\exp(-i\beta_n z)\,, \quad n = 1, 2, \ldots, N \tag{6.5-7}$$

where β_n is the propagation constant and n is the mode number. An incident wave $\mathbf{E}_\mathrm{I}$ can be expressed as a sum over the N modes at the input ($z = 0$)

$$\mathbf{E}_\mathrm{I} = \sum_n A_n \mathbf{E}_n \tag{6.5-8}$$

where the A_n are the mode amplitudes. The output ($z = L$) field $\mathbf{E}_\mathrm{II}$ can be written

$$\mathbf{E}_\mathrm{II} = (E_{\mathrm{II}x}, E_{\mathrm{II}y}) \tag{6.5-9}$$

where $E_{\mathrm{II}x}$ and $E_{\mathrm{II}y}$ are the x and y components. If modal scrambling is complete, then the energy is distributed equally among the x and y polarizations

$$\int |E_{\mathrm{II}x}|^2\, dx\, dy = \int |E_{\mathrm{II}y}|^2\, dx\, dy \tag{6.5-10}$$

Let this field be incident on a phase conjugator which has unit reflectivity and a polarizer along the x axis in front of it. The reflected field $\mathbf{E}_{\mathrm{III}}$ is given by

$$\mathbf{E}_{\mathrm{III}} = (E^*_{\mathrm{II}x}, 0) \tag{6.5-11}$$

This field can be written as the sum of two components; a component $\mathbf{E}^*_{\mathrm{II}}$, that is, the phase conjugate of the light exiting from the fiber at $z = L$ before reaching the polarizer and a component $\mathbf{E}^*_{\mathrm{II}\perp}$, that is orthogonal to the conjugate in the sense that $\int \mathbf{E}_{\mathrm{II}} \cdot \mathbf{E}^*_{\mathrm{II}\perp}\, dx\, dy = 0$,

$$\mathbf{E}_{\mathrm{III}} = a\mathbf{E}^*_{\mathrm{II}} + b\mathbf{E}^*_{\mathrm{II}\perp} \tag{6.5-12}$$

where a and b are constants. Scalar multiplying both sides of Eq. (6.5-12) by $\mathbf{E}_{\mathrm{II}}$, integrating over x and y, and using Eqs. (6.5-10) and (6.5-12), we obtain the results $a = 1/2$, $b = 1/2$, and $E_{\mathrm{II}\perp} = (E_{\mathrm{II}x}, -E_{\mathrm{II}y})$. Using these results, $\mathbf{E}_{\mathrm{III}}$ becomes

$$\mathbf{E}_{\mathrm{III}} = 1/2\mathbf{E}^*_{\mathrm{II}} + 1/2\mathbf{E}^*_{\mathrm{II}\perp} \tag{6.5-13}$$

This field propagates back through the fiber. If the fiber is linear and lossless, then the conjugate component $1/2\mathbf{E}^*_{\mathrm{II}}$ generates $1/2\mathbf{E}^*_{\mathrm{I}}$ by time reversal and the orthogonal component $1/2\mathbf{E}^*_{\mathrm{II}\perp}$ generates $1/2\mathbf{E}^*_{\mathrm{I}\perp}$, where $\mathbf{E}^*_{\mathrm{I}\perp}$ is orthogonal to $\mathbf{E}^*_{\mathrm{I}}$ in the sense that $\int \mathbf{E}_{\mathrm{I}} \cdot \mathbf{E}^*_{\mathrm{I}\perp}\, dx\, dy = 0$. Since we assume complete modal scrambling, the energy in $\mathbf{E}^*_{\mathrm{I}\perp}$ is distributed equally among the x and y polarizations. The returning field at the fiber input becomes

$$\mathbf{E}_{\mathrm{IV}} = 1/2\mathbf{E}^*_{\mathrm{I}} + 1/2\mathbf{E}^*_{\mathrm{I}\perp} \tag{6.5-14}$$

The second term, $1/2\mathbf{E}^*_{\mathrm{I}\perp}$, represents a noise background and cannot be ignored. This is a result of the fact that only one polarization component is phase conjugated. Note that when only one polarization component is conjugated at the output of the fiber, only 1/4 of the energy ($|1/2\mathbf{E}^*_{\mathrm{I}}|^2$) is recovered in the reconstruction of the input field. The other 1/4 of the energy ($|1/2\mathbf{E}^*_{\mathrm{I}\perp}|^2$) is randomly distributed among the spatial and polarization modes. Therefore, $(1/4 + 1/2 \times 1/4) = 3/8$ of the original energy returns with the polarization of the input and $(1/2 \times 1/4) = 1/8$ returns orthogonal to it. Half of the original energy is lost at the polarizer. The phase conjugation of only one polarization component does not eliminate any of the spatial information about the image. It does reduce the signal to noise ratio because the restored image is superposed with a noise background. The restoration of image information using both PPPC and NPPPC has been demonstrated using a step-index multimode fiber [31–33].

PROBLEMS

6.1 Frequency Shift and Momentum Mismatch in Phase Conjugation

a. Show that momentum mismatch is related to the frequency shift by

$$\Delta k = (\mathbf{k}_4 + \mathbf{k}_1)_z = \frac{2\delta}{c} n \cos\theta$$

b. Let $\omega_2 = \omega_3 = \omega$, $\omega_1 = \omega + \delta$, $\omega_4 = \omega - \delta$ and $\mathbf{k}_2 + \mathbf{k}_3 = 0$, show that the coupled equations are

$$\frac{d}{dz} A_1 = -\frac{1}{2} \Gamma(A_1 A_2^* + A_3 A_4^* e^{i\Delta k z}) A_2 / I_0$$

$$\frac{d}{dz} A_2 = \frac{1}{2} \Gamma^*(A_1^* A_2 + A_3^* A_4 e^{-i\Delta k z}) A_1 / I_0$$

$$\frac{d}{dz} A_3 = \frac{1}{2} \Gamma(A_1 A_2^* e^{-i\Delta k z} + A_3 A_4^*) A_4 / I_0$$

$$\frac{d}{dz} A_4 = -\frac{1}{2} \Gamma^*(A_1^* A_2 e^{i\Delta k z} + A_3^* A_4) A_3 / I_0$$

where Δk is the momentum mismatch.

6.2 Conical Scattering in MPPC

a. Let $A_2(x, y) \exp[i\omega_2 t - i\mathbf{k}_2 \cdot \mathbf{r}]$ be a solution to Eq. (6.4-42). Show that

$$\nabla^2 A_2 - 2i\left(k_{2x} \frac{\partial}{\partial x} + k_{2y} \frac{\partial}{\partial y} + k_{2z} \frac{\partial}{\partial z}\right) A_2 = 0$$

b. We now rotate the vector $\mathbf{k}_2$ around an axis parallel to $\mathbf{k}_1 - \mathbf{k}_3$ and obtain $A_2(x, y) \exp[i\omega_2 t - i\mathbf{k}_2' \cdot \mathbf{r}]$. Show that this new wave is a solution of Eq. (6.4-42) only if

$$(\mathbf{k}_2' \cdot \nabla) A_2 = (\mathbf{k}_2 \cdot \nabla) A_2$$

c. Since $A_2(x, y)$ is an arbitrary wave function, $A_2(x, y) \exp[i\omega t - \mathbf{k}_2' \cdot \mathbf{r}]$, in general, is not a solution of the wave equation.

6.3 Show that the argument used in the paragraph after Eq. (6.3-12) is valid only when $I_1 + I_2 \simeq$ constant. This is not the case in phase conjugation via backward stimulated scattering. Show that a background intensity $I_B \gg I_1, I_2$ may be responsible for the high gain for phase conjugate waves.

6.4 Derive Eqs. (6.3-29) and (6.3-39).

6.5 Discuss the reflectivity of a self-pumped ring phase conjugator when R

is greater than unity (i.e., gain is present in the cavity). What about the threshold condition Eq. (6.3-38) when $R > 1$.

6.6 Discuss the solutions of Eq. (6.3-36) when ΓL is complex.

6.7 Show that Eq. (6.4-25) is equivalent to Eq. (6.3-38).

REFERENCES

1. Sec, for example, A. Yariv and P. Yeh, Optical Waves in Crystals (Wiley, 1984) Chap. 13.
2. See, for example, A. Yariv, "Phase Conjugate Optics and Real-Time Holography," IEEE J. Quantum Electron. **14**, 650–660 (1978).
3. J. Feinberg, "Self-Pumped Continuous-Wave Conjugator Using Internal Reflections," Opt. Lett. **7**, 486 (1982); J. Feinberg, "Continuous-Wave Self-Pumped Phase Conjugator with Wide Field of View," Opt. Lett. **8**, 480 (1983).
4. M. D. Ewbank and P. Yeh, "Frequency Shifts of Self-Pumped Phase Conjugators," Proc. SPIE **613**, 11 (1986).
5. J. Feinberg, "Asymmetric Self-Defocusing of an Optical Beam from the Photorefractive Effect," J. Opt. Soc. Am. **72**, 46 (1982).
6. T. Y. Chang and R. W. Hellwarth, "Optical Phase Conjugation by Backscattering in the Photorefractive Crystal Barium Titanate," Opt. Lett. **10**, 108–410 (1985).
7. B. Ya. Zel'dovich, V. I. Popovichev, V. V. Ragul'skii, and F. S. Faizullov, "Connection Between the Wavefronts of the Reflected and Exciting Light in SBS," ZhETF Pis. Red. **15**, 160 (1972).
8. M. Cronin-Golomb, B. Fischer, J. O. White, and A. Yariv, "Theory and Applications of Four-Wave Mixing in Photorefractive Media," IEEE J. Quantum Electron. **20**, 12–30 (1984).
9. See, for example, P. Yeh, T. Y. Chang, and M. D. Ewbank, "Model for Mutually Pumped Phase Conjugation," J. Opt. Soc. Am. **B5**, 1743 (1988), and references therein.
10. S. Weiss, S. Sternklar, and B. Fischer, "Double Phase Conjugate Mirror: Analysis, Demonstration and Applications," Opt. Lett. **12**, 114–116 (1987).
11. R. W. Eason and A. M. C. Smout, "Bistability and Noncommutative Behavior of Multiple-Beam Self-Pushing and Self-Pumping in $BaTiO_3$," Opt. Lett. **12**, 51–53 (1987); A. M. C. Smout and R. W. Eason, "Analysis of Mutually Incoherent Beam Coupling in $BaTiO_3$," Opt. Lett. **12**, 498–500 (1987).
12. M. D. Ewbank, "Incoherent Beam Sharing in Photorefractive Holograms," Topical Meetings on Photorefractive Materials, Effects, and Devices. Technical Digest Vol. 17 (Optical Society of America, Washington DC, 1987), pp. 179–182.
13. D. Wang, Z. Zhang, Y. Zhu, S. Zhang, and P. Ye, "Observations on the Coupling Channel of Two Mutually Incoherent Beams Without Internal Reflection in $BaTiO_3$," Opt. Comm. **73**, 495–500 (1989).
14. E. J. Sharp, W. W. Clark III, M. J. Miller, G. L. Wood, B. Monson, G. J.

Salamo, and R. R. Neurganokar, "Double Phase Conjugation in Tungsten Bronze Crystals," Appl. Opt. **29**, 743–749 (1990).

15. M. D. Ewbank, "Mechanism for Photorefractive Phase Configuration Using Incoherent Beams," Opt. Lett. **13**, 47 (1988).
16. P. Yeh, "Photorefractive Phase Conjugators," Proc. IEEE **80**, 436 (1992).
17. P. Yeh, "Coupled Mode Theory of Hologram Sharing Model of MPPC's," Appl. Opt. **28**, 1961–1964 (1989).
18. F. Laeri, T. Tschudi, and J. Albers, "Coherent CW Image Amplifier and Oscillator Using Two-Wave Mixing in a $BaTiO_3$ Crystal," Opt. Comm. **47**, 387 (1983).
19. R. Ulrich and M. Johnson, "Fiber-Ring Interferometer: Polarization Analysis," Opt. Lett. **4**, 152 (1979).
20. J. Feinberg and R. W. Hellwarth, "Phase-Conjugating Mirror with Continuous Wave Gain," Opt. Lett. **5**, 519 (1980).
21. P. Yeh, "Scalar Phase Conjugator for Polarization Correction," Opt. Comm. **51**, 195 (1984).
22. See, for example, A. Yariv, and P. Yeh, Optical Waves in Crystals (Wiley, New York, 1984).
23. N. G. Basov, V. F. Efimkov, I. G. Zuberev, A. V. Kotov, S. I. Mikhailov, and M. G. Smirnov, "Inversion of Wavefront in SMBS of the Depolarized Pump," JETP Lett. **4**, 197 (1978).
24. G. Martin, L. K. Lam, and R. W. Hellwarth, "Generation of a Time-Reversed Replica of a Nonuniformly Polarized Image-Bearing Optical Beam," Opt. Lett. **5**, 185 (1980).
25. J. Feinberg, "Self-Pumped Continuous-Wave Conjugator Using Internal Reflections," Opt. Lett. **7**, 486 (1982).
26. W. B. Whitten and J. M. Ramsey, "Self-Scanning of a Dye Laser Due to Feedback from a $BaTiO_3$ Phase-Conjugate Reflector," Opt. Lett. **9**, 44 (1985).
27. J. Feinberg and G. D. Bacher, "Self-Scanning of a Continuous-Wave Dye Laser Having a Phase-Conjugating Resonator Cavity," Opt. Lett. **9**, 420 (1984).
28. M. Ewbank and P. Yeh, "Frequency Shift of Self-Pumped Phase Conjugator," SPIE Proc. **613**, 59 (1986).
29. M. D. Ewbank, P. Yeh, M. Khoshnevisan, and J. Feinberg, "Time Reversal by an Interferometer with Coupled Phase-Conjugate Reflectors," Opt. Lett. **10**, 282 (1985).
30. I. McMichael, M. Khoshnevisan, and P. Yeh, "Polarization-Preserving Phase Conjugator," Opt. Lett. **11**, 525–527 (1986).
31. I. McMichael, P. Yeh, and P. Beckwith, "Correction of Polarization and Modal Scrambling in Multimode Fibers by Phase Conjugation," Opt. Lett. **12**, 507–509 (1987).
32. K. Kyuma, A. Yariv, and S. Kwong, "Polarization Recovery in Phase Conjugation by Modal Dispersal," Appl. Phys. Lett. **49**, 617 (1986).
33. A. Yariv, Y. Tomita, and K. Kyuma, "Theoretical Model for Dispersal of Polarization Information and its Recovery by Phase Conjugation," Opt. Lett. **11**, 809 (1986).

7

Diffraction Properties of Gratings in Photorefractive Media

We consider the diffraction properties of a volume index grating present inside a photorefractive medium. This grating can be either photo-induced by a pair of coherent interfering beams, or simply a fixed (or permanent) grating. When this grating is read by an incident beam at the Bragg angle, strong diffraction occurs. The incident beam and the diffracted beam will not only affect the existing grating, but also induce a new photorefractive grating. Thus the diffraction of light in photorefractive media is different from that of a simple Bragg diffraction.

7.1 DIFFRACTION PROPERTIES OF PHOTOREFRACTIVE GRATINGS

In two-wave mixing, two coherent beams interfere inside a photorefractive medium and produce a volume index grating. The depth of index modulation is a function of position as a result of the energy coupling. In optical phase conjugation via four-wave mixing, a third beam is incident at the Bragg angle from the other side ($z = L$), and a fourth beam is generated. Here in this section, we consider the situation when the third beam is incident from the front ($z = 0$) side at the Bragg angle and a diffracted beam is generated. We assume that the third beam is incoherent with the two beams that were interfering and forming the index grating. It seems that there are two pairs of beams each undergoing TWM independently. They are actually coupled through the common grating which has two components, one from each pair of beams.

Referring to Figure 7.1, we consider the interaction of these two pairs of beams in a photorefractive medium. The electric fields of these waves are written

$$E_j = A_j \exp[i(\omega_j t - \mathbf{k}_j \cdot \mathbf{r})] \tag{7.1-1}$$

with

$$\omega_1 = \omega_2 \tag{7.1-2}$$

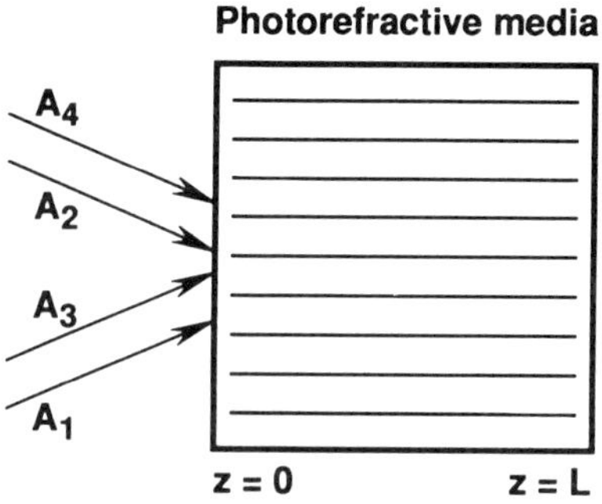

Figure 7.1. Multi-beam mixing in photorefractive media. Beam A_1 is coherent with beam A_2, and beam A_3 is coherent with beam A_4. The index grating consists of two contributions (i.e., $A_1^* A_2$ and $A_3^* A_4$) with a common wave vector.

$$\omega_3 = \omega_4 \tag{7.1-3}$$

$$\mathbf{k}_2 - \mathbf{k}_1 = \mathbf{k}_4 - \mathbf{k}_3 = \mathbf{K} \tag{7.1-4}$$

where A_j ($j = 1, \ldots, 4$) is the amplitude of the jth wave, ω_j, $\mathbf{k}_j$ are the frequency and wave vector of the jth wave, respectively. We assume that the frequency difference between these two pairs of beams $(\omega_3 - \omega_1)$ is much larger than the inverse of the response time of the medium, that is, $(\omega_3 - \omega_1) \gg 1/\tau$. This presents the formation of grating between A_1 and A_4. Similarly there is no grating formed between A_2 and A_3. Thus $(A_1 A_2^* + A_3 A_4^*)$ is the only grating that can be formed by these two pairs of beams.

The coupled equations for these two pairs of beams are very similar to those obtained in four-wave mixing using one-grating approximation. They are written

$$\frac{d}{dz} A_1 = -\frac{1}{2} \Gamma(A_1 A_2^* + A_3 A_4^*) A_2 / I_0 \tag{7.1-5}$$

$$\frac{d}{dz} A_2 = \frac{1}{2} \Gamma^*(A_1^* A_2 + A_3^* A_4) A_1 / I_0 \tag{7.1-6}$$

$$\frac{d}{dz} A_3 = -\frac{1}{2} \Gamma(A_1 A_2^* + A_3 A_4^*) A_4 / I_0 \tag{7.1-7}$$

$$\frac{d}{dz} A_4 = \frac{1}{2} \Gamma^*(A_1^* A_2 + A_3^* A_4) A_3 / I_0 \tag{7.1-8}$$

By comparing with Eqs. (4.5-6)–(4.5-9), we note that the signs of the coupled equations for A_3 and A_4 are reversed. This is due to the opposite direction of propagation. In arriving at these equation, we have ignored the small difference between the frequencies (i.e., $\omega_1 \simeq \omega_3$) and wave vectors (i.e., $k_1 \simeq k_3$), leading to the same coupling constants for all the beam pairs. For typical values of 1 μm grating period and 1 mm thickness of the photorefractive grating, the frequencies of the various beam pairs can differ by a few tens of gigaHertz and their angles differ by a few milliradians for the Bragg phase-matching condition to be still satisfied, so that assuming the same coupling constant for all the beam pairs seems reasonable.

The coupled equations can be solved analytically using a change of variable similar to the one used in the treatment of cross-polarization two-wave mixing. Before we proceed with the integration, we note that there are several constants of integration. These are

$$I_1 + I_2 = |A_1|^2 + |A_2|^2 = c_1 \tag{7.1-9}$$

$$I_3 + I_4 = |A_3|^2 + |A_4|^2 = c_2 \tag{7.1-10}$$

$$A_1^* A_3 + A_2^* A_4 = b \tag{7.1-11}$$

$$A_1 A_4 - A_2 A_3 = c \tag{7.1-12}$$

where $c_1 c_2$, b, and c are constants.

We now define two new variables,

$$f = \frac{A_1}{A_3}, \qquad g = \frac{A_2}{A_4} \tag{7.1-13}$$

Taking the derivative of f and g with respect to z and using the coupled equations, we obtain

$$\frac{df}{dz} = \frac{1}{2}\Gamma[bf^2 - \sigma f - b^*]/I_0 \tag{7.1-14}$$

$$\frac{dg}{dz} = -\frac{1}{2}\Gamma^*[bg^2 - \sigma g - b^*]/I_0 \tag{7.1-15}$$

where σ is a constant given by

$$\sigma = c_1 - c_2 \tag{7.1-16}$$

These two differential equations can be integrated by using the formula Eq. (4.5-35). The results are

$$\tanh^{-1}\frac{2bf - \sigma}{s} = -\frac{s\Gamma}{4I_0} z + D \tag{7.1-17}$$

$$\tanh^{-1}\frac{2bg - \sigma}{s} = \frac{s\Gamma^*}{4I_0} z + D' \tag{7.1-18}$$

where D and D' are constants of integration and s is given by

$$s^2 = \sigma^2 + 4|b|^2 \tag{7.1-19}$$

These two constants can be determined by the boundary conditions at $z = 0$ (i.e., the incident amplitudes $A_1(0)$, $A_2(0)$, $A_3(0)$, and $A_4(0)$. If we define

$$f_0 = \frac{A_1(0)}{A_3(0)}, \qquad g_0 = \frac{A_2(0)}{A_4(0)} \tag{7.1-20}$$

then f and g can be written

$$f(z) = \frac{sf_0 - (\sigma f_0 + 2b^*)\tanh \kappa z}{s + (\sigma - 2bf_0)\tanh \kappa z} \tag{7.1-21}$$

$$g(z) = \frac{sg_0 + (\sigma g_0 + 2b^*)\tanh \kappa^* z}{s - (\sigma - 2bg_0)\tanh \kappa^* z} \tag{7.1-22}$$

with

$$\kappa = \frac{s\Gamma}{4I_0} \tag{7.1-23}$$

Now with f and g expressed in closed form as functions of z, all the beam intensities can be obtained. According to the definition and the constants of integration, we have

$$I_1 = I_3|f|^2 \tag{7.1-24}$$

$$I_2 = I_4|g|^2 \tag{7.1-25}$$

$$I_1 + I_2 = c_1 \tag{7.1-26}$$

$$I_3 + I_4 = c_2 \tag{7.1-27}$$

These equations can be solved and the intensities are

$$I_1(z) = \frac{c_2|fg|^2 - c_1|f|^2}{|g|^2 - |f|^2} \tag{7.1-28}$$

$$I_2(z) = \frac{c_1|g|^2 - c_2|gf|^2}{|g|^2 - |f|^2} \tag{7.1-29}$$

$$I_3(z) = \frac{c_2|g|^2 - c_1}{|g|^2 - |f|^2} \tag{7.1-30}$$

$$I_4(z) = \frac{c_1 - c_2|f|^2}{|g|^2 - |f|^2} \tag{7.1-31}$$

The above equations are general solutions of forward four-wave mixing. For the special case of interest, we have a boundary condition of $A_2(0) = 0$. In other words, beam 3 and beam 4 are undergoing two-wave mixing and the hologram written by them is read out by beam 1. The reading and writing happen simultaneously. The diffraction efficiency is defined as

$$\eta = \left|\frac{A_2(L)}{A_1(0)}\right|^2 = \frac{I_2(L)}{I_1(0)} \tag{7.1-32}$$

where $I_1(0)$ is the intensity of the read beam, and $I_2(L)$ is the intensity of the diffracted beam. The intensity $I_2(L)$ can be obtained from Eq. (7.1-29).

The same problem can also be solved by using the method of grating integral (Section 2.7) for the case when Γ is real. We define

$$G = \frac{A_1 A_2^* + A_3 A_4^*}{I_0} \tag{7.1-33}$$

where we note that I_0 is a constant. This is the grating interference function which represents the amplitude of the index grating. By using the definition of G, the coupled equations are written

$$\frac{d}{dz} A_1 = -\frac{1}{2} \Gamma G A_2 \tag{7.1-34}$$

$$\frac{d}{dz} A_2 = \frac{1}{2} \Gamma^* G^* A_1 \tag{7.1-35}$$

$$\frac{d}{dz} A_3 = -\frac{1}{2} \Gamma G A_4 \tag{7.1-36}$$

$$\frac{d}{dz} A_4 = \frac{1}{2} \Gamma^* G^* A_3 \tag{7.1-37}$$

Using the coupled equations, we obtain

$$\frac{d}{dz} G = \frac{1}{2} \Gamma G (I_1 - I_2 + I_3 - I_4)/I_0 \tag{7.1-38}$$

Since all the intensities are real, we conclude that the phase of G is a constant provided Γ is real. In fact, G can be written

$$\ln G = \ln G(0) + \int_0^z \frac{1}{2I_0} \Gamma (I_1 - I_2 + I_3 - I_4)\, dz' \tag{7.1-39}$$

where $G(0)$ is the value of G at $z = 0$. We note that the second term in the above equation is always real. Thus, the phases of G and $G(0)$ are identical. If we further define

$$F = \frac{I_1 - I_2 + I_3 - I_4}{I_0} \tag{7.1-40}$$

we obtain

$$\frac{d}{dz} |G|^2 = \frac{1}{2} (\Gamma + \Gamma^*) |G|^2 F \tag{7.1-41}$$

$$\frac{d}{dz} F = -(\Gamma + \Gamma^*) |G|^2 \tag{7.1-42}$$

These two equations can be easily integrated. First, we note that

$$F^2 + 4|G|^2 = C^2 \tag{7.1-43}$$

where C is a constant determined by the boundary conditions at $z = 0$,

$$C^2 = F^2(0) + 4|G(0)|^2 \tag{7.1-44}$$

By eliminating $|G|^2$ from Eqs. (7.1-42) and (7.1-43), we obtain

$$\frac{d}{dz} F = -\frac{\gamma}{2}(C^2 - F^2) \tag{7.1-45}$$

where γ is the real part of Γ. An integration of the above equation yields

$$F(z) = -C \tanh\left[\frac{C}{2}\gamma z - D\right] \tag{7.1-46}$$

where D is a constant given by

$$D = \tanh^{-1} \frac{F(0)}{C} \tag{7.1-47}$$

By substituting Eq. (7.1-46) for F in Eq. (7.1-43), we obtain

$$|G| = \frac{C}{2\cosh\left[\frac{C}{2}\gamma z - D\right]} \tag{7.1-48}$$

Once $|G|$ is obtained, the diffraction efficiency can be evaluated by using the method of grating integral described in Section 2.7. With the boundary condition $A_2(0) = 0$ and that Γ is real, the diffraction efficiency is given, according to Eq. (2.7-9), by

$$\eta = \sin^2 \int_0^L \frac{1}{2}\gamma|G|\, dz \tag{7.1-49}$$

The integral in the above equation can be obtained by using the following formula:

$$\int \frac{dx}{\cosh x} = 2\tan^{-1} e^x \tag{7.1-50}$$

Using Eq. (7.1-48) for $|G|$, we obtain an expression for the diffraction efficiency

$$\eta = \sin^2\left[\tan^{-1} \exp\left(\frac{C}{2}\gamma L - D\right) - \tan^{-1} e^{-D}\right] \tag{7.1-51}$$

Similar expressions were derived earlier in Chapter 4 (see Eqs. (4.7-32) and (4.8-30)). This expression is useful in evaluating the diffraction efficiency of a "soft" grating which is formed by two-beam interference in a photorefractive medium. It is important to note that the diffraction efficiency depends on γL as well as the constants C and D which are functions of the input intensities at $z = 0$. To illustrate the use of this equation, we consider the following examples.

Example Consider the recording of a photorefractive grating in a crystal with $\gamma L = 10$ by using a pair of beams with intensities $I_3(0) = I_4(0) = 1\ \text{W/cm}^2$. Let the read beam be $I_1(0) = 1\ \text{W/cm}^2$ and $I_2(0) = 0$. The constants are $F(0) = 0.33$, $C = 0.75$, $D = 0.48$, according to Eqs. (7.1-40), (7.1-44), and (7.1-47). Substituting these constants in Eq. (7.1-51), we obtain a diffraction efficiency of $\eta = 0.72$.

Example Consider a similar example with a read beam intensity of $I_1(0) = 1\ \text{mW/cm}^2$. These constants are $F(0) = 0.5 \times 10^{-3}$, $C = 0.67$, $D = 7.5 \times 10^{-4}$, according to Eqs. (7.1-40), (7.1-44), and (7.1-47). Substituting these constants in Eq. (7.1-51), we obtain a diffraction efficiency of $\eta = 0.46$.

Example Consider a similar example with a read beam incident from the other port. The intensities are $I_1(0) = 0$ and $I_2(0) = 1\ \text{W/cm}^2$. The constants are $F(0) = -0.33$, $C = 0.75$, $D = -0.48$, according to Eqs. (7.1-40), (7.1-44), and (7.1-47). Substituting these constants in Eq. (7.1-51), we obtain a diffraction efficiency of $\eta = 0.28$.

Based on these examples, we find that the diffraction efficiency of a "soft" grating depends on the intensity of the read beam. By examining the expression for the diffraction efficiency, we find that maximum diffraction efficiency of 100% is achievable with a large coupling constant γL such that

$$\exp\left[\frac{C}{2}\gamma L\right] \gg \exp(D) \gg 1 \tag{7.1-52}$$

To achieve the condition of $\exp(D) \gg 1$, the constant $F(0)/C$ has to be close to unity, according to Eq. (7.1-47). This requires that the grating interference function $G(0)$ at $z = 0$ be relatively small, according to Eq. (7.1-44). Physically, a small interference function means a large beam ratio (i.e., $I_3(0) \gg I_4(0)$ or $I_3(0) \ll I_4(0)$). This is illustrated in the following example.

Example Consider the recording of a photorefractive grating in a crystal with $\gamma L = 10$ by using a pair of beams with intensities $I_{30} = 1\ \text{W/cm}^2$, $I_{40} = 1\ \text{mW/cm}^2$. Let the read beam be $I_{10} = 1\ \text{mW/cm}^2$ and $I_{20} = 0$. The constants are $F(0) = 0.998$, $C = 0.999998$, $D = 3.45$, according to Eqs. (7.1-40), (7.1-44), and (7.1-47). This leads to $\exp(\frac{1}{2}C\gamma L) = 148.41$, $\exp(D) = 31.65$, and a diffraction efficiency of $\eta = 0.94$.

If we examine Eq. (7.1-51), we find that the diffraction efficiency is an increasing function of γL. In the limit when γL is very large such that $(\frac{1}{2}C\gamma L - D) \gg 1$, we have

$$\eta = \cos^2[\tan^{-1} e^{-D}] = \tfrac{1}{2}(1 + \tanh D) \tag{7.1-53}$$

Using Eqs. (7.1-47), (7.1-44), (7.1-40), and the boundary condition of $I_{20} = 0$, we obtain

$$\eta = \frac{1}{2}\left[1 + \frac{I_{10} + (I_{30} - I_{40})}{[(I_{10} + I_{30} - I_{40})^2 + 4I_{30}I_{40}]^{1/2}}\right] \tag{7.1-54}$$

where $I_{10} = I_1(0)$, $I_{30} = I_3(0)$, $I_{40} = I_4(0)$ are intensities at $z = 0$. For equal intensity of the write beams, the expression reduces to

$$\eta = \frac{1}{2}\left[1 + \frac{I_{10}}{(I_{10}^2 + 4I_{30}^2)^{1/2}}\right] \tag{7.1-55}$$

where $I_{30} = I_{40}$ is the write intensity. Although these expressions show explicit intensity dependence, they are valid only when $(\frac{1}{2}C\gamma L - D) \gg 1$ which may not be valid in some ranges of intensity.

To illustrate the nonreciprocal diffraction, let us consider the case of equal intensity of the write beams. The constant D can be written

$$D = \tanh^{-1} \frac{(I_{10} - I_{20})}{[(I_{10} - I_{20})^2 + 4I_{30}I_{40}]^{1/2}} \tag{7.1-56}$$

For reading with beam 1 (i.e., $I_{20} = 0$), the constant D is written

$$D = \tanh^{-1} \frac{I}{(I^2 + 4I_w^2)^{1/2}} \tag{7.1-57}$$

where I_w is the write intensity and I is the read intensity. On the other hand, if the reading beam is beam 2 with $I_{10} = 0$, then

$$D = \tanh^{-1} \frac{-I}{(I^2 + 4I_w^2)^{1/2}} \tag{7.1-58}$$

We note that the constant D becomes negative. This leads to a different diffraction coefficient. The nonreciprocity is a result of the relative phase between $A_3A_4^*$ and the self-induced grating $A_1A_2^*$. Depending on the port of incidence, the self-induced grating $A_1A_2^*$ will be either exactly in phase or exactly π out of phase with $A_3A_4^*$. For equal intensity of the write beams, changing the port of incidence is equivalent to reversing the sign of γ.

Figure 7.2 is a plot of the diffraction efficiency as a function of the read beam intensity, as given by Eq. (7.1-51). The intensity of the write beams is

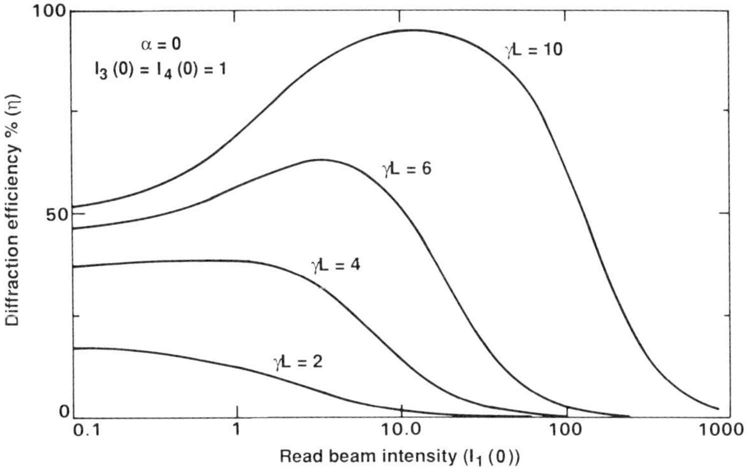

Figure 7.2. Diffraction efficiency η as a function of read beam intensity $I_1(0)$ for various coupling strengths γL. The normalized, equal intensities of the write beam is unity: $I_3(0) = I_4(0) = 1$.

taken to be unity, $I_3(0) = I_4(0) = 1$, and the absorption losses are assumed to be negligible, $\alpha = 0$. For small coupling strengths ($\frac{1}{2}\gamma L \approx 1$) and weak read beam ($I_1(0) < I_3(0)$, $I_4(0)$), the diffraction efficiency is independent of the read beam intensity. Increase of the read beam intensity above the write beam intensities serves to erase the grating being read and the diffraction efficiency falls to zero. For strong coupling, the behavior of η is different; the diffraction efficiency is ~50% for weak read beams, and increases as the read beam becomes stronger than the write beams, reaches a broad maximum and eventually falls to zero when $I_1(0) \gg I_3(0)$, $I_4(0)$.

To understand the nonlinear behavior of the diffraction efficiency with the read beam intensity that occurs at strong coupling, we have plotted the intensities of the four beams as a function of the distance propagated in the medium in Figures 7.3(a–d). The absorption losses are again assumed to be negligible ($\alpha = 0$), and the coupling is assumed to be strong ($\gamma = 10\,\text{cm}^{-1}$, $L = 1$ cm). As in Figure 7.2, the intensity of the write beams (solid curves) is taken to be unity, and the intensities of the read beam and diffracted beam (dashed curves) are normalized by the read beam intensity at the input, $I_1(0)$, with $I_2(0) = 0$. For these input intensities, grating $A_3A_4^*$ has its maximum amplitude near the input plane, while grating $A_1A_2^*$ is zero. For strong coupling, there can be a complete energy transfer between the write beams, and when this happens, the grating formed by the write beams reduces to zero. This particular propagation distance is a sensitive function of the read beam intensity. For a weak read beam which does not perturb the grating written by beams 3 and 4, total energy transfer requires long interaction lengths, as is typical of two-beam coupling in photorefractive media, and is shown in Figure 7.3(a) for $I_1(0) = 0.01$. As scattering of read beam I_1 off grating $A_3A_4^*$ generates beam I_2, grating $A_1A_2^*$ slowly builds up. An examination of the curves reveals that near the exit plane $z = L$, grating $A_3A_4^*$ is zero and only grating $A_1A_2^*$ is finite, but it is very weak due to the cross-washout factor I_0 in the denominator. Hence the maximum diffraction efficiency of the read beam is 50%. As the intensity of the read beam

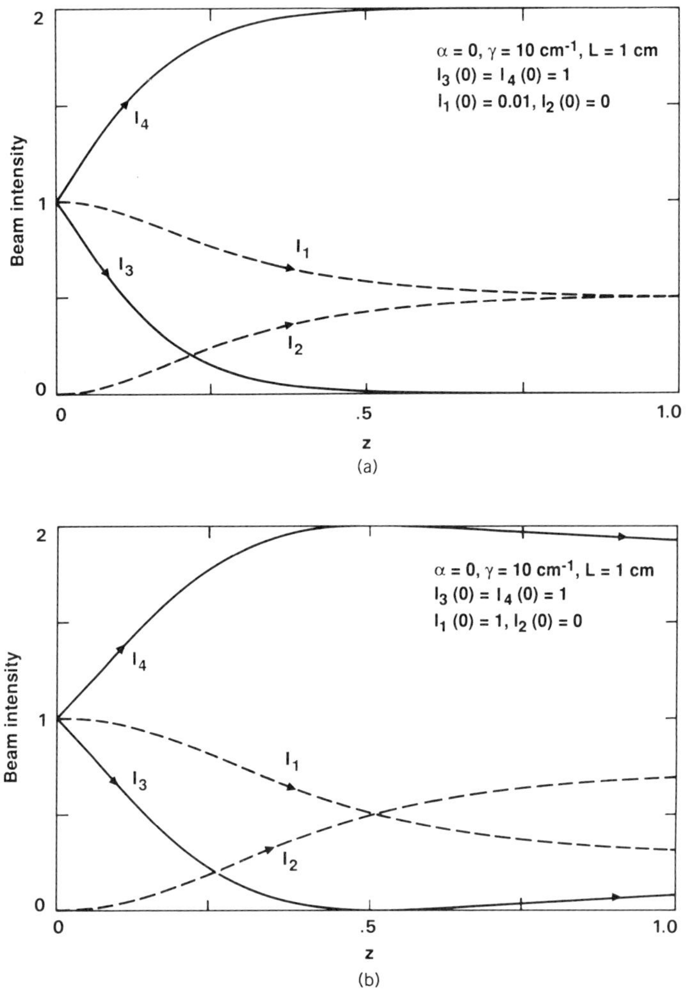

Figure 7.3. Intensities $I(z)$ of the four beams as a function of the depth z within the medium. The absorption coefficient α is taken to be zero, the coupling coefficient γ is equal to $10\,\text{cm}^{-1}$, and the interaction length L is 1 cm. The normalized, equal intensities of the write beams (shown as solid curves) is unity: $I_3(0) = I_4(0) = 1$, and $I_2(0) = 0$. The read beam $I_1(0)$ has the following intensity corresponding to each figure: (a) 0.01, (b) 1, (c) 10, (d) 100.

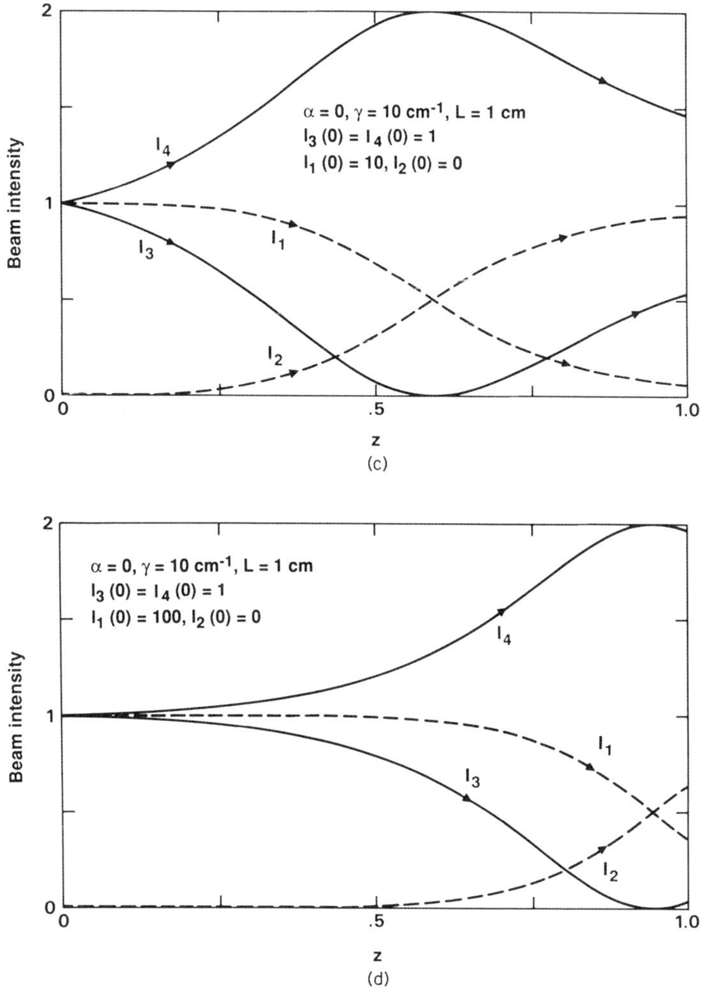

Figure 7.3. (*Continued*)

increases, grating $A_1A_2^*$ can build up to a magnitude comparable to grating $A_3A_4^*$ and since the two gratings are in phase, the overall shared grating strength increases, leading to a stronger coupling of the beams in a given pair. The propagation distance for total depletion of $I_3(0)$ decreases as $I_1(0)$ is increased, being approximately half the interaction length when all the three input beams have equal intensities (Figure 7.3(b)). Beyond this distance, scattering of I_4 off grating $A_1A_2^*$ begins to restore the depleted beam I_3, and grating $A_3A_4^*$ is rebuilt. Although these two gratings are π out of phase, the grating integral is an increasing function of distance. This leads to a larger diffraction efficiency. Since scattering of I_4 increases with the intensity of the read beam, grating $A_3A_4^*$ rebuilds to a bigger amplitude with increasing I_3, hence higher η (Figure 7.3(c)). However, if $I_1(0) \gg I_3(0)$, $I_4(0)$ (Figure 7.3(d)), then grating $A_3A_4^*$ is very small near the input plane, and grating $A_1A_2^*$ is zero. Consequently, depletion of I_3 occurs very slowly, almost near the end of the interaction region, so that grating $A_3A_4^*$ no longer has a chance to rebuild. Once again only a single grating, $A_1A_2^*$, in the overall shared grating near the exit plane, leading to a decrease in η. For still higher read beam intensities, grating $A_3A_4^*$ becomes very small near the input plane so that diffraction of the read beam becomes negligible throughout the interaction region and η reduces to zero. The nonlinear behavior of the diffraction efficiency with read beam intensity arises because of the photorefractive nature of the index gratings; the grating depends on the local beam ratios and not on the absolute intensities of the beams.

Figure 7.4 is the same set of curves as in Figure 7.2 but for negative coupling constant γ, which corresponds to readout from the other port provided $I_3(0) = I_4(0)$. In this case, the two gratings are out of phase, so that there is no enhancement of the diffraction efficiency for any intensity of the read beam. As predicted by Eq. (7.1-55), the maximum diffraction efficiency is only 50% at large coupling strengths.

Figure 7.5 is a plot of the diffraction efficiency η versus the normalized coupling strength γL for negligible absorption and equal intensity of write beams $I_3(0) = I_4(0) = 1$. The asymmetry in η is negligible for a weak read beam and becomes appreciable only when $I_1(0) > I_3(0)$, $I_4(0)$.

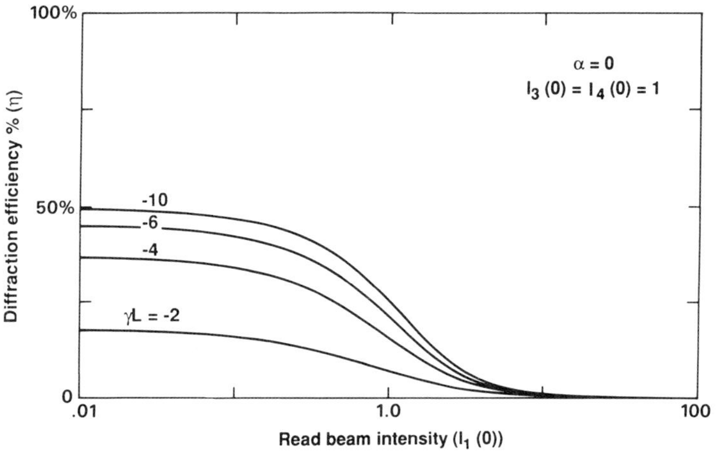

Figure 7.4. Diffraction efficiency η as a function of read beam intensity $I_1(0)$ for various negative coupling strengths γL. The normalized equal intensities of the write beams is unity: $I_3(0) = I_4(0) = 1$.

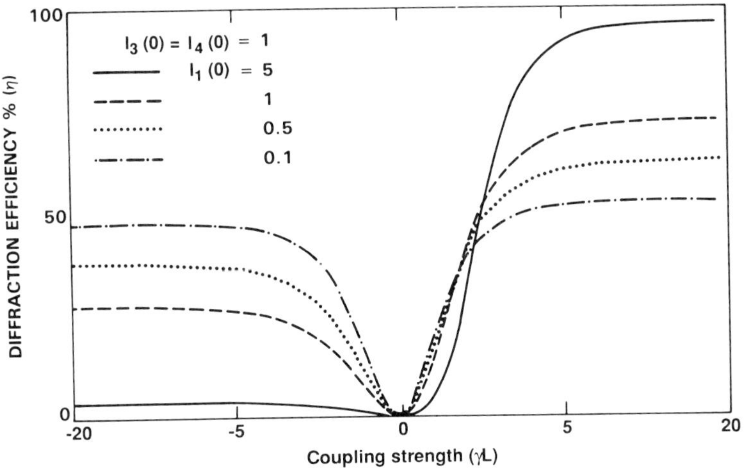

Figure 7.5. Diffraction efficiency η as a function of coupling strength γL for various read beam intensities $I_1(0)$. The normalized, equal intensities of the write beams is unity: $I_3(0) = I_4(0) = 1$, and absorptive losses are assumed to be negligible ($\alpha = 0$).

7.2 FIXED GRATINGS

According to the discussions in the previous chapters, we know that photorefractive materials such as $BaTiO_3$, $LiNbO_3$, $Sr_{1-x}Ba_xNb_2O_6$, BSO, are, by far, the most efficient media for the recording of dynamic holograms. In these media, information can be stored, retrieved, and erased by the illumination of light. In addition to the holographic properties, energy coupling occurs between the recording beams and also between the reading beam and the scattered beam. This is the two-wave mixing discussed in Chapter 4 (TWM) [1]. The unique properties of beam coupling and high diffraction efficiency can be employed for many applications in optical computing [2] including information storage [3], image processing [4, 5], reconfigurable optical interconnections [6, 7], and implementation of neural networks [8–10]. Some of these applications will be addressed later in the book.

While the holograms in photorefractive media can be manipulated dynamically, the stored holograms are partially erased by the reading beam each time the information is retrieved. There are situations where it is desirable to have photorefractive holograms that can be fixed by physical processes. These "fixed" holograms are not subject to erasure during read-out [11–15]. For example, it is desirable to fix the photorefractive holograms at the end of the learning cycles in optical perceptrons [16–19]. After the fixing, a trained perceptron will perform pattern classification without memory fading. It is important to note that energy coupling between the reading beam and the scattered beam occurs due to not only the fixed grating but also the photorefractive grating formed by these two beams. The diffraction is therefore different from that of a simple Bragg scattering. Previously, Bragg diffraction from fixed gratings and photorefractive beam coupling have been treated separately [20, 21]. In this section, we investigate the diffraction properties of these fixed holograms in photorefractive media taking into account the effect of the photo-induced index gratings.

Consider a fixed volume grating inside a photorefractive medium with a well defined grating wave vector of **K**. A plane wave is incident upon the fixed volume grating along a direction which satisfies the Bragg condition (see Figure 7.6). As a result of the Bragg scattering, a plane wave along a new direction is generated. These two plane waves can be written as

$$E_j = A_j \exp[i(\omega t - \mathbf{k}_j \cdot \mathbf{r})] \tag{7.2-1}$$

where $j = 1, 2$, ω is their common frequency, $\mathbf{k}_j$ is the corresponding wave vector. Conservation of momentum requires that

$$\mathbf{k}_2 = \mathbf{k}_1 \pm \mathbf{K} \tag{7.2-2}$$

It is important to note that the photorefractive grating induced by the interference intensity pattern, formed by the incident beam and the scattered beam, has exactly the same grating wave vector **K**. The photo-induced grating is either in phase or 180° out of phase with the fixed grating (assuming a phase grating) [22] depending on the direction of incidence. This leads to a strengthened or weakened grating and thus nonreciprocal

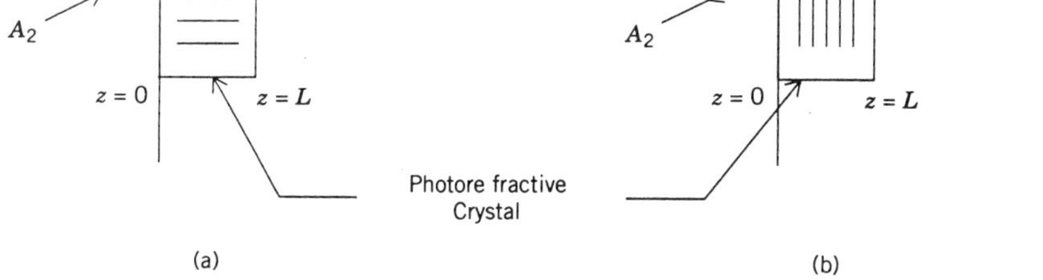

Figure 7.6. Schematic drawings of the incident and the scattered beams in the case of (a) transmission gratings and (b) reflection gratings.

diffraction. The nonreciprocity has been observed in thick color center holograms [23]. A similar effect has also been analyzed and observed in Kerr media [24, 25].

The total index grating can be written as

$$n = n_0 + n_p \cos \mathbf{K} \cdot \mathbf{r} - in_a(1 + \cos \mathbf{K} \cdot \mathbf{r}) + \left[\frac{n_1}{2} e^{i\phi} \frac{A_1^* A_2}{I_0} e^{-i\mathbf{K} \cdot \mathbf{r}} + \text{c.c.} \right] \tag{7.2-3}$$

where n_0 is the index of refraction in the absence of gratings, n_p and n_a represent amplitude of the fixed phase and absorption gratings respectively, n_1 represents the amplitude of the index grating formed by the incident wave and the diffracted wave via photorefractive effect, ϕ is the spatial phase shift between the index grating and the interference pattern, and I_0 is the sum of intensities of the two waves

$$I_0 = |A_1|^2 + |A_2|^2 \tag{7.2-4}$$

All these amplitudes n_p, n_a, n_1 are real numbers. In photorefractive crystals where diffusion is the dominant mechanism for charge transportation, ϕ is either $\pi/2$ or $-\pi/2$ depending on the direction of the c-axis.

The coupled mode equations can be derived by substituting Eq. (7.2-3) for the index of refraction into the wave equation. For simplicity, we assume the grating wave vector is along the x-direction, the angle between the two waves is 2θ, and the z-component of $\mathbf{k}_j$ is β_j. Using the slow-varying amplitude approximation,

$$\left| \frac{d^2}{dz^2} A_j \right| \ll \left| \beta_j \frac{d}{dz} A_j \right| \tag{7.2-5}$$

and the equality

$$|\mathbf{k}_j|^2 = \frac{\omega^2}{c^2} n_0^2 \tag{7.2-6}$$

we can write the coupled mode equations as

$$2i\beta_1 \frac{d}{dz} A_1 = \frac{\omega^2}{c^2} \left[n_0 n_1 e^{-i\phi} \frac{|A_2|^2}{I_0} A_1 + (n_0 n_p - in_0 n_a) A_2 - 2in_0 n_a A_1 \right] \tag{7.2-7}$$

$$2i\beta_2 \frac{d}{dz} A_2 = \frac{\omega^2}{c^2} \left[n_0 n_1 e^{i\phi} \frac{|A_1|^2}{I_0} A_2 + (n_0 n_p - in_0 n_a) A_1 - 2in_0 n_a A_2 \right] \tag{7.2-8}$$

where c is the speed of light in a vacuum. These two equations will be used in the following for various interaction configurations.

According to Eq. (7.2-3), fixed gratings are represented by the second and third terms. The second term $n_p \cos \mathbf{K} \cdot \mathbf{r}$ is a pure phase grating, while the third term $-in_a(1 + \cos \mathbf{K} \cdot \mathbf{r})$ represents an absorption grating. The fourth term, which is dependent on the product of the two-wave amplitudes, represents the photo-induced grating. The coupling of these two waves depends on their directions of propagation. The diffraction properties of fixed gratings in photorefractive media are thus divided into the following categories.

7.2.1 Transmission Phase Gratings ($n_a = 0$, $\beta_1 \beta_2 > 0$)

For transmission gratings, the z-component of the two-wave vectors are of the same sign. Without loss of generality, we take

$$\beta_1 = \beta_2 = \frac{2\pi}{\lambda} n_0 \cos \theta \tag{7.2-9}$$

Using Eq. (7.2-9), the coupled mode equations can be written in the form

$$\frac{d}{dz} A_1 = -\frac{\Gamma}{2} \frac{|A_2|^2}{I_0} A_1 - i\left(\kappa - i\frac{\alpha}{4}\right) A_2 - \frac{\alpha}{2} A_1 \tag{7.2-10}$$

$$\frac{d}{dz} A_2 = \frac{\Gamma^*}{2} \frac{|A_1|^2}{I_0} A_2 - i\left(\kappa - i\frac{\alpha}{4}\right) A_1 - \frac{\alpha}{2} A_2 \tag{7.2-11}$$

where Γ is the complex photorefractive coupling coefficient for TWM, $(\kappa - i\alpha/4)$ is the complex Bragg coupling coefficient, and α is the absorption constant due to the absorption grating,

$$\Gamma = i \frac{2\pi n_1}{\lambda \cos \theta} e^{-i\phi} \tag{7.2-12}$$

$$\kappa = \frac{\pi n_p}{\lambda \cos \theta} \tag{7.2-13}$$

and

$$\alpha = \frac{4\pi n_a}{\lambda \cos \theta} \tag{7.2-14}$$

These coupled equations (7.2-10) and (7.2-11) can be written in terms of intensities and phases of the two waves. We write

$$A_1 = \sqrt{I_1} e^{-i\psi_1} \tag{7.2-15}$$

$$A_2 = \sqrt{I_2} e^{-i\psi_2} \tag{7.2-16}$$

and assume that the medium is lossless, that is, pure phase grating with $n_a = \alpha = 0$. Equations (7.2-10) and (7.2-11) can be transferred into

$$\frac{d}{dz} I_1 = -\gamma \frac{I_1 I_2}{I_1 + I_2} - 2\kappa \sin \Delta\psi \sqrt{I_1 I_2} \tag{7.2-17}$$

$$\frac{d}{dz} I_2 = \gamma \frac{I_1 I_2}{I_1 + I_2} + 2\kappa \sin \Delta\psi \sqrt{I_1 I_2} \tag{7.2-18}$$

and

$$\frac{d}{dz} \psi_1 = \beta \frac{I_2}{I_1 + I_2} + \kappa \cos \Delta\psi \sqrt{I_2 / I_1} \tag{7.2-19}$$

$$\frac{d}{dz} \psi_2 = \beta \frac{I_1}{I_1 + I_2} + \kappa \cos \Delta\psi \sqrt{I_1 / I_2} \tag{7.2-20}$$

where

$$\Delta\psi = \psi_2 - \psi_1 \tag{7.2-21}$$

and

$$\gamma = \frac{2\pi n_1}{\lambda \cos \theta} \sin \phi \tag{7.2-22}$$

$$\beta = \frac{\pi n_1}{\lambda \cos \theta} \cos \phi \tag{7.2-23}$$

Notice that γ and β are related to Γ by $\Gamma = \gamma + i2\beta$. In photorefractive crystals which operate by pure diffusion, $\beta = 0$. Since wave A_2 is generated by Bragg scattering of wave A_1 from the fixed grating, the phase of A_2 is automatically locked to that of A_1. Thus the possible solutions to Eq. (7.2-19) and Eq. (7.2-20) are

$$\Delta\psi = \frac{\pi}{2} \tag{7.2-24}$$

and

$$\Delta\psi = -\frac{\pi}{2} \tag{7.2-25}$$

Using the boundary condition $A_2(0) = 0$, we find that $\Delta\psi = \pi/2$ is a proper solution by examining the phase of A_2 in Eq. (7.2-11). Substituting $\Delta\psi = \pi/2$ into Eq. (7.2-17) and Eq. (7.2-18), we obtain

$$\frac{d}{dz} I_1 = -\gamma \frac{I_1 I_2}{I_1 + I_2} - 2\kappa \sqrt{I_1 I_2} \tag{7.2-26}$$

$$\frac{d}{dz} I_2 = \gamma \frac{I_1 I_2}{I_1 + I_2} + 2\kappa \sqrt{I_1 I_2} \tag{7.2-27}$$

Since $I_1 + I_2$ is a constant, Eqs. (7.2-26) and (7.2-27) are mathematically identical to those of nonlinear optical Bragg scattering and can be solved exactly [24]. The solutions, subject to the boundary condition $A_2(0) = 0$, are written

$$I_1(z) = I_0 \cos^2 u \tag{7.2-28}$$

$$I_2(z) = I_0 \sin^2 u \tag{7.2-29}$$

where I_0 is the incident intensity, and u is given by [24]

$$\tan u = \frac{\tan[\kappa z(1-b^2)^{1/2}]}{(1-b^2)^{1/2} - b \tan[\kappa z(1-b^2)^{1/2}]} \tag{7.2-30}$$

with

$$b = \frac{\gamma}{4\kappa} \tag{7.2-31}$$

The intensity of the diffracted wave as a function of z has been investigated in previous work [24] for various values of b. The parameter b is a measure of the strength of the photorefractive grating relative to the fixed grating. Equation (7.2-30) is valid for all values of b. When the magnitude of b becomes greater than 1 (i.e., $|b| > 1$), $(1-b^2)^{1/2}$ becomes $i(b^2-1)^{1/2}$ and $\tan \kappa z(1-b^2)^{1/2}$ becomes $i \tanh \kappa z(b^2-1)^{1/2}$. We also note that κ is a positive number as defined earlier. The photorefractive constant γ can be either positive or negative, depending on the orientation of the crystal.

We now examine the intensity variation with respect to z for various values of b. For $b > 1$, $I_2(z)$ reaches its maximum value I_0 (100% energy transfer) at distance z such that $\tanh[\kappa z(b^2-1)^{1/2}] = (b^2-1)^{1/2}/b$. Beyond this point, the intensity $I_2(z)$ decreases and reaches its asymptotic value of $I/\{2b[b-(b^2-1)^{1/2}]\}$ which becomes I_0 when b approaches infinity.

For $b = 1$, $I_2(z)$ reaches its maximum value I_0 at $z = 1/\kappa$. Beyond this point, the intensity $I_2(z)$ decreases and reaches its asymptotic value of $I_0/2$ at $z = \infty$.

For $0 \leq b < 1$, $I_2(z)$ is a periodic function of z with maximum value I_0 at points when $\kappa z(1-b^2)^{1/2} = (1-b^2)^{1/2}/b$. The minimum value of $I_2(z)$ is zero which occurs when $\tan[\kappa z(1-b^2)^{1/2}] = 0$. Notice that a maximum or minimum occurs when $I_1 I_2 = 0$.

For $-1 < b < 0$, $I_2(z)$ is also a periodic function of z with maximum value I_0 at points when $\tan[\kappa z(1-b^2)^{1/2}] = (1-b^2)^{1/2}/b$. Compared with the case $0 \leq b < 1$, we notice that it takes longer interaction length for I_2 to reach its maximum value because of the negative photorefractive coupling. The minimum value of $I_2(z)$ is zero which also occurs when $\tan[\kappa z(1-b^2)^{1/2}] = 0$.

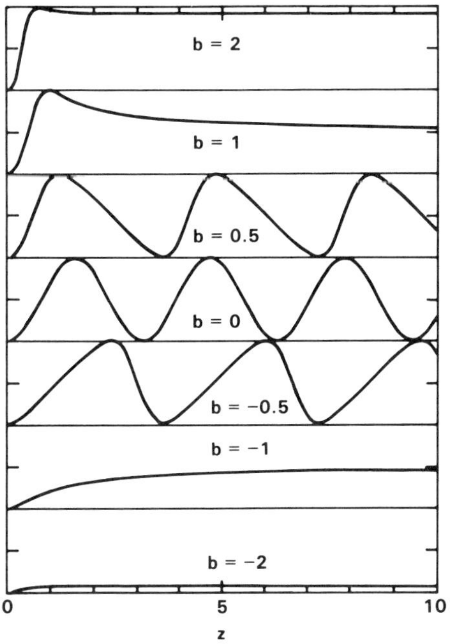

Figure 7.7. I_2 as a function of z for various values of b.

For $b = -1$, $I_2(z)$ is a monotonically increasing function of z with an asymptotic value of $I_2(z) = I_0/2$ at $z = \infty$. For $b < -1$, $I_2(z)$ is also a monotonically increasing function of z with an asymptotic value of $I_2(z) = I_0/[2b^2 - 2b(b^2 - 1)^{1/2}]$ at $z = \infty$. Figure 7.7 plots the intensity of $I_2(z)$ as a function z at various values of b.

The diffraction efficiency is a nonlinear function of b. For a crystal with interaction length L, the diffraction efficiency is

$$\eta = \frac{I_2(L)}{I_0} = \sin^2 u(z = L) \tag{7.2-32}$$

We notice that η approaches 100% at large b. When $b \gg 1$ and $\kappa L b \gg 1$, the asymptotic form of η can be written as

$$\eta = 1 - \frac{1}{4b^2}(1 - 4b^2 e^{-2\kappa L b})^2 = 1 - \frac{1}{4b^2}(1 - 8b^2 e^{-2\kappa L b}) \tag{7.2-33}$$

The asymptotic form for $b \to -\infty$ can be written as

$$\eta = \frac{1}{4b^2}\left(1 + \frac{1}{4b^2} - 2e^{-2\kappa L|b|}\right) \tag{7.2-34}$$

At $b = 0$, $u = \kappa z$, according to Eq. (7.2-30), the diffraction efficiency reduces to $\eta = \sin^2 \kappa L$. Figure 7.8 depicts the diffraction efficiency as a

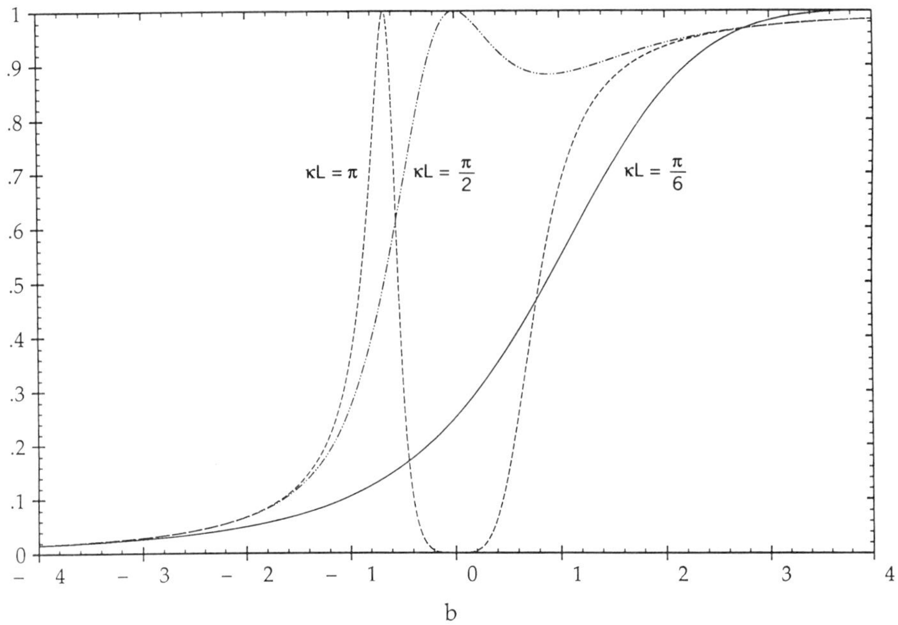

Figure 7.8. Diffraction efficiency as a function of b for transmission phase gratings with different values of κL. Solid curve: $\kappa L = \pi/6$. Double dot curve: $\kappa L = \pi/2$. Dashed curve: $\kappa L = \pi$.

function of b for different values of κL. For $b > 0$, the photorefractive grating is in phase with the fixed grating. When κ is small, the presence of the photo-induced grating strengthens the grating and the diffraction efficiency increases as a function of b. When $\kappa L = \pi/2$, the diffraction efficiency is 1 at $b = 0$ (no photo-induced grating). The diffraction efficiency decreases as b increases from $b = 0$, reaching a minimum at $b \approx 0.89$, and then increases to 1 when b is very large. When $\kappa L > \pi$, the diffraction efficiency will oscillate between 0 and 100% within $|b| < 1$. This oscillatory behavior can also be seen in Fig. 2 of Ref. [24], since I_1 and I_2 are periodic functions of z for $-1 < b < 1$. The sign of the parameter b depends on the direction of incidence. Thus if b is positive for an incident angle of θ, it will be negative when the beam is incident at an angle of $-\theta$. The different values of η for $+b$ and $-b$ reflect nonreciprocal diffractions due to TWM. In other words, the diffraction efficiency of fixed gratings in photorefractive media depends on the direction of incidence at Bragg angles.

Figure 7.9 shows the grating amplitude as a function of z. When $b > 0$, for example, $b = 1$, the photo-induced grating is in phase with the fixed grating, therefore, the grating is strengthened. When $b < 0$, for example, $b = -1$, the photo-induced grating is out of phase with the fixed grating, therefore, the grating is weakened. From the curve for $b = 10$, we notice

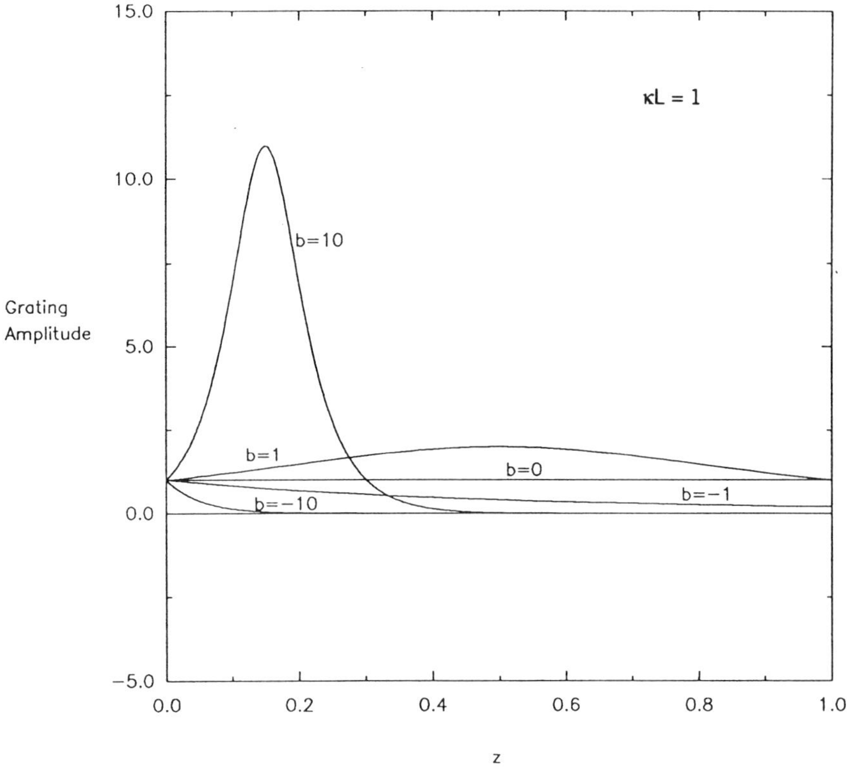

Figure 7.9. Grating amplitudes as functions of z for $b = 10, 1, 0, -1, -10$, in the case of transmission phase gratings.

that at the beginning, the photo-induced grating is in phase with the fixed grating, therefore, the grating is strengthened. At the point in space where all the incident power is converted into diffracted beam, that is, $I_2 = I_0$ and $I_1 = 0$, the photo-induced grating disappears. After this point, the photo-induced grating is out of phase with the fixed grating. Physically, this is equivalent to the case when the beam is incident from the other input port (i.e., at $-\theta$). When the photo-induced grating cancels the fixed grating, there is no more diffraction and thus the grating amplitude stays 0. The curve for $b = -10$ is the same as the part of that for $b = 10$ after the point where $I_1 = 0$.

7.2.2 Transmission Absorption Gratings ($n_p = 0$, $\beta_1\beta_2 > 0$)

Nonreciprocal diffractions can also occur in the case of absorption gratings. When $\alpha \neq 0$, the coupled mode equations Eq. (7.2-10) and Eq. (7.2-11) cannot be easily solved analytically, however, nonreciprocal behavior can still be seen in the numerical solutions. We can solve the coupled mode equations, Eqs. (7.2-10) and (7.2-11), using the fifth-order Runge–Kutta method with adaptive step size. Figure 7.10 shows the intensities of the incident and diffracted beams as functions of z for various values of $b = \gamma/\alpha$, where the fixed grating is a pure absorption grating ($\kappa = 0$). Similar to the phase gratings, diffraction from one input port ($b > 0$) is enhanced and that from the other input port ($b < 0$) is suppressed, as compared with the case when TWM disappears ($b = 0$). The diffraction efficiency as a function of b is plotted in Figure 7.11 for various values of αL. We note that the nonreciprocity in absorption gratings is quite strong.

For fixed gratings with both phase and absorption components, the diffraction property is a combination of that of the pure phase and pure absorption gratings. As an example, we give the numerical solutions for $\kappa = \cos(\pi/30)$ and $\alpha = 4\sin(\pi/30)$. Figure 7.12 shows the intensities of the incident and diffracted beams as functions of z for various values of

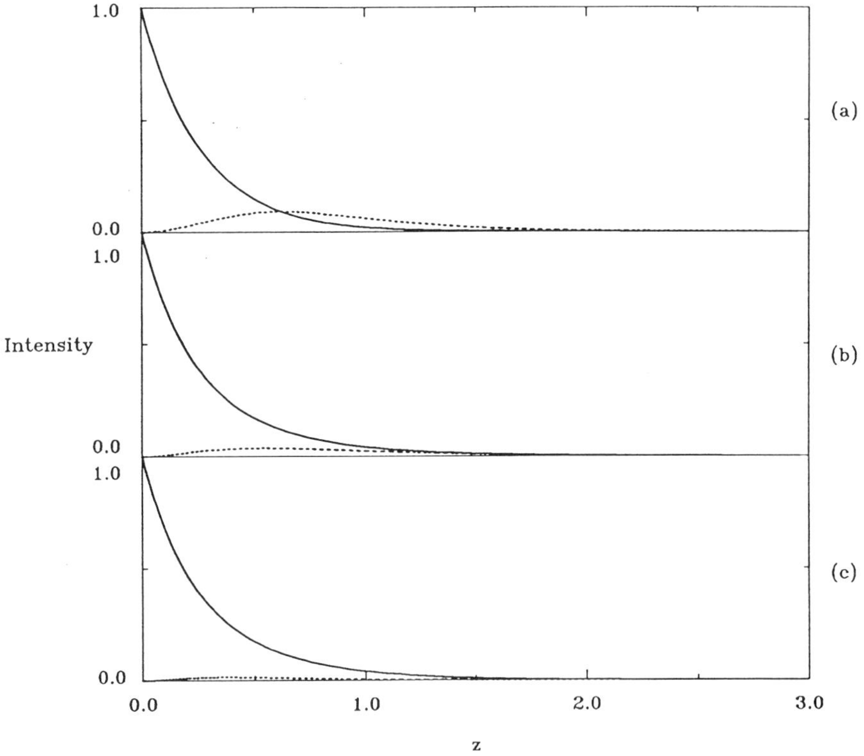

Figure 7.10. Intensities of the incident and diffracted beams as functions of z for transmission absorption gratings ($\kappa = 0$) with various values of b. Solid curve: I_1. Dotted curve: I_2. (a) $b = 1$, (b) $b = 0$, (c) $b = -1$.

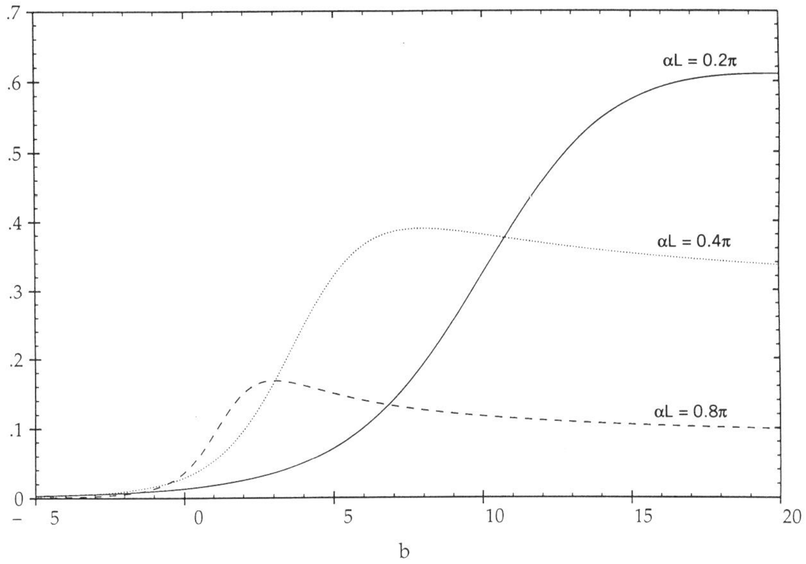

Figure 7.11. Diffraction efficiency as a function of b for transmission absorption gratings ($\kappa = 0$) with various values of αL. Solid curve: $\alpha L = 0.2\pi$. Dotted curve: $\alpha L = 0.4\pi$. Dashed curve: $\alpha L = 0.8\pi$.

$b = \gamma/[(4\kappa)^2 + \alpha^2]^{1/2}$, and Figure 7.13 shows the diffraction efficiency as a function of b for various values of L. In the numerical calculations, the units of κ, α and z are chosen arbitrarily, since κz and αz are dimensionless. Again, we note that the diffraction is nonreciprocal.

7.2.3 Reflection Phase Gratings ($n_a = 0$, $\beta_1\beta_2 > 0$)

For reflection gratings, the z-components of the two wave vectors are of opposite sign. They are given by

$$\beta_1 = -\beta_2 = \frac{2\pi}{\lambda} n_0 \cos\theta \tag{7.2-35}$$

Using Eqs. (7.2-7) and (7.2-8) the coupled mode equations can be written in the form

$$\frac{d}{dz} A_1 = -\frac{\Gamma}{2} \frac{|A_2|^2}{I_0} A_1 - i\left(\kappa - i\frac{\alpha}{4}\right) A_2 - \frac{\alpha}{2} A_1 \tag{7.2-36}$$

$$\frac{d}{dz} A_2 = -\frac{\Gamma^*}{2} \frac{|A_1|^2}{I_0} A_2 + i\left(\kappa - i\frac{\alpha}{4}\right) A_1 + \frac{\alpha}{2} A_2 \tag{7.2-37}$$

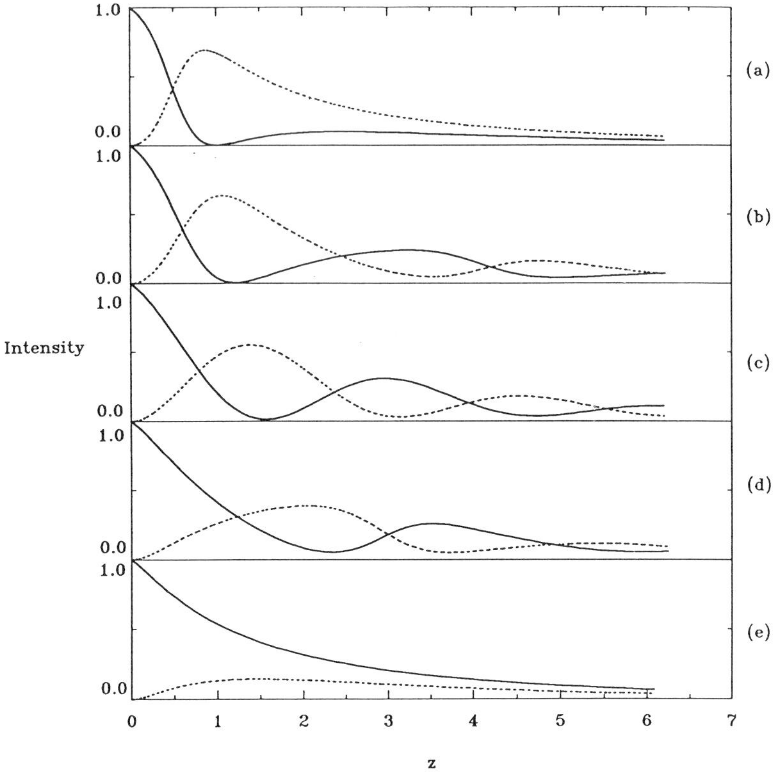

Figure 7.12. Intensities of the incident and diffracted beams as functions of z for fixed gratings with both phase and absorption components ($\kappa = \cos(\pi/30) = 0.9945$, $\alpha = 4\sin(\pi/30) = 0.4181$) and various values of b. Solid curve: I_1. Dotted curve: I_2. (a) $b = 1$, (b) $b = 0.5$, (c) $b = 0$, (d) $b = -0.5$, (e) $b = -1$. The units of κ, α, and z are chosen arbitrarily, since κz and αz are dimensionless.

Following similar steps as in transmission phase gratings, the coupled equations for the intensities of the two waves can be found as

$$\frac{d}{dz} I_1 = -\gamma \frac{I_1 I_2}{I_1 + I_2} - 2\kappa\sqrt{I_1 I_2} \tag{7.2-38}$$

$$\frac{d}{dz} I_2 = -\gamma \frac{I_1 I_2}{I_1 + I_2} - 2\kappa\sqrt{I_1 I_2} \tag{7.2-39}$$

In deriving Eqs. (7.2-38) and (7.2-39), we have used $\phi = \pi/2$ and $\Delta\psi = \pi/2$, which is one of the two solutions ($\Delta\psi = \pi/2$ or $\Delta\psi = -\pi/2$) for the equations of phase change.

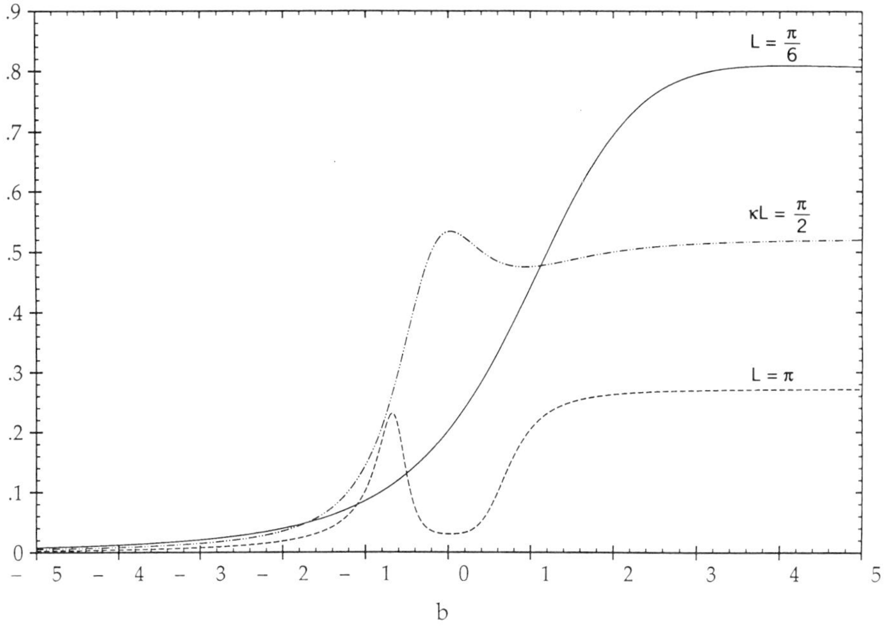

Figure 7.13. Diffraction efficiency as a function of b for fixed gratings with both phase and absorption components ($\kappa = \cos(\pi/30) = 0.9954$, $\alpha = 4\sin(\pi/30) = 0.4181$) and various values of L. Solid curve: $L = \pi/6$. Double dot curve: $L = \pi/2$. Dashed curve: $L = \pi$. The units of κ, α, and L are chosen arbitrarily, since κz and αz are dimensionless.

Equations (7.2-38) and (7.2-39) can be solved analytically. The solutions with the boundary condition $I_2 = 0$ at $z = L$ are given by

$$I_1(z) = C \cosh^2 u \qquad (7.2\text{-}40)$$

$$I_2(z) = C \sinh^2 u \qquad (7.2\text{-}41)$$

where C is the transmitted intensity at $z = L$, and u is related to z by

$$\frac{e^{2u}}{[\cosh(2u) + b\sinh(2u)]^b} = \exp[-2\kappa(1 - b^2)(z - L)] \qquad (7.2\text{-}42)$$

where b was given by Eq. (7.2-31). Equation (7.2-42) must be handled with care when $|b| = 1$. The solution for u as a function of z when $b = 1$ can be written

$$u = 2\kappa(L - z) - \tfrac{1}{4} + \tfrac{1}{4}e^{-4u} \qquad (7.2\text{-}43)$$

whereas when $b = -1$,

$$u = 2\kappa(L - z) + \tfrac{1}{4} - \tfrac{1}{4}e^{4u} \tag{7.2-44}$$

Figure 7.14 shows the intensities as functions of z for $b = 1$, $b = 0$, and $b = -1$. We note that backward coupling of energy is enhanced for the case when $b > 0$.

The diffraction efficiency is given by the expression

$$\eta = \frac{I_2(0)}{I_1(0)} = \tanh^2 u(z = 0) \tag{7.2-45}$$

Given a κL, $u(0)$ can be solved from Eq. (7.2-42). For example, when $b = 1$, $u(0)$ is given, according to Eq. (7.2-43), by

$$u = 2\kappa L - \tfrac{1}{4} + \tfrac{1}{4}e^{-4u} \tag{7.2-46}$$

We note that when $\kappa L > 1/8$, $u(0)$ is positive, and e^{-4u} becomes very small when $2\kappa L - 1/4 > 1/4$. Therefore when $\kappa L > 1/4$, $u(0) \approx 2\kappa L - 1/4$, and

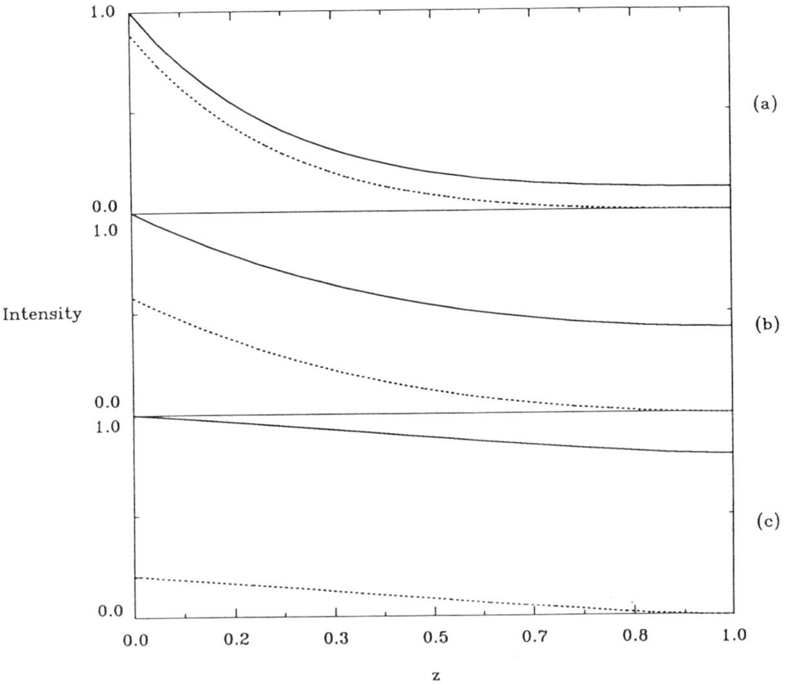

Figure 7.14. Intensities of incident (solid curve) and diffracted (dotted curve) beams as functions of z for reflection phase gratings ($\kappa = 1$, $\alpha = 0$). (a) $b = 1$, (b) $b = 0$, (c) $b = -1$.

$\eta \approx \tanh^2(2\kappa L - 1/4)$. The diffraction efficiency approaches 100% exponentially as $(2\kappa L - 1/4) \gg 1$, that is, $\kappa L \gg 5/8$. Figure 7.15(a) shows the diffraction efficiency as a function of κL for $b = 0$, ± 1, and also for $b = \pm 2$. When $b = 0$, $\eta = \tanh^2 \kappa L$. When $b \gg 1$ and $\kappa L b \gg 1$, the asymptotic expression for the diffraction efficiency is

$$\eta - 1 - 2be^{-2\kappa L b} \tag{7.2-47}$$

We notice that the diffraction efficiency approaches 100% exponentially as b increases. When $b \to -\infty$, the asymptotic expression for the diffraction efficiency is

$$\eta = \frac{1}{4b^2}\left(1 - 2e^{-2\kappa L|b|}\right) \tag{7.2-48}$$

As $2\kappa L|b| \gg 1$, $\eta = 1/4b^2$ for all κL. Figure 7.15(b) shows the diffraction efficiency as a function of b for various values of κL. Comparison of Figures 7.15 and 7.8 reveals that the diffraction is nonreciprocal for both transmission and reflection gratings. The oscillatory behavior in Figure 7.8 does not occur in Figure 7.15 because the intensities in Eqs. (7.2-40) and (7.2-41) are not periodic functions of z.

Figure 7.16 shows the grating amplitude as a function of z. Similar to co-directional diffractions, the photo-induced grating is in phase with the fixed grating when $b > 0$ and out of phase when $b < 0$, therefore, strengthens or weakens the grating, respectively. We notice from Figure 7.16 that the grating amplitude for $b = 1$ and that for $b = -1$ are not symmetric with respect to the value corresponding to $b = 0$. This is because the photo-induced grating is proportional to $|A_1 A_2|/(I_1 + I_2)$ which depends on the sign of b as well as its absolute value, as shown in Figure 7.14. This asymmetric behavior also occurs in the transmission phase gratings, as shown in Figure 7.9.

7.2.4 Reflection Absorption Gratings ($n_p = 0$, $\beta_1\beta_2 < 0$)

For absorption gratings, closed form solutions of the coupled equations are not easily available. The diffraction properties of these gratings can be studied using numerical techniques. We can solve the coupled mode equations, Eqs. (7.2-36) and (7.2-37), numerically with nonzero absorption constant. Since the photo-induced grating depends only on the modulation depth, Eqs. (7.2-36) and (7.2-37) are scaling invariant. In other words, if A_1 and A_2 are solutions to Eqs. (7.2-36) and (7.2-37), cA_1 and cA_2 are also solutions to these two equations, where c is a complex constant. Therefore, we can reduce the 2-point boundary value problem to a 1-point one. Numerical integration starts from $z = L$, where $A_2 = 0$ and A_1 is set to be 1. As we go from $z = L$ to $z = 0$, complex $A_1(z)$ and $A_2(z)$ are obtained. Then

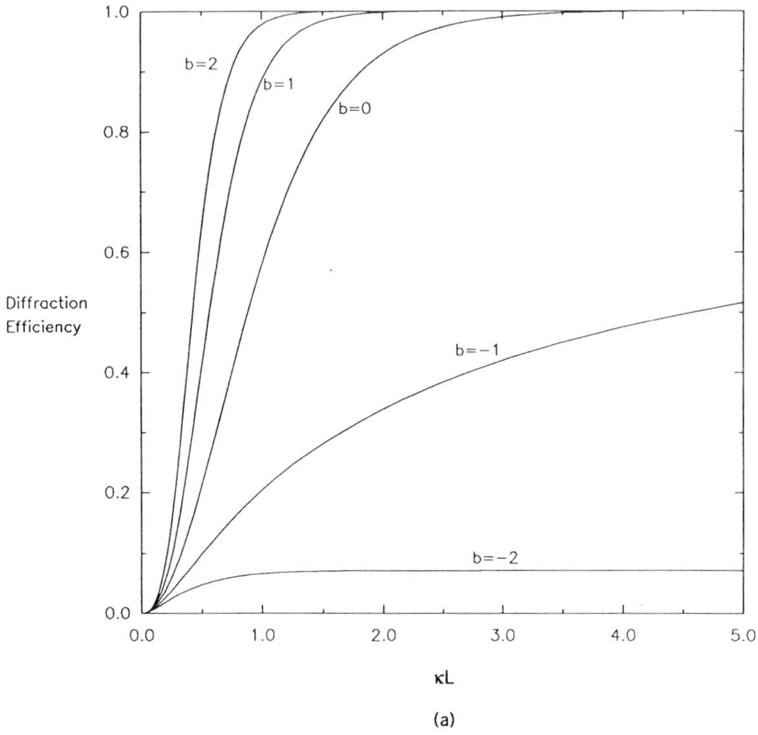

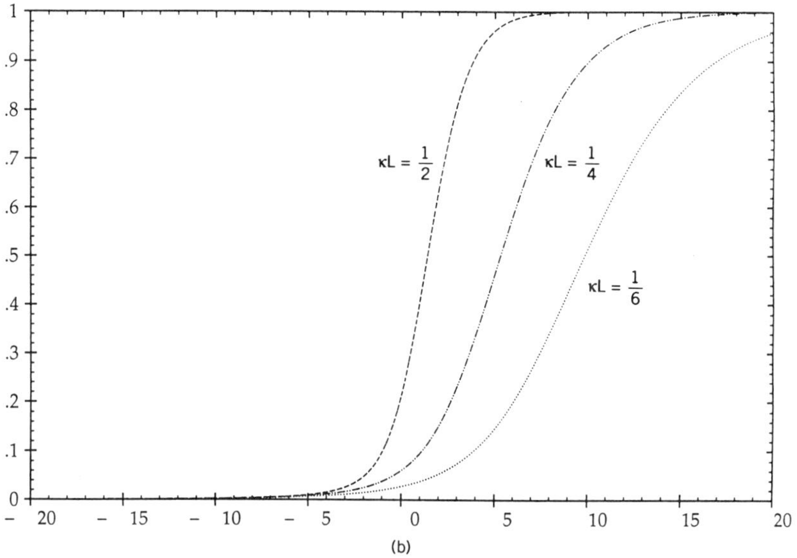

Figure 7.15. (a) Diffraction efficiency as a function of κL for $b = 0, \pm 1, \pm 2$. (b) Diffraction efficiency as a function of b for reflection phase gratings with various values of κL. Dotted curve: $\kappa L = 1/6$. Double dot curve: $\kappa L = 1/4$. Dashed curve: $\kappa L = 1/2$.

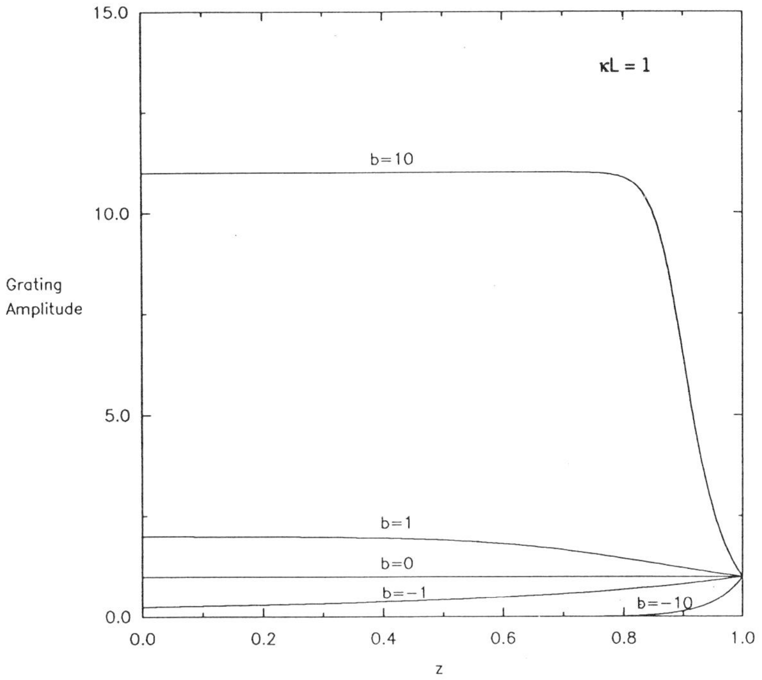

Figure 7.16. Grating amplitudes as functions of z for $b = 10, 1, 0, -1, -10$, in the case of reflection phase gratings.

the intensities $I_1(z)$ and $I_2(z)$ are normalized by taking the normalized $I_1(0)$ to be 1.

In the case of pure absorption grating ($\kappa = 0$), Figure 7.17 shows the intensities as functions of z. We notice that I_2 grows from 0 as the diffracted beam propagates from $z = L$ to $z = 0$. The growth is slower than that in the case of phase gratings due to the absorption. The incident beam I_1 decays as it propagates from $z = 0$ to $z = L$ due to both energy exchange with I_2 and absorption. Figure 7.18 shows the diffraction efficiency as a function of b for various values of αL. The diffraction property is similar to that in the absence of absorption (Figure 7.15), except that due to the loss in the medium, the diffraction efficiency cannot reach 100%. The nonreciprocity, in the case of absorption gratings, can be explained by the phase shift between the photo-induced phase grating and the interference pattern. The fixed absorption grating provides a source to generate the second beam when the incident beam satisfies the Bragg condition.

In summary, we have analyzed the diffraction properties of fixed gratings in photorefractive media with the consideration of TWM between the incident and the diffracted waves. Both co-directional and contra-directional

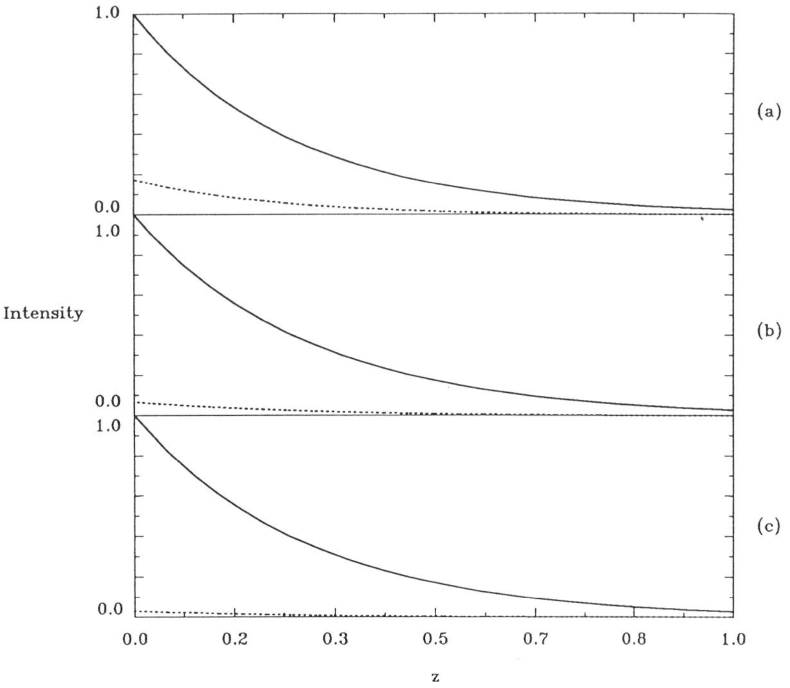

Figure 7.17. Intensities of incident (solid curve) and diffracted (dotted curve) beams as functions of z for reflection absorption gratings ($\kappa = 0$, $\alpha = 4$). (a) $b = 1$, (b) $b = 0$, (c) $b = -1$.

diffractions have been solved including the presence of both phase and absorption gratings. The results indicate that fixed gratings in photorefractive media exhibit strong optical nonreciprocity which cannot be explained without considering TWM. Since the photorefractive grating depends on the modulation depth instead of the intensities, there is no optical nonlinearity which appeared in the Kerr media [24].

In this section, we have considered only diffraction properties of fixed gratings with exact Bragg match and the photorefractive media are assumed to be in the regime of pure diffusion. Bragg mismatch ($\Delta\kappa \neq 0$) as well as arbitrary phase shift ($\phi \neq \pi/2$) are important issues that will affect the diffraction properties of fixed gratings in photorefractive media. These subjects are beyond the scope of the book.

7.3 RECIPROCITY IN PHOTOREFRACTIVE WAVE MIXING

Although we discussed nonreciprocal diffraction in this chapter, the nonreciprocity is a result of different gratings due to different writing conditions. Given a grating, the diffraction should be reciprocal. In this section

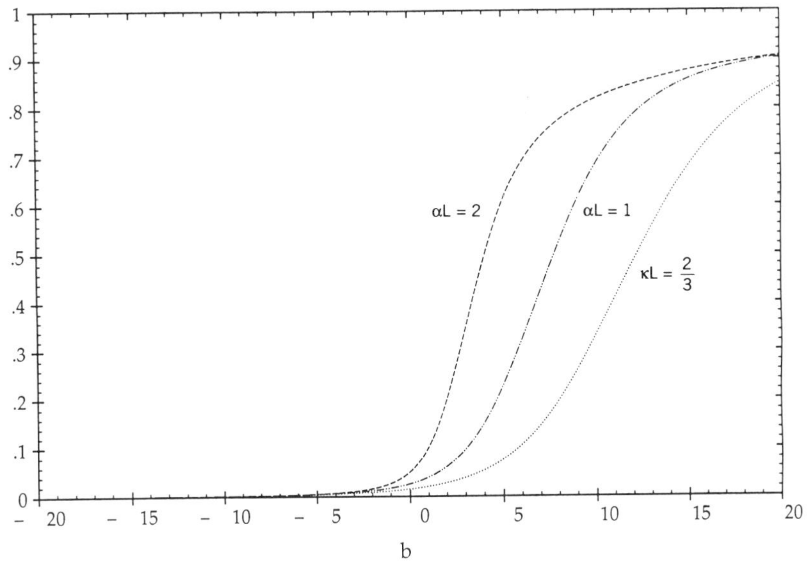

Figure 7.18. Diffraction efficiency as a function of b for reflection absorption gratings ($\kappa = 0$) with various values of αL. Dotted curve: $\alpha L = 2/3$. Double dot curve: $\alpha L = 1$. Dashed curve: $\alpha L = 2$.

we consider the validity of the reciprocity theorem as applied to photorefractive wave mixing. Some constants of integration in four-wave-mixing are physically interpreted as results of reciprocity.

Reciprocity is a fundamental law in both quantum physics [26, 27] and classical physics [28, 29]. In electromagnetic scattering, the reciprocity theorem can be proven with a similar approach as in [26]. For electromagnetic scattering by a static dielectric medium, the reciprocity relationship is written

$$S(\mathbf{k}, \mathbf{k}') = S(-\mathbf{k}', -\mathbf{k}) \tag{7.3-1}$$

where $S(\mathbf{k}, \mathbf{k}')$ is the amplitude scattering coefficient for incident plane wave with wave vector $\mathbf{k}$ and scattered plane wave with wave vector $\mathbf{k}'$. We note that $(-\mathbf{k}', -\mathbf{k})$ is the reversed scattering process of the direct scattering $(\mathbf{k}, \mathbf{k}')$, and the above reciprocity relationship is valid even when the medium has absorption loss.

In photorefractive wave mixing, however, since the index of refraction of the medium may depend on the incidence condition, application of the reciprocity theorem must be handled with care. Nonreciprocal diffractions have been reported in photorefractive two-wave mixing [30–33]. The seeming failure of the reciprocity law is a result of the fact that the photo-induced index variation when the incident wave is $\mathbf{k}_1$ and the diffracted wave is $\mathbf{k}_2$, is different from that when the incident wave is $-\mathbf{k}_2$ and the diffracted wave is $-\mathbf{k}_1$. As stated above, the reciprocity theorem is valid only when the index of refraction is not changing with time. In photorefractive four-wave mixing (FWM), four waves are present in the medium simultaneously and form two counterpropagating pairs, as shown in Figure 7.19. At steady state in degenerate FWM, the photo-induced index variation does not change with time. Therefore, we can apply the reciprocity theorem.

It is known that in photorefractive FWM via transmission grating as shown in Figure 7.19(a), there are several constants of integration (see Eqs. (4.5-11)–(4.5-13)). These are [34]

$$|A_1|^2 + |A_2|^2 = c_1 \tag{7.3-2}$$

$$|A_3|^2 + |A_4|^2 = c_2 \tag{7.3-3}$$

$$A_1A_4 + A_2A_3 = c \tag{7.3-4}$$

$$A_1A_3^* - A_2A_4^* = d \tag{7.3-5}$$

where c_1, c_2, c, and d are constants. The first two are results of conservation of energy. The other two constants have never been explained in terms of the existing principles of physics. In the discussion that follows, we point out that these two constants of integration are results of the reciprocity theorem.

Referring to Figure 7.19(a), the steady state relationship between the output and the input amplitudes of the four waves can be written formally as

$$\begin{bmatrix} A_{1L} \\ A_{2L} \\ A_{30} \\ A_{40} \end{bmatrix} = S \begin{bmatrix} A_{10} \\ A_{20} \\ A_{3L} \\ A_{4L} \end{bmatrix} \tag{7.3-6}$$

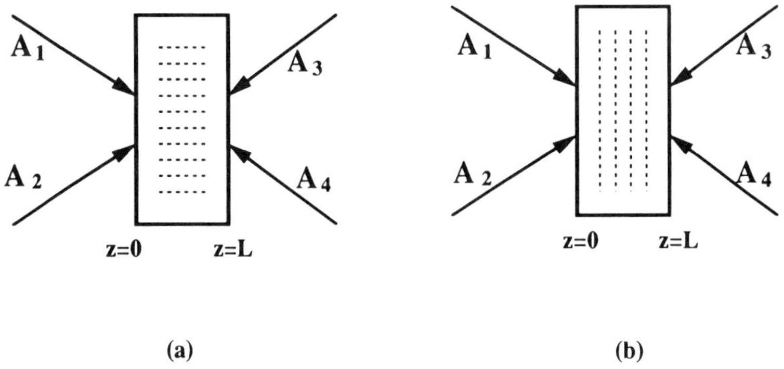

Figure 7.19. FWM in a photorefractive medium. (a) Transmission grating, (b) reflection grating.

where A_{i0} and A_{iL} are the amplitudes of the ith wave at $z = 0$ and $z = L$, respectively, and S is the scattering matrix. It is important to note that S depends on the steady state index grating $\Delta n(\mathbf{r})$ which depends on the incident amplitudes. According to the Bragg condition for the transmission grating, S can be written as

$$\begin{bmatrix} S_{11} & S_{12} & 0 & 0 \\ S_{21} & S_{22} & 0 & 0 \\ 0 & 0 & S_{33} & S_{34} \\ 0 & 0 & S_{43} & S_{44} \end{bmatrix} \tag{7.3-7}$$

Since the four waves are diffracted by the same steady state grating, we can use the reciprocity theorem, Eq. (7.3-1), and obtain

$$S_{11} = S_{44}, \qquad S_{12} = S_{34}, \qquad S_{21} = S_{43}, \qquad S_{22} = S_{33} \tag{7.3-8}$$

Substituting the above reciprocity relations into Eq. (7.3-6) and calculating $A_1A_4 + A_2A_3$ at $z = L$, we obtain

$$\begin{aligned} A_{1L}A_{4L} + A_{2L}A_{3L} &= (S_{11}A_{10} + S_{12}A_{20})A_{4L} + (S_{21}A_{10} + S_{22}A_{20})A_{3L} \\ &= A_{10}(S_{43}A_{3L} + S_{44}A_{4L}) + A_{20}(S_{33}A_{3L} + S_{34}A_{4L}) \\ &= A_{10}A_{40} + A_{20}A_{30} \end{aligned} \tag{7.3-9}$$

We note that the above derivation can be carried out at any z between 0 and L. Therefore, $A_1A_4 + A_2A_3$ is a constant of integration.

As an example, we apply the reciprocity theorem to mutually pumped phase conjugation (MPPC) with one interactive region [9]. At steady state, there exists a static index grating $\Delta n(\mathbf{r})$ in the medium. Referring to Figure 7.20, A_1 and A_3 can be viewed as scattered beams when the index grating is read by beams A_2 and A_4. Let the diffraction coefficients be defined as

$$\eta_{21} = \frac{A_{1L}}{A_{20}} \tag{7.3-10}$$

$$\eta_{43} = \frac{A_{30}}{A_{4L}} \tag{7.3-11}$$

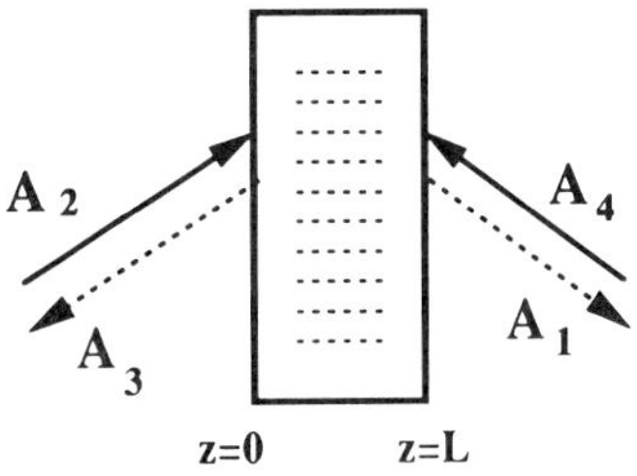

Figure 7.20. MPPC where A_2 and A_4 are pump beams, A_1 and A_3 are phase conjugate beams.

These two diffraction coefficients are identical according to the principle of reciprocity. It follows that $A_{1L}A_{4L} = A_{20}A_{30}$ and

$$\left|\frac{\rho_0}{\rho_L}\right|^2 = q^2 \tag{7.3-12}$$

where $|\rho_0|^2 = I_{30}/I_{20}$ and $|\rho_L|^2 = I_{1L}/I_{4L}$, $I_i = |A_i|^2$ is the intensity of the ith plane wave, and $q = I_{4L}/I_{20}$ is the pump intensity ratio. In addition, we obtain

$$|\rho_0 \rho_L|^2 = \eta^2 \leq 1 \tag{7.3-13}$$

where $\eta \equiv |\eta_{21}|^2 = |\eta_{43}|^2$. Notice that Eqs. (7.3-12) and (7.3-13) are general results for both real and complex coupling constants. Similar results can be derived for other configurations of MPPC [35].

We can also find the reciprocity relationship for FWM with the reflection grating, as shown in Figure 7.19(b). The scattering matrix and the reciprocity relations for the reflection grating are written as

$$\begin{bmatrix} S_{11} & 0 & S_{13} & 0 \\ 0 & S_{22} & 0 & S_{24} \\ S_{31} & 0 & S_{33} & 0 \\ 0 & S_{42} & 0 & S_{44} \end{bmatrix} \tag{7.3-14}$$

and

$$S_{11} = S_{44}\,, \qquad S_{13} = S_{24}\,, \qquad S_{31} = S_{42}\,, \qquad S_{33} = S_{22} \tag{7.3-15}$$

respectively. Following similar steps as in Eq. (7.3-9), we obtain

$$\begin{aligned} A_{10}A_{40} + A_{2L}A_{3L} &= A_{10}(S_{42}A_{20} + S_{44}A_{4L}) + (S_{22}A_{20} + S_{24}A_{4L})A_{3L} \\ &= (S_{11}A_{10} + S_{13}A_{3L})A_{4L} + A_{20}(S_{31}A_{10} + S_{33}A_{3L}) \\ &= A_{1L}A_{4L} + A_{20}A_{30} \end{aligned} \tag{7.3-16}$$

Rearranging the two sides of the above equation leads to

$$A_1A_4 - A_2A_3 = \text{constant} \tag{7.3-17}$$

Notice that there is a sign difference between Eqs. (7.3-4) and (7.3-17). However, both equations are directly related to the reciprocity theorem.

When the reciprocity theorem is combined with other properties of photorefractive diffraction, other constants and integration can be found. From the coupled wave equations of FWM with the transmission grating, by examining the differential scattering matrix and using the properties of matrix multiplication, we can derive the following relations

$$S_{11} = S_{22}^*, \qquad S_{12} = S_{21}^* \tag{7.3-18}$$

Following similar steps as in Eq. (7.3-9), we obtain Eq. (7.3-5), that is,

$$A_1 A_3^* - A_2 A_4^* = \text{constant of integration} \tag{7.3-19}$$

For the case of reflection grating, if the medium does not have absorption loss (in other words, the S matrix is unitary), similar relations as those in Eq. (7.3-18) can be found

$$S_{11} = S_{33}, \qquad \frac{S_{13}}{S_{11}} = -\left(\frac{S_{31}}{S_{11}}\right)^* \tag{7.3-20}$$

Following similar steps as in Eq. (7.3-16), we obtain [36]

$$A_1 A_2^* - A_3 A_4^* = \text{constant of integration} \tag{7.3-21}$$

Notice that Eq. (7.3-19) for transmission gratings is valid even when there is absorption loss, whereas Eq. (7.3-21) for reflection gratings is only true when there is no absorption loss.

In conclusion, we have discussed the reciprocity theorem in photorefractive wave mixing. We have also pointed out that some constants of integration in photorefractive FWM are direct consequences of the reciprocity theorem. These general principles of physics can also be applied to FWM in other media, for instance, Kerr media. In addition, the results can be extended to cover nondegenerate FWM, provided the velocity of the moving index grating is much smaller compared to the speed of light.

PROBLEMS

7.1 Interference Function $G(z)$

a. Derive Eq. (7.1-38).

b. Show that

$$|G(z)| = \frac{C|G(0)| \operatorname{sech}\left(\frac{C}{2}\gamma z\right)}{C + (C^2 - 4|G(0)|^2)^{1/2} \tanh\left(\frac{C}{2}\gamma z\right)}$$

c. Show that the constants defined in Eqs. (7.1-19), (7.1-43) are identical, that is, $s^2 = C^2$.

7.2 Nonreciprocal Diffraction. The diffraction efficiency of a photorefractive grating is given by Eq. (7.1-51). Let η_1 be the diffraction efficiency for incidence at port 1 (i.e., $A_2(0) = 0$) and η_2 be the diffraction efficiency for incidence at port 2 (i.e., $A_1(0) = 0$).

a. Show that

$$\eta_1(\gamma) = \eta_2(-\gamma; 3 \leftrightarrow 4)$$

where $3 \leftrightarrow 4$ indicates that beams 3 and 4 are exchanged.

b. Show that for equal intensity of write beams (i.e., $I_3(0) = I_4(0)$)

$$\eta_1(\gamma) = \eta_2(-\gamma)$$

7.3 Use the method of grating integral to derive Eqs. (7.2-28) and (7.2-29).

7.4 Derive Eq. (7.2-30).

7.5 Show that the diffraction efficiency of a "soft" grating approaches zero when the read intensity becomes very large (i.e., $I_{10} \gg I_{30}, I_{40}$), according to Eq. (7.1-51).

REFERENCES

1. See, for example, P. Yeh, "Two-Wave Mixing in Nonlinear Media," IEEE J. Quantum Electron. **25**, 484 (1989).
2. See, for example, P. Yeh, A. E. Chiou, J. Hong, P. Beckwith, T. Chang, and M. Khoshnevisan, "Photorefractive Nonlinear Optics and Optical Computing," Opt. Eng. **28**, 328 (1989).
3. P. J. van Heerden, "Theory of Optical Information Storage in Solids," Appl. Opt. **2**, 393 (1963).
4. A. E. T. Chiou and P. Yeh, "Parallel Image Subtraction Using Phase-Conjugate Michelson Interferometer," Opt. Lett. **11**, 306 (1989).
5. T. Y. Chang, P. H. Beckwith, and P. Yeh, "Real-Time Optical Image Subtraction Using Dynamic Holographic Interference in Photorefractive Media," Opt. Lett. **13**, 586 (1988).
6. P. Yeh, A. E. T. Chiou, and J. Hong, "Optical Interconnection Using Photorefractive Dynamic Holograms," Appl. Opt. **27**, 2093 (1988).
7. A. Chiou and P. Yeh, "Energy Efficiency of Optical Interconnection Using Photorefractive Holograms," Appl. Opt. **29**, 1111 (1990).
8. H. Lee, X. G. Gu, and D. Psaltis, "Volume Holographic Interconnections with Maximal Capacity and Minimal Cross Talk," J. Appl. Phys. **65**, 2191 (1989).
9. D. Psaltis, X. G. Gu, and D. Brady, "Holographic Implementations of Neural Networks," in An Introduction to Neural and Electronic Networks, S. F. Zornetzer, J. L. Davis and C. Lau, Eds. (Academic Press, New York, 1990).
10. D. Psaltis, D. Brady, X. G. Gu, and S. Lin, "Holography in Artificial Neural Networks," Nature **343**, 325 (1990).

11. D. L. Staebler and J. J. Amodei, Ferroelectrics **3**, 107 (1972).
12. F. Micheron and G. Bismuth, "Electrical Control of Fixation and Erasure of Holographic Patterns in Ferroelectric Materials," Appl. Phys. Lett. **20**, 79 (1971).
13. F. Micheron, C. Mayeux, and J. C. Trotier, "Electric Control in Photo-Ferroelectric Materials for Optical Storage," Appl. Opt. **13**, 784 (1974).
14. D. von der Linde, A. M. Glass, and K. F. Rodgers, "High Sensitivity Optical Recording in KTN by Two-Photon Absorption," Appl. Phys. Lett. **26**, 22 (1975).
15. A. Delboulbe, C. Fromont, J. P. Herriau, S. Mallick, and J. P. Huignard, "Quasi-Nondestructive Readout of Holographically Stored Information in Photorefractive $Bi_{12}SiO_{20}$ Crystals," Appl. Phys. Lett. 55, **713** (1989).
16. J. Hong, S. Campbell, and P. Yeh, "Optical Pattern Classifier with Perceptron Learning," Appl. Opt. **29**, 3019 (1990).
17. J. Hong, S. Campbell, and P. Yeh, "Optical Learning Machines for Pattern Classification," OSA Annual Meeting Tech. Digest, WJ3, Orlando, FL, 1989.
18. D. Psaltis, D. Brady, and K. Wagner, "Adaptive Optical Networks Using Photorefractive Crystals," Appl. Opt. **27**, 1752 (1988).
19. E. G. Paek, J. Wullert, and J. S. Patel, "Holographic Implementation of a Learning Machine Based on a Multi-Category Perceptron Algorithm," Opt. Lett. **14**, 1303 (1989).
20. H. Kogelnik, "Coupled-Wave Theory for Thick Holographic Gratings," Bell Syst. Tech. J. **48**, 2909 (1969).
21. R. Saxena, F. Vachss, I. McMichael, and P. Yeh, "Diffraction Properties of Multiple-Beam Photorefractive Gratings," J. Opt. Soc. Am. **B7**, 1210 (1990).
22. T. Chang and P. Yeh, "Dark rings from photorefractive conical diffraction in a $BaTiO_3$ crystal," Proc. SPIE **739**, 109 (1987).
23. N. A. Vaines, S. L. Clapham, and R. W. Eason, "Multiplexed Permanent and Real Time Holographic Recording in Photorefractive BSO," Appl. Opt. **28**, 4381 (1989).
24. P. Yeh and M. Khoshnevisan, "Nonlinear Optical Bragg Scattering in Kerr Media," J. Opt. Soc. Am. **B4**, 1954 (1987).
25. F. Vachss, I. McMichael, M. Khoshnevisan, and P. Yeh, "Enhanced Acousto-Optic Diffraction in Electrostrictive Media," J. Opt. Soc. Am. **B7**, 859 (1990).
26. See, for example, L. I. Schiff, Quantum Mechanics, 3rd edition (McGraw-Hill, New York, 1968), pp. 114–135.
27. See, for example, L. D. Landau and E. M. Lifshitz, Quantum Mechanics, 3rd edition (Pergamon Press, Oxford, 1977), pp. 508–512, 587–603.
28. M. Born and E. Wolf, Principles of Optics, 6th edition (with corrections) (Pergamon Press, Oxford, 1980), pp. 378–382.
29. See, for example, P. Yeh, Optical Waves in Layered Media (Wiley, New York, 1988), pp. 112–114.
30. P. Yeh, "Electromagnetic Propagation in a Photorefractive Layered Medium," J. Opt. Soc. Am. **73**, 1268 (1983).
31. M. Z. Zha and P. Gunter, "Nonreciprocal Optical Transmission Through Photorefractive $KNbO_3$:Mn," Opt. Lett. **10**, 184 (1985).

32. K. R. MacDonald, J. Feinberg, M. Z. Zha, and P. Gunter, "Asymmetric Transmission Through a Photorefractive Crystal of Barium Titanate," Opt. Comm. **50**, 146 (1984).
33. P. Yeh, "Two-Wave Mixing in Nonlinear Media," IEEE J. Quantum Electron. **25**, 484 (1989).
34. See, for example, M. Cronin-Golomb, B. Fischer, J. O. White, and A. Yariv, "Theory and Applications of Four-Wave Mixing in Photorefractive Media," IEEE J. Quantum Electron. **20**, 12 (1984).
35. See, for example, P. Yeh, T. Y. Chang, and M. D. Ewbank, "A Model for Mutually Pumped Phase Conjugators," J. Opt. Soc. Am. **B5**, 1743 (1988), and references therein.
36. N. V. Kukhtarev and T. I. Semenets, "Phase Conjugation by Reflection Gratings in Electro-Optic Crystals," Appl. Phys. **B41**, 259 (1986).

8

Volume Holograms and Planar Holograms

We know by now that photorefractive materials can be used as holographic recording media. When the complex amplitude of an object is recorded in the medium by means of interference, a photo-induced periodic index variation Δn is produced. It has been demonstrated that from such a recorded interference pattern (the so-called hologram, meaning "whole information" or "total recording"), a three-dimensional image of the original object can be reconstructed, even if the recording medium is a thin film. In this chapter, we discuss some of the fundamental properties of volume holograms and planar holograms.

8.1 GRATINGS AND HOLOGRAMS

In the holographic recording, the reference beam is usually a plane wave. In the extreme case when the object beam is also a plane wave, the interference pattern is a pure sinusoidal function of position. Such a pure sinusoidal variation of the index of refraction is called a grating. Since a plane wave (or a grating) carries no spatial information, a hologram must consist of several gratings in order to record spatial information of a general object. This can be explained by examining the recording process. A general object which contains spatial information can always be represented as a linear superposition of several plane waves. By interfering with the reference plane wave, each plane wave component of the object beam will form a grating with the reference beam. Thus the recorded hologram will consist of several gratings each with a different wave vector.

In the early days of holography, the interference pattern between the object and reference beam was recorded in thin photographic film. The result is a planar hologram. With photorefractive media, we can record volume holograms. Here in this section, we will examine the fundamental difference between a volume and a planar hologram. Consider a transmission grating with an index of refraction given by

$$n = \begin{cases} n_0 + n_1 \cos \mathbf{K} \cdot \mathbf{r} , & |z| < \dfrac{L}{2} \\ n_0 , & \text{otherwise} \end{cases} \tag{8.1-1}$$

where n_0, n_1 are constants, $\mathbf{K}$ is the grating wave vector and L is the thickness of the grating. In Chapter 2, we defined a dimensionless parameter Q,

$$Q = \frac{2\pi\lambda L}{n_0 \Lambda^2} \tag{8.1-2}$$

where Λ is the grating period

$$\Lambda = \frac{2\pi}{|K|} \tag{8.1-3}$$

A grating is thick, or called a volume grating if $Q > 1$. On the other hand if $Q < 1$, then the grating is thin, or called a planar grating.

Example: Gratings in Liquid Crystals Nonlinear optical gratings can also be formed in liquid crystals by two-beam interference. Typical thicknesses of liquid crystals are of the order of 10 μm. Let the grating period be $\Lambda = 5$ μm, $\lambda = 0.5$ μm, $n_0 = 1.5$. The parameter is

$$Q = 0.84$$

and the grating is planar.

Example: Gratings in SBN Crystals Consider a thin plate of SBN crystal of thickness $L = 1$ mm. Let the grating period be $\Lambda = 5$ μm, $\lambda = 0.5$ μm, $n_0 = 2.5$. The parameter is

$$Q = 50$$

and the grating is a volume grating.

In a planar grating, the thickness (or interaction length) is small such that

$$L < \frac{n_0 \Lambda^2}{2\pi\lambda} \tag{8.1-4}$$

The finite thickness leads to an uncertainty in K_z by

$$\Delta K_z = \frac{2\pi}{L} \tag{8.1-5}$$

In the extreme case when L is near zero, the uncertainty in K_z becomes infinite. Such an uncertainty in K_z is useful in terms of relaxing the Bragg condition for diffraction. In volume gratings, efficient diffraction requires that

$$\mathbf{k}' = \mathbf{k} + \mathbf{K} \tag{8.1-6}$$

where $\mathbf{k}$ is the incident wave vector and $\mathbf{k}'$ is the wave vector of the diffracted beam. Equation (8.1-6) is known as the Bragg condition which is precisely the conservation of momentum. In addition, the magnitude of the wave vectors must obey

$$|\mathbf{k}'| = |\mathbf{k}| = \frac{\omega}{c} n_0 \tag{8.1-7}$$

which is a result of conservation of energy. These two conditions lead to a unique direction of incidence for efficient diffraction. In planar gratings, the conservation of momentum is given by

$$k'_{\|} = k_{\|} + K_{\|} \tag{8.1-8}$$

where the symbol $\|$ indicates a component parallel to the plane of grating (i.e., xy-plane). The uncertainty in K_z allows the efficient diffraction over a continuous range of incidence.

We now investigate the recording of holograms and the reconstruction of the original object beam. An interesting question people often ask is: Do volume holograms record more information than planar holograms as far as reconstructing the original object is concerned? The answer is no and the proof is as follows. Let the object wave be written

$$E = A(\mathbf{r})e^{i(\omega t - kz)} \tag{8.1-9}$$

where $A(\mathbf{r})$ is the amplitude and the wave propagates along the $+z$ direction. The propagation of the wave in free space is governed by the following equation:

$$\nabla_{\perp}^2 A - 2ik \frac{\partial}{\partial z} A = 0 \tag{8.1-10}$$

Since this is a first-order differential equation in z, a unique determination of the wave requires only the knowledge of $A(\mathbf{r})$ at any one plane $z =$ constant. Thus if we can reconstruct E at a given plane $z =$ constant by using a planar hologram, then the object wave is generated everywhere.

Let us now consider the recording of the object wave in a planar medium at $z = 0$. Let the reference beam be written

$$E_1 = A_1 \exp[i(\omega t - \mathbf{k}_1 \cdot \mathbf{r})] \tag{8.1-11}$$

where A_1 is a constant, $\mathbf{k}_1$ is the wave vector of the reference beam. As a result of the interference and the recording, a hologram is formed with Δn given by

$$\Delta n = n_1 A_1^* A \exp[i(k_{1x}x + k_{1y}y)] + \text{c.c.} \quad \text{at } z = 0 \qquad (8.1\text{-}12)$$

where n_1 is a constant. If now the hologram is illuminated with the same reference beam (8.1-11), the light transmitted by the hologram is

$$E_2 = E_1 \exp\left(-i \frac{2\pi}{\lambda} \Delta n\, d\right) \qquad (8.1\text{-}13)$$

where d is the thickness of the hologram.

For thin holograms, $2\pi\, \Delta n\, d/\lambda \ll 1$ and the transmitted wave can be written

$$E_2 \simeq E_1 - i \frac{2\pi}{\lambda} d\, \Delta n\, E_1 \qquad (8.1\text{-}14)$$

which consists of an undiffracted portion E_1 and a diffracted one (the second term on the right side). Using Eqs. (8.1-12) and (8.1-13), the diffracted wave at $z = 0$ can be written

$$E_3 = -i \frac{2\pi}{\lambda} d\, \Delta n\, E_1 = -i \frac{2\pi}{\lambda} dn_1 |A_1|^2 A e^{i\omega t} \qquad (8.1\text{-}15)$$

which is exactly proportional to the object wave E at $z = 0$. According to the above discussion, E_3 will be exactly proportional to E everywhere for $z > d$.

We have shown that planar holograms are adequate for the reconstruction of object waves. What is the purpose of using volume holograms? Bragg selectivity and high diffraction efficiency are important issues associated with volume holograms. We have mentioned earlier that diffraction (or read out) from thick holograms requires a precise alignment of the reading beam. In other words, the reading beam has to be almost exactly parallel to the reference beam in order to reconstruct the object wave. The thickness of the hologram not only imposes the stringent Bragg condition, but also provides a high diffraction efficiency. The diffraction efficiency of planar holograms are relatively small because of the thin interaction length.

Despite the low diffraction efficiency, planar holograms offer some unique properties in optical signal processing. First of all, planar holograms can be read by a reading beam incident over a wide range of angles. Suppose that the reading beam is given by

$$E_1' = A_1 \exp[i(\omega t - \mathbf{k}_1' \cdot \mathbf{r})] \qquad (8.1\text{-}16)$$

where $\mathbf{k}_1'$ is the wave vector. The diffracted wave at $z = 0$, will be given by

$$E_3' = -i\,\frac{2\pi}{\lambda}\,dn_1|A_1|^2 A \exp(-i\,\Delta\mathbf{k}_\parallel \cdot \mathbf{r})\,, \quad z=0 \tag{8.1-17}$$

where

$$\Delta\mathbf{k}_\parallel = (\mathbf{k}_1' - \mathbf{k}_1)_\parallel \tag{8.1-18}$$

where the symbol $\|$ indicates a component parallel to the plane $z = 0$. We note that within a factor of $\exp[-i\,\Delta\mathbf{k}_\parallel \cdot \mathbf{r}]$, the diffracted wave is proportional to the object wave. This phase factor represents a tilt of the reconstructed object wave. If we compare the Fourier transform of E_3 and E_3', we find that the Fourier transform of E_3' is shifted in space relative to that of E_3 by an amount proportional to $\Delta\mathbf{k}_\parallel$. This is the only difference. Thus we find that planar holograms are useful in optical information processing. In thick holograms, the misalignment of the readout beam will lead to no diffraction at all.

We consider now the inverse problem of holography. Given an arbitrary Δn, can we find a pair of beams whose interference pattern is proportional to Δn? Let the index modulation of a planar hologram be written

$$\Delta n = n_1 f(x, y)e^{-i(\alpha x + \beta y)} \tag{8.1-19}$$

where n_1 is a constant, α and β are components of grating wave vectors and $f(x, y)$ is an arbitrary function. There are an infinite number of pairs of beams whose interference pattern is proportional to Δn. An obvious example will be the pair given by Eqs. (8.1-9) and (8.1-11) with $A(x, y, z=0) = f(x, y)$ and $\mathbf{k}_1 = (\alpha, \beta, k_{1z})$. The amplitude $A(\mathbf{r})$ in Eq. (8.1-9) can be arbitrarily specified at the plane $z = 0$. It is important to note that $(|\alpha|^2 + |\beta|^2)$ should be less than the wave number of the beam.

The inverse problem for volume holograms is not as simple. Let the index modulation of a volume hologram be written

$$\Delta n = n_1 f(x, y, z)e^{-i\mathbf{K}\cdot\mathbf{r}} \tag{8.1-20}$$

where n_1 is a constant, $\mathbf{K}$ is the grating wave vector and $f(x, y, z)$ is an arbitrary function. Such a volume hologram may not be produced by a pair of beams. It looks as though a pair of beams with electric field

$$E_j = A_j e^{i(\omega t - \mathbf{k}_j\cdot\mathbf{r})}\,, \quad j = 1, 2 \tag{8.1-21}$$

can produce an interference pattern proportional to Δn provided

$$A_1^* A_2 = f(x, y, z) \tag{8.1-22}$$

and

$$\mathbf{k}_2 - \mathbf{k}_1 = \mathbf{K} \tag{8.1-23}$$

The reason that this may not be possible is because A_1 and A_2, which satisfy Eq. (8.1-22), may not be the solutions of the wave equation.

The volume hologram described by Eq. (8.1-20) can, however, be produced by a series of two-beam interferometric exposure. This is obvious because Eq. (8.1-20) can always be represented in terms of a summation of Fourier components:

$$\Delta n = \sum_{\mathbf{K}'} A(\mathbf{K}')e^{-i\mathbf{K}'\cdot\mathbf{r}} \tag{8.1-24}$$

Each of the grating components in Eq. (8.1-24) can be produced by using the interference of two plane waves, provided

$$\mathbf{K}' = \mathbf{k}_2 - \mathbf{k}_1 \tag{8.1-25}$$

where $\mathbf{k}_1$, $\mathbf{k}_2$ are the wave vectors of the plane waves. There are an infinite number of solutions to the above equation. In other words, the plane wave pair that can produce a grating of $\mathbf{K}'$ is not unique even if the frequency of the waves is fixed. This is a result of K-space degeneracy which is the subject of the next section.

8.2 *K*-SPACE DEGENERACY

Degeneracy arises when a grating is Bragg matched by more than one pair of incident-diffracted plane waves. The degeneracy can be easily understood by examining the triangle formed by the three vectors $\mathbf{K}'$, $\mathbf{k}_2$, $\mathbf{k}_1$ in Eq. (8.1-25). We now rotate the triangle by keeping the vector $\mathbf{K}'$ fixed. For each angle of rotation, a new set of wave vectors $\mathbf{k}_2'$, $\mathbf{k}_1'$ are obtained. These new wave vectors also satisfy the same Bragg condition, Eq. (8.1-25). In this section, we consider the degeneracy condition in a uniaxial crystal, as in the case of most photorefractive crystals. In the following discussion, we will consider the ordinary and extraordinary waves separately, assuming that there is no polarization change during diffraction.

According to Eq. (1.3-13), the normal surface of a uniaxial crystal can be separated into a sphere,

$$\frac{k_x^2}{n_o^2} + \frac{k_y^2}{n_o^2} + \frac{k_z^2}{n_o^2} = k_0^2 \tag{8.2-1}$$

and an ellipsoid,

$$\frac{k_x^2}{n_e^2} + \frac{k_y^2}{n_e^2} + \frac{k_z^2}{n_o^2} = k_0^2 \tag{8.2-2}$$

where $\mathbf{k} = (k_x, k_y, k_z)$ is the wave vector, $k_0 = \omega/c$, n_e and n_o are the extraordinary and ordinary indices of refraction, respectively. We first examine the diffraction of extraordinary waves and then extend the result to that of ordinary waves, since the sphere can be regarded as a special case of an ellipsoid with $n_e = n_o$. The coordinate system is chosen such that the optical axis is in the z-direction.

Consider a grating recorded in a photorefractive crystal by two-beam interference. When two different plane waves $\exp(-i\mathbf{k}_i \cdot \mathbf{r})$ and $\exp(-i\mathbf{k}_d \cdot \mathbf{r})$ form an interference pattern $\cos[(\mathbf{k}_d - \mathbf{k}_i)\cdot \mathbf{r}]$ inside the crystal, the resulting dielectric phase grating will have a first order sinusoidal component $\cos(\mathbf{K}\cdot\mathbf{r} + \phi)$, where $\mathbf{K}$ is the grating wave vector and ϕ is the phase shift between the grating and the interference pattern. The relation $\mathbf{K} = \mathbf{k}_d - \mathbf{k}_i$ is exactly the Bragg condition.

All pairs of degenerate incident-diffracted wave vectors matched by the same wave vector $\mathbf{K}$ are located on two ellipses, called degeneracy ellipses, on the normal surface. To find the two degeneracy ellipses in the $\mathbf{k}$-space, consider a pair of incident-diffracted plane waves $(\mathbf{k}_i, \mathbf{k}_d)$. These two wave vectors, confined by the normal surface Eq. (8.2-2), must satisfy the following conditions:

$$\frac{k_{ix}^2}{n_e^2} + \frac{k_{iy}^2}{n_e^2} + \frac{k_{iz}^2}{n_o^2} = k_0^2 \tag{8.2-3}$$

$$\frac{k_{dx}^2}{n_e^2} + \frac{k_{dy}^2}{n_e^2} + \frac{k_{dz}^2}{n_o^2} = k_0^2 \tag{8.2-4}$$

where k_{ix}, k_{iy}, k_{iz}, k_{dx}, k_{dy}, k_{dz} are components of the wave vectors and

$$\mathbf{K} = \mathbf{k}_d - \mathbf{k}_i \tag{8.2-5}$$

Equation (8.2-5) is exactly the Bragg condition. Substituting $\mathbf{k}_d$ by $\mathbf{K} + \mathbf{k}_i$ in Eq. (8.2-4), we obtain the ellipse of the incident wave which is the intersection of the following two surfaces

$$\begin{cases} \dfrac{k_{ix}^2}{n_e^2} + \dfrac{k_{iy}^2}{n_e^2} + \dfrac{k_{iz}^2}{n_o^2} = k_0^2 \\[2ex] \dfrac{(2k_{ix} + K_x)K_x}{n_e^2} + \dfrac{(2k_{iy} + K_y)K_y}{n_e^2} + \dfrac{(2k_{iz} + K_z)K_z}{n_o^2} = 0 \end{cases} \tag{8.2-6}$$

Similarly, the equations for the ellipse of the diffracted wave are

$$\begin{cases} \dfrac{k_{dx}^2}{n_e^2} + \dfrac{k_{dy}^2}{n_e^2} + \dfrac{k_{dz}^2}{n_o^2} = k_0^2 \\[2ex] \dfrac{(2k_{dx} - K_x)K_x}{n_e^2} + \dfrac{(2k_{dy} - K_y)K_y}{n_e^2} + \dfrac{(2k_{dz} - K_z)K_z}{n_o^2} = 0 \end{cases} \tag{8.2-7}$$

Given a grating wave vector **K**, these equations yield an infinite pair of solutions $(\mathbf{k}_i, \mathbf{k}_d)$. These pairs of wave vectors are confined in two curves. It can be recognized that these two curves are two ellipses resulting from the intersection of the normal surface with two parallel planes. The direction of the vector normal to these two planes, $\hat{\mathbf{K}}$, is given by

$$\hat{\mathbf{K}} = \begin{bmatrix} K_x/n_e^2 \\ K_y/n_e^2 \\ K_z/n_o^2 \end{bmatrix} \tag{8.2-8}$$

which, in general, is not parallel to the vector **K**. The distance between the origin of the **K**-space and these two planes is given by

$$d = \frac{\dfrac{1}{2}\left(\dfrac{K_x^2}{n_e^2} + \dfrac{K_y^2}{n_e^2} + \dfrac{K_z^2}{n_o^2}\right)}{\left(\dfrac{K_x^2}{n_e^4} + \dfrac{K_y^2}{n_e^4} + \dfrac{K_z^2}{n_o^4}\right)^{1/2}} \tag{8.2-9}$$

These two degenerate ellipses are shown in Figure 8.1.

The degeneracy ellipses can be used to locate all pairs of incident-diffracted wave vectors which are Bragg matched by a given grating wave vector **K**. A pair of incident-diffracted wave vectors can be found by picking up one point on one ellipse, for instance $\mathbf{k}_1'$; then adding the vector **K** to $\mathbf{k}_1'$, resulting in a point on the second ellipse, $\mathbf{k}_d' = \mathbf{k}_i' + \mathbf{K}$. For each point on one degeneracy ellipse, there exists one and only one point on the other degeneracy ellipse such that these two points can be Bragg matched by the given grating wave vector **K**.

For ordinary waves, degeneracy ellipses become degeneracy circles. The two circles are perpendicular to the grating wave vector **K**, since the $\hat{\mathbf{K}}$

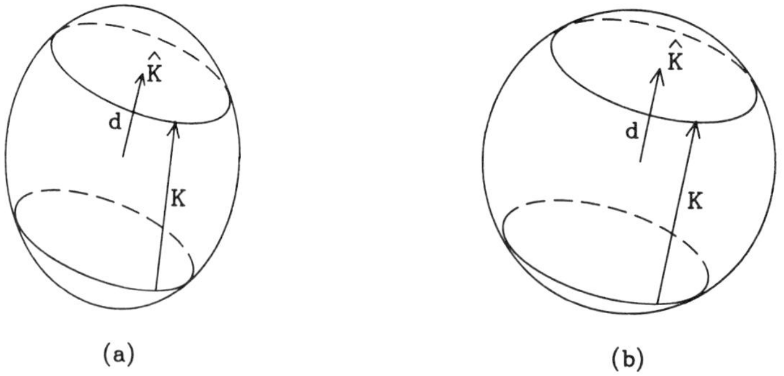

Figure 8.1. **K**-Space degeneracy curves for the grating wave vector **K**: (a) degeneracy ellipses for extraordinary waves; (b) degeneracy circles for ordinary waves.

vector is parallel to the grating wave vector **K** for $n_e = n_o$. The distance between the plane of each circle and the origin is half the magnitude of the grating wave vector, as given by Eq. (8.2-9). The degeneracy ellipses for anisotropic Bragg scattering can be obtained in a similar way (see Problem 8.5).

8.3 STORAGE CAPACITY

Photorefractive materials such as $BaTiO_3$, $LiNbO_3$, SBN can be employed for the storage of optical images. Two-dimensional image information can be recorded by using either planar or volume holograms in these media. In what follows, we consider the storage capacity of these two types of holograms.

8.3.1 Storage Capacity of Volume Holograms

The storage capacity of a photorefractive crystal is the maximum number of distinguishable gratings that can be stored. The value can be obtained from the **K**-space representation of gratings.

We first introduce a grating space, called **K**-space, in which each point **K** represents, mathematically, a set of sinusoidal fringes extended to infinity in the real space. With the help of the Fourier transform, the spatial superposition of multiple gratings in a photorefractive crystal is described as the distribution of points in the **K**-space.

To evaluate the storage capacity, we need to consider all the gratings formed by all possible pairs of wave vectors. Since the wave vectors are confined on the normal surface, the grating wave vectors **K** consist of all the vectors formed by all possible pairs of points on the normal surface. Figure 8.2 shows a grating wave vector **K** and a normal surface centered at N depicted in the **K**-space. The origin of wave vectors, $\mathbf{k}_i$ and $\mathbf{k}_d$, is at the center of the normal surface. The origin of the grating wave vector, **K**, is at the origin of the **K**-space, that is, $K_x = K_y = K_z = 0$. The Bragg condition $\mathbf{K} = \mathbf{k}_d - \mathbf{k}_i$ requires that the three vectors **K**, $\mathbf{k}_i$, and $\mathbf{k}_d$ form a triangle. Therefore, the center of the normal surface is displaced to the point $\mathbf{K} = -\mathbf{k}_i$.

In what follows, we want to show that all these grating vectors **K** occupy a space confined by an ellipsoid in K-space. This ellipsoid of grating wave vectors is exactly twice the size of the normal surface in linear dimensions. In other words, the principal axes are $2n_e k_0$, $2n_e k_0$, $2n_o k_0$ in K-space. Referring to Figure 8.2, we consider gratings formed by a given incident plane wave with wave vector $\mathbf{k}_i$ and an arbitrary plane wave of the same frequency ω. We place the normal surface of the plane waves of frequency ω at $\mathbf{K} = -\mathbf{k}_i$ (Point N in K space). Consider another point P on this normal surface. The vector from point N to point P represents a possible diffracted

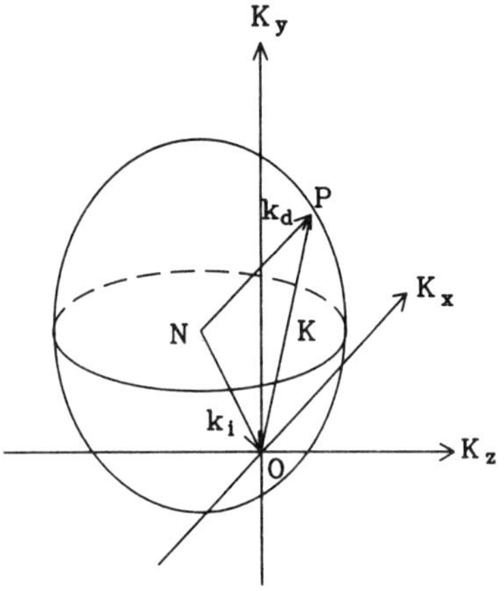

Figure 8.2. The **K**-space of grating wave vectors and the normal surface. For a given incident wave vector $\mathbf{k}_i$, all the possible wave vectors $\mathbf{k}_d$ of the diffracted beam are confined on the normal surface (i.e., NP). We place the center of the normal surface N at $K = -\mathbf{k}_i$, so that the origin of the **K**-space O is at the tip of the vector $\mathbf{k}_i$. The normal surface centered at N thus defines all the possible grating wave vectors **K** for a given incident wave vector $\mathbf{k}_i$.

wave vector $\mathbf{k}_d$. The vector from point O, which is the origin of the **K**-space, to point P is $-\mathbf{k}_i + \mathbf{k}_d$. Thus, point P represents a grating wave vector $\mathbf{K} = \mathbf{k}_d - \mathbf{k}_i$. Since point P can be any point on the normal surface, all possible grating wave vectors are distributed on this two-dimensional normal surface. Another set of grating wave vectors can be found by using another incident plane wave. Having a different incident wave vector $\mathbf{k}_i'$ is equivalent to displacing the center of the normal surface to a different point N', which is given by $-\mathbf{k}_i'$, in the **K**-space. All possible grating wave vectors $\mathbf{K}' = \mathbf{k}_d' - \mathbf{k}_i'$ will be distributed on the two-dimensional normal surface centered at point N'. The two normal surfaces (centered at N and N', respectively) have exactly the same orientation, which is determined by the given orientation of the crystal.

All possible locations of the centers of the normal surfaces are confined to an ellipsoid described by Eq. (8.2-3), since all the $-\mathbf{k}_i$ are confined to the normal surface described by Eq. (8.2-3). This ellipsoid is centered at the origin of the **K**-space and will be called the center ellipsoid (see Figure 8.3). By using all possible incident and diffracted plane waves, we can find all possible gratings distributed in the three-dimensional **K**-space. All possible grating wave vectors can be found by continuously moving the normal surface in the **K**-space. Since the center of the normal surface is restricted to the center ellipsoid, all possible gratings are distributed inside a bounded grating space.

The boundary of the grating space is an ellipsoid with its axis twice that of the normal surface ellipsoid. The equation for that boundary is

$$\frac{K_x^2}{(2n_e)^2} + \frac{K_y^2}{(2n_e)^2} + \frac{K_z^2}{(2n_o)^2} = k_0^2 \qquad (8.3\text{-}1)$$

This is called the grating ellipsoid. Only half of the gratings inside the above

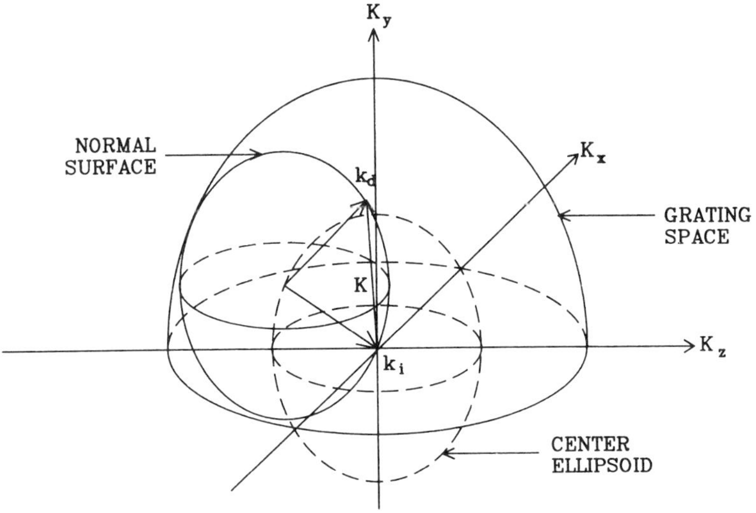

Figure 8.3. The ellipsoid of grating wave vectors in **K**-space. The normal surface centered at N in Figure 8.2 defines all the possible grating wave vectors **K** for a given incident wave vector $\mathbf{k}_i$. Since N can be any arbitrary point on the center ellipsoid (dashed ellipsoid in figure) to account for all possible incident waves, the grating wave vectors **K** are thus confined by the outer ellipsoid.

bounded **K**-space are necessary to describe independent gratings, because for sinusoidal gratings, wave vectors **K** and $-\mathbf{K}$ represent the same fringe spacing and orientation. In the following discussion, the top half of the **K**-space bounded by Eq. (8.3-1) will be used. Figure 8.3 shows the relationship between the normal surface, the center ellipsoid and the grating ellipsoid. The volume of the grating space, that is, the upper half of the ellipsoid bounded by Eq. (8.3-1), can be evaluated as

$$V_K = \frac{1}{2}\frac{4\pi}{3}(2n_e k_0)(2n_e k_0)(2n_o k_0) \tag{8.3-2}$$

that is,

$$V_K = \frac{16\pi}{3} n_e^2 n_o k_0^3 \tag{8.3-3}$$

where we recall that $k_0 = 2\pi/\lambda$.

Now we calculate the **K**-space volume of each grating, which is related to the dimensions of the recording crystal. The uncertainty principle provides an estimate of this **K**-space volume of a grating, which will also be called the uncertainty volume of a grating.

Suppose the crystal has a rectangular shape with its sides L_x, L_y and L_z along the x-, y- and z-directions. The rectangular crystal is chosen for simplicity. Other crystal geometrics can be analyzed similarly. The uncertainty values of a grating wave vector are related to the sides of the crystal, L_i by $\delta K_i L_i \approx 2\pi$, where $i = x, y, z$. These uncertainty values are explicitly written as

$$\delta K_x \approx \frac{2\pi}{L_x} \tag{8.3-4}$$

$$\delta K_y \approx \frac{2\pi}{L_y} \tag{8.3-5}$$

$$\delta K_z \approx \frac{2\pi}{L_z} \tag{8.3-6}$$

The uncertainty volume, v_g, of a grating is obtained by a simple multiplication of these three uncertainty values, that is,

$$v_g = \frac{(2\pi)^3}{L_x L_y L_z} \tag{8.3-7}$$

The strength of each grating inside the uncertainty volume can be calculated by using the Fourier transform of the dielectric modulation of a typical grating of finite dimension,

$$\Delta\epsilon(x, y, z) = \mathrm{rect}(x/L_x)\,\mathrm{rect}(y/L_y)\,\mathrm{rect}(z/L_z)(\Delta\epsilon_0)\exp(-i\mathbf{K}_g \cdot \mathbf{r}) \tag{8.3-8}$$

where $\Delta\epsilon_0$ is a constant representing the amplitude of the grating, $\mathrm{rect}(x/L_x)\,\mathrm{rect}(y/K_y)\,\mathrm{rect}(z/L_z)$ represents the rectangular crystal, and $\exp(-i\mathbf{K}_g \cdot \mathbf{r})$ represents the grating with wave vector $\mathbf{K}_g$. The Fourier transform of $\Delta\epsilon(x, y, z)$ is

$$\begin{aligned}\Delta\tilde{\epsilon}(K_x, K_y, K_z) &= \iiint \Delta\epsilon(x, y, z)\exp[-i(K_x x + K_y y + K_z z)]\,dx\,dy\,dz \\ &= \Delta\epsilon_0 L_x L_y L_z\,\mathrm{sinc}\left(\frac{K_x - K_{gx}}{2\pi/L_x}\right)\mathrm{sinc}\left(\frac{K_y - K_{gy}}{2\pi/L_y}\right) \\ &\quad \times \mathrm{sinc}\left(\frac{K_z - K_{gz}}{2\pi/L_z}\right)\end{aligned} \tag{8.3-9}$$

The strength of each grating component $\mathbf{K}$ is represented by $\Delta\tilde{\epsilon}(K_x, K_y, K_z)$. Equation (8.3-9) indicates that the grating component $\mathbf{K} = \mathbf{K}_g$ has the maximum grating strength. The grating components with $K_x = K_{gx} \pm 2\pi/L_x$, or $K_y = K_{gy} \pm 2\pi/L_y$, or $K_z = K_{gz} \pm 2\pi/L_z$ have zero grating strength. The uncertainty box specified by $|K_x - K_{gx}| \le \pi/L_x$, $|K_y - K_{gy}| \le \pi/L_y$, and $|K_z - K_{gz}| \le \pi/L_z$ includes all grating components with significant strength. The center of this box is at the point $\mathbf{K}_g$, and the sides are $2\pi/L_x$, $2\pi/L_y$, and $2\pi/L_z$ along the x-, y-, and z-directions, respectively. For any grating component outside the box, its grating strength will be reduced by sinc(1/2) or more, according to Eq. (8.3-9). The volume of this box, as given by Eq. (8.3-7), is also called the **K**-space volume of the nominal grating $\mathbf{K}_g$. Two gratings written in a finite crystal are distinguishable only if their uncertainty volumes do not overlap.

The storage capacity of a photorefractive crystal is thus defined as the maximum number of distinguishable gratings that can be contained by the grating space. It is a theoretical upper limit based upon the uncertainty consideration. To reach this upper limit, gratings have to be stacked in the **K**-space, so that two adjacent gratings are barely distinguishable. Since the **K**-space volume of each grating is independent of the grating wave vector, as shown in Eq. (8.3-7), the total number of distinguishable gratings is the ratio between the volume of the grating space and the **K**-space volume of each grating, that is, $C = V_K/v_g$. According to Eqs. (8.3-3) and (8.3-7),

$$C = \frac{16\pi}{3} n_e^2 n_o \frac{V_{\text{crystal}}}{\lambda^3} \tag{8.3-10}$$

where $V_{\text{crystal}} = L_x \times L_y \times L_z$ is the volume of the crystal and λ is the wavelength of light in vacuum.

Example: Storage Limit of $LiNbO_3$ Consider a $LiNbO_3$ crystal of $1 \times 1 \times 1\ \text{cm}^3$ in size. Using $\lambda = 0.5\ \mu\text{m}$ as the light source for the recording of holograms, the indices of refraction are $n_e = 2.29$ and $n_o = 2.20$. According to Eq. (8.3-10), the maximum number of gratings that can be stored is $C = 1.55 \times 10^{15}$. This is an enormous amount of storage.

8.3.2 Storage Capacity of Planar Holograms

Due to the lack of information in the third dimension, the storage capacity of a planar holographic medium will be smaller than that of a volume medium with comparable linear size. The **K**-space discussion can be extended to that of planar holograms. Figure 8.4 shows the **K**-space geometry when the recording medium becomes very thin. The uncertainty value $\delta K_z = 2\pi/L_z$ increases as the thickness L_z decreases. When the thickness L_z becomes comparable with the wavelength λ, the diffraction occurs at any angle of incidence. Figure 8.5 shows that when the incident plane wave $\mathbf{k}_i'$ is significantly deviated from the Bragg angle, which is determined by the nominal grating wave vector $\mathbf{K}_g$, the wave vector $\mathbf{k}_i'$ can still be Bragg matched by a grating wave vector $\mathbf{K}$ inside the uncertainty volume to produce a diffracted plane wave $\mathbf{k}_d'$. From Eq. (8.3-9), it can be noticed that the grating strength of $\mathbf{K}$ is close to that of $\mathbf{K}_g$ as long as $|K_z - K_{gz}| \ll 2\pi/L_z$.

The storage capacity of the two-dimensional recording medium is also defined as the maximum number of distinguishable gratings that can be

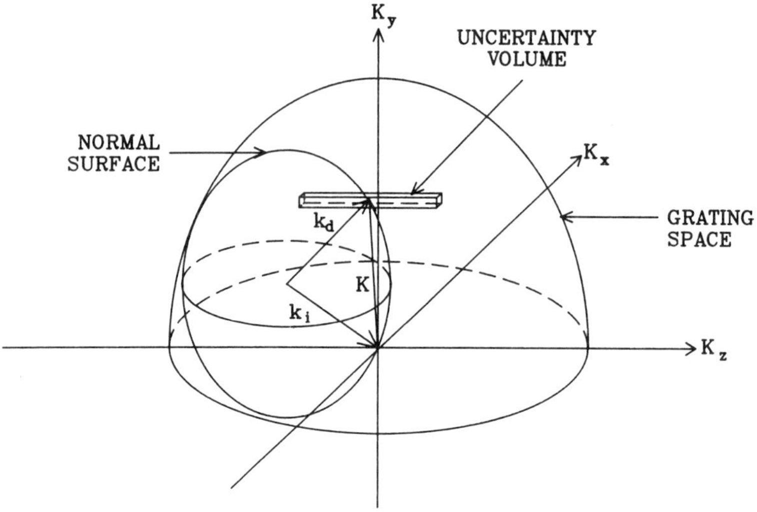

Figure 8.4. The **K**-space geometry for a thin recording medium. Notice that the uncertainty volume is an elongated rectangular prism along z-direction which is normal to the plane of the recording medium.

contained in the grating space. To derive an expression for the capacity, we assume that the medium is isotropic, and the magnitude of any wave vector is k. Therefore, the bounded grating space (K-space) is the top half of the sphere,

$$K_x^2 + K_y^2 + K_z^2 = (2k)^2 \tag{8.3-11}$$

For a planar hologram, k is comparable with the uncertainty value $\delta K_z = 2\pi/L_z$. The dimension of the uncertainty volume along the K_z-direction is comparable with the dimension of the grating space. Different

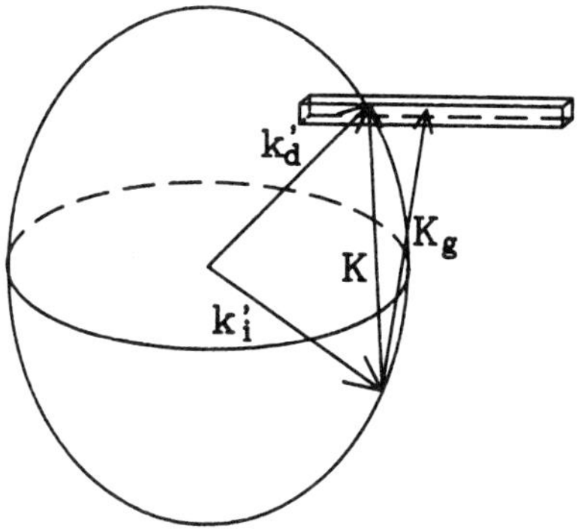

Figure 8.5. The elongated uncertainty volume in K-space leads to a relaxation of the Bragg diffraction condition. The figure indicates that significant diffraction occurs even when the Bragg condition is not satisfied (i.e., $\mathbf{k}_i'$, $\mathbf{k}_d'$ and $\mathbf{K}_g$ do not form a triangle).

gratings are represented by nonoverlapping uncertainty volumes separated in the K_x or K_y direction. The largest cross section of the grating space perpendicular to the K_z-direction is the circle in the (K_x, K_y)-plane. The area of this circle is $\pi(2k)^2$. The cross section of the uncertainty volume perpendicular to the K_z-direction is $\delta K_x \, \delta K_y$. Therefore, the storage capacity is given by

$$C = \frac{\pi(2K)^2}{\delta K_x \, \delta K_y} \tag{8.3-12}$$

where $\delta K_x = 2\pi/L_x$, $\delta K_y = 2\pi/L_y$. Using Eqs. (8.3-4), (8.3-5), and $k = 2\pi n/\lambda$, we obtain

$$C = 4\pi n^2 \frac{A}{\lambda^2} \tag{8.3-13}$$

where n is the refractive index of the material and $A = L_x L_y$ is the area of the planar holographic medium.

Example: Storage Capacity of a Planar Hologram Consider the storage of holograms in a thin plate of BSO crystal with a size of $1 \times 1\ \text{cm}^2$. Using a laser radiation with a wavelength of $\lambda = 0.5\ \mu\text{m}$, the storage capacity is $C = 3.24 \times 10^{10}$. We assume an index of refraction of $n = 2.54$.

8.4 FANNING HOLOGRAMS AND CONICAL DIFFRACTIONS

When a laser beam of finite transverse cross section passes through a photorefractive crystal, beam fanning often occurs [1]. The fanned beam originates from the scattering of the incident beam due to defects, imperfections, surface roughness of the crystal. The scattered light and the incident beam overlap and undergo two-wave mixing. Along certain directions, the scattered light gets amplified. This leads to beam fanning. Because of the dependence of coupling constant Γ on angles (θ, ϕ), the fanned light has a unique angular distribution. In most photorefractive crystals, the fanned light has the same frequency as that of the incident beam. Inside the photorefractive crystal, the incident beam and the fanned light together form an interference pattern and create a photo-induced hologram. Such a fanning hologram consists of a multitude of gratings with a wide range of grating wavevectors and orientations because one of the writing beams (the fanned beam) spans a wide solid angle in space.

In addition, conical diffractions are observed most of the time when fanning occurs. Such conical diffractions, similar to Laue patterns in X-ray scattering from crystal powders, can be explained in terms of the Bragg diffraction of light by the hologram formed by the incident beam and the fanned light. The hologram consists of a multitude of gratings because the fanned light spans a wide solid angle in space. When such a multitude of gratings is read out by a laser beam, only a subset of these gratings matches the Bragg diffraction condition with the readout beam. This gives rise to the conical diffractions [2, 3].

We first consider the situation of beam fanning. Let $\mathbf{k}_1$ be the wave vectors of the incident beam. The incident beams scatter off defects of impurities into a broad range of wave vectors $\mathbf{k}_f$. These scattered beams interfere with the incident beam and create a fanning hologram with grating wave vectors given by the following equation:

$$\mathbf{K} = \mathbf{k}_f - \mathbf{k}_1 \tag{8.4-1}$$

Note that both $\mathbf{K}$ and $\mathbf{k}_f$ span a wide solid angle in momentum space. The only constraint is that both $\mathbf{k}_1$ and $\mathbf{k}_f$ must lie on the normal surface (i.e., surface of constant frequency in momentum space $\omega(\mathbf{k}) = \text{constant}$) [4]. In uniaxial crystals such as $BaTiO_3$, SBN, $LiNbO_3$, $LiTaO_3$, etc., the normal surface consists of a sphere and an ellipsoid (see Figure 1.3). The sphere represents the ordinary wave, whereas the ellipsoid represents the extra-ordinary wave. In isotropic crystals such as GaAs, CdTe, the normal surface is a sphere regardless of the polarization states. When the hologram formed by the incident beam and the fanned beam is read out by a laser beam, the scattering is described by the following equation:

$$\mathbf{k}_s = \mathbf{k}_2 + N\mathbf{K} \tag{8.4-2}$$

where $\mathbf{k}_s$ represents the wave vectors of the scattered light, $\mathbf{k}_2$ is the wave vector of the readout beam and N is an integer. When the integer N is ± 1, the scattering is due to the fundamental component of the hologram, whereas $|N| > 1$ corresponds to the scattering due to higher harmonics of the gratings.

By eliminating $\mathbf{K}$ from Eqs. (8.4-1) and (8.4-2) for the cases of $N = \pm 1$, we obtain

$$\mathbf{k}_s = \mathbf{k}_f + (\mathbf{k}_2 - \mathbf{k}_1) \tag{8.4-3}$$

and

$$\mathbf{k}_s = -\mathbf{k}_f + (\mathbf{k}_2 + \mathbf{k}_1) \tag{8.4-4}$$

respectively. These two equations determine the direction of the scattered light. Since $\mathbf{k}_f$ spans over a wide solid angle in momentum space, these equations yield infinite solutions for $\mathbf{k}_s$. These infinite number of wave vectors form two hollow cones. The shape of such cones can be easily visualized by using a graphic method. For example, the solutions of Eq.

(8.4-4) can be obtained by drawing a normal surface centered at the origin representing the scattered light and a normal surface centered at $\mathbf{k}_2 + \mathbf{k}_1$ representing the fanned light. The intersection of these two normal surfaces gives the solutions. Since most normal surfaces are either a sphere or ellipsoid, the intersection usually has the shape of a ring. Figure 8.6 shows the graphic method. Such a ring and the center of the normal surface of the scattered lights determine the cone. The same method can be applied to Eq. (8.4-3) to obtain another cone. In most photorefractive crystals, the polarization state of the fanned light is the same as that of the incident beam. The same polarization state ensures that an intensity interference pattern can be formed to induce a fanning hologram. The hologram is a multitude of index gratings which can be either scalar or birefringent depending on the tensor property of the electro-optic coefficients involved. For the case of scalar holograms, the scattered light has the same polarization state as that of the readout beam. For birefringent holograms, the scattered light can have a polarization state which is different from that of the readout beam. Thus, in uniaxial photorefractive crystals, for each set of birefringent fanning hologram and polarized readout beam (i.e., either ordinary or extraordinary), there are, in principle, four conical diffractions. These four cones degenerate into two in isotropic crystals. Due to Fresnel refraction and the specific crystal cut, some or part of the cones may be difficult to observe. The graphic method will be used in the following to explain some of the rings observed.

To illustrate the use of the graphic method, we consider conical diffractions in $BaTiO_3$ crystals. Generally speaking, the incident beam and the readout beam can be in different polarization states, even different colors. Here we will limit ourselves to the case when both beams are of the same frequency and same polarization state.

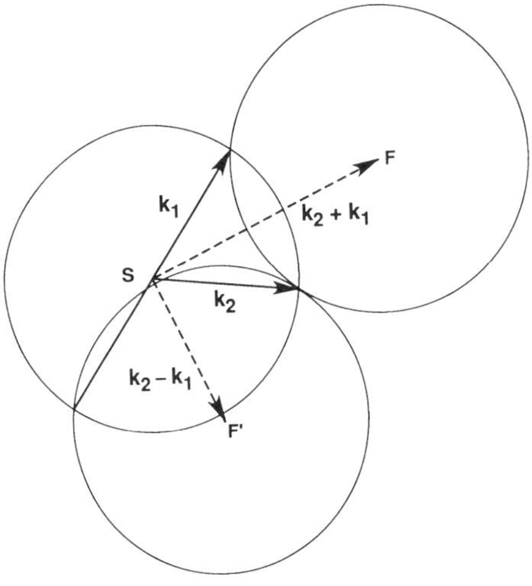

Figure 8.6. Graphic method for conical diffusion. S is the center of the normal surface for scattered (diffracted) light. F and F' are centers of the normal surfaces for fanned light. The intersection with the normal surface centered at S defines the rings of conical diffraction.

Case 1. Both the incident beam and the readout beam are assumed to be ordinary waves. Since the fanned light has the same polarization state as the incident beam, both normal surfaces centered at F and F' are exact spheres. The ordinary conical diffractions are obtained by drawing a sphere representing the ordinary normal surface of the scattered light. Two cones are obtained as shown in Figure 8.6. One of the cones is defined by the two wave vectors $\mathbf{k}_1$ and $\mathbf{k}_2$. The other cone is defined by the vectors $\mathbf{k}_2$ and $-\mathbf{k}_1$. For conical diffraction with extraordinary polarization, we draw an ellipsoid of revolution centered at S. The intersections with the two spheres (F and F') yield the cones of extraordinary polarized light.

Case 2. Both the incident beam and the readout beam are assumed to be extraordinary waves. In this case we draw ellipsoids centered at F and F' representing normal surfaces of the fanned light. We then draw the normal surface (a sphere and an ellipsoid) of the scattered light at S. Four cones are obtained, in general. Two of them are ordinary and two are extraordinary.

Case 3. We consider a unique case when the readout beam is the incident beam itself and is polarized as an extraordinary wave. In this case, the fanned light is also extraordinary. The fanning hologram, however, is in general birefringent. Using the graphic method, we draw two ellipsoids of revolution centered at F and F'. Let $\mathbf{k}_2 = \mathbf{k}_1 = \mathbf{k}_e$, where $\mathbf{k}_e$ represents the wave vector of the incident beam. The center F is located at $2\mathbf{k}_e$ from the origin S. The other center F' is exactly on top of S. We now draw the normal surface representing the scattered light. The normal surface consists of a sphere and an ellipsoid. Because of the unique situation ($\mathbf{k}_2 = \mathbf{k}_1$), a nontrivial solution for conical diffraction occurs only for the ordinary waves. As shown in Figure 8.7, the ellipsoid F only interests the ellipsoid portion of the normal surface S at one point (i.e., at $\mathbf{k}_e$). The sphere portion of the

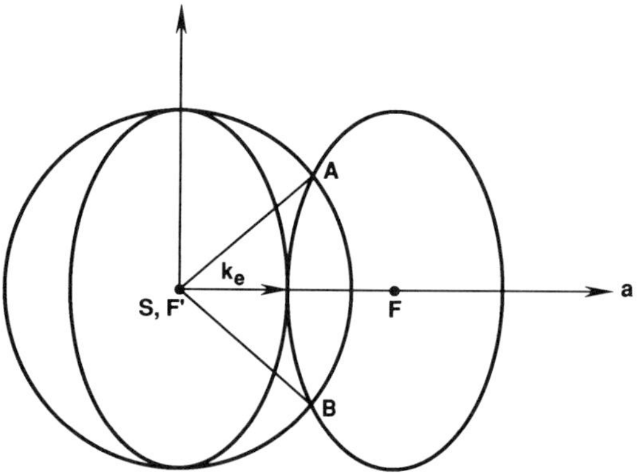

Figure 8.7. Anisotropic conical diffraction. The figure shows the intersection of the ellipsoid centered at F with the normal surface centered at S. We assume the incident wave vector $\mathbf{k}_e$ is perpendicular to the c-axis.

normal surface S is physically larger than the ellipsoid portion. We recall that $BaTiO_3$ is a negative uniaxial crystal with $n_o > n_e$. The intersection of the sphere with the ellipsoid of revolution centered at F leads to a ring of ordinary conical diffraction, according to Eq. (8.4-4). The angular position and shape of the diffraction cone depend on the angle between the wave vector $\mathbf{k}_e$ and the c-axis of the crystal. For normal incidence (i.e., $\mathbf{k}_e$ perpendicular to the c-axis), the internal half-apex angle of the cone in the plane containing the incident beam and the x-axis is given by [3]

$$\cos \alpha = \frac{2n_e}{n_o + n_e} \tag{8.4-5}$$

and for scattering in the plane containing the incident beam but normal to the c-axis is given by [3]

$$\cos \alpha = \frac{3n_e^2 + n_o^2}{4n_o n_e} \tag{8.4-6}$$

The intensity distribution of the conical diffraction is determined by the dielectric tensor of the fanning hologram. A detailed calculation would require the evaluation of

$$\left\langle \mathbf{p}' \middle| \Delta\left(\frac{1}{n^2}\right) \middle| \mathbf{p} \right\rangle$$

where $\Delta(1/n^2)$ is given by Eq. (3.3-14) and $\mathbf{p}'$, $\mathbf{p}$ are the polarization vectors of the scattered and incident beam, respectively.

Dark rings have also been observed when a $BaTiO_3$ crystal is illuminated with two laser beams. These rings originate from one beam reading the fanning holograms created by the other beams. In the above discussion, we ignore the fanning hologram created by the reading beam. When both fanning holograms are present, the intersection of these two holograms in K-space consists of several rings. The intersection will either strengthen or weaken the index gratings. Under the appropriate conditions, the index gratings of the intersected region can be destroyed by mutual cancellation. This leads to dark rings. For details, interested readers are referred to Ref. [5].

PROBLEMS

8.1 Recording of Multiple Holograms. Consider a set of multiple holograms given by

$$n(r) = \sum_{m=1}^{N} n_m \exp(-i\mathbf{K}_m \cdot \mathbf{r})$$

where n_m are constants and $\mathbf{K}_m$ are grating wave vectors. Using two-beam interference, the holograms can be recorded in sequence.

a. Let the beam intensity be fixed so that the erasure rate is $1/\tau$. Show that at the end of the Nth recording, the grating amplitudes are

$$n_m = n_{m0} \exp(-T_m/\tau)$$

where n_{m0} is the grating amplitude of the mth grating when it was recorded, and T_m is

$$T_m = \sum_{j=m+1}^{N} t_j$$

where t_j is the recording time for the jth grating.

b. Show that the dynamic range of the grating amplitude Δn must obey the following condition

$$\Delta n > e^{T/\tau} n_1$$

where

$$T = \sum_{j=2}^{N} t_j$$

and n_1 is the first grating amplitude to be recorded.

8.2 Simultaneous Recording of N Gratings

a. Show that, by using N mutually incoherent laser sources, the multiple gratings in Problem 8.1 can be recorded simultaneously.

b. Show that the recorded grating amplitude can be written

$$n_m = \frac{2A_m B_m^*}{\sum_j [|A_j|^2 + |B_j|^2]} \Delta n_s$$

where Δn_s is the saturated index modulation, A_j, B_j are amplitudes of the jth writing beams.

c. Show that

$$n_m \le \frac{1}{N} \Delta n_s$$

for all $m = 1, 2, \ldots, N$. In other words, the amplitude of each grating is less than $1/N$ of the saturated index modulation. Assuming $|A_j|^2 + |B_j|^2 = C$ for $j = 1, 2, 3, \ldots, N$.

8.3 Multiple Gratings and Dynamic Range of Δn. Consider a photorefractive space charge density given by

$$\rho(x) = \rho_1 \sin(Kx + \phi_1)$$

where K is the grating wave vector and ρ_1, ϕ_1 are constants. The magnitude of ρ_1 is limited by the dopant density in the medium. The same limitation applies to multiple space-charge gratings

$$\rho(x) = \sum_{m=1}^{N} \rho_1 \sin(K_m x + \phi_m)$$

a. If all K_m are distinct and ϕ_m are random, then $\rho(x)$ at any point x is a sum of N random numbers which are limited by a circle of radius ρ_1 in the complex plane. Show that

$$\langle \rho(x) \rangle = 0 , \qquad \langle |\rho(x)|^2 \rangle = N|\rho_1|^2$$

b. Show that, statistically, the magnitude of $\rho(x)$ is around $\sqrt{N}\rho_1$.

c. Show that if $\rho_{\max}$ is the maximum achievable space-charge density, then the maximum number of gratings is given by

$$N = \left| \frac{\rho_{\max}}{\rho_{\min}} \right|^2$$

where $\rho_{\min}$ is the minimum detectable space-charge density.

8.4 Compressible Holographic Media. If the holographic recording medium is compressible, then the storage capacity can be increased by diluting the material which leads to a larger volume. The increase in the geometrical limit is accompanied by a decrease in the space-charge density. Discuss the effect on the number of gratings that can be stored in the medium.

8.5 *K*-Space Degeneracy for Anisotropic Bragg Scattering. Many photorefractive crystals exhibit birefringent gratings which couple ordinary waves with extraordinary waves. In other words, the incident and the diffracted waves can be of different polarization states. Derive the degeneracy curves.

8.6 Show that there is no conical diffraction for the case when $\mathbf{k}_1 = \mathbf{k}_2$ and the incident beam is polarized as an ordinary wave.

REFERENCES

1. J. Feinberg, "Asymmetric Self-Defocusing of an Optical Beam from the Photorefractive Effect," J. Opt. Sci. Am. **73**, 46 (1982).

2. D. A. Temple and C. Warde, "Anisotropic Scattering in Photorefractive Crystals," J. Opt. Soc. Am. **B3**, 337 (1986).
3. M. D. Ewbank, P. Yeh and J. Feinberg, "Photorefractive Conical Diffraction in $BaTiO_3$," Opt. Comm. **59**, 423–428 (1986).
4. See, for example, A. Yariv and P. Yeh, Optical Waves in Crystals (Wiley, New York, 1984).
5. T. Y. Chang and P. Yeh, "Dark Rings in Photorefractive Conical Diffractions," Proc. SPIE **739**, 109–116 (1987).

9

Phase Conjugate Interferometry

In classical optics, interferometers consist of mirrors and beams splitters. There are many different interferometers, each designed for the purpose of specific applications. These include Fabry–Perot interferometers, Michelson interferometers, Sagnac interferometers, etc. With phase conjugate mirrors available, we can construct interferometers using these new mirrors. In this chapter, we discuss properties of some of the phase conjugate interferometers.

9.1 INTRODUCTION

Before we proceed with the discussion of phase conjugate interferometers, we review briefly the properties of dielectric beam splitters. Referring to Figure 9.1, we consider reflection and transmission properties of a planar dielectric beam splitter. Let r and t be reflection and transmission coefficient, respectively, for light incident from the left side. We assume that the beam splitter consists of an arbitrary number of dielectric layers and does not necessarily possess any left-right symmetry. Let r' and t' be the corresponding coefficients when the light is incident from the right side. It has been known for some time that these four coefficients are related by the so-called Stokes relationships [1].

$$t^*t' + r^*r = 1 \tag{9.1-1}$$

$$t^*r' + r^*t = 0 \tag{9.1-2}$$

These equations are results of time reversal symmetry and are derived as follows. Consider the incidence of an electromagnetic field represented by E, as shown in Figure 9.1(a). The amplitude of the reflected wave and transmitted wave are rE and tE, respectively. We now imagine that the time is reversed so that r^*E^* and t^*E^* retrace their paths and recombine at the same beam splitter. Here the complex conjugation is a result of a time reversal operation (see Problem 9.1). If the properties of the dielectric beam splitter are invariant under the time reversal operation, then these two incident waves will undergo reflection and transmission with the coefficients as described in Figures 9.1(a, b). In addition, these two incident waves (see

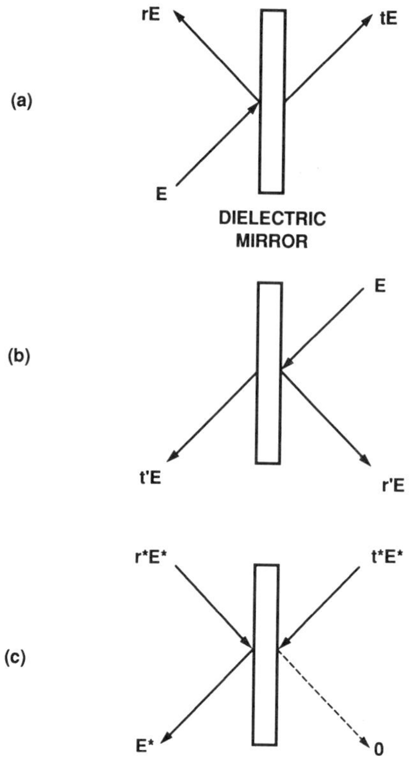

Figure 9.1. Stokes treatment of the principle of reversibility.

Figure 9.1(c)) should recombine and reconstruct the time reversed version of the incident wave E. Thus we obtain

$$t't^*E^* + rr^*E^* = E^* \tag{9.1-3}$$

and

$$r't^*E^* + tr^*E^* = 0 \tag{9.1-4}$$

Since E^* is arbitrary, these two equations are exactly the Stokes equations (9.1-1) and (9.1-2). It is important to note that Stokes equations are results of the time reversal symmetry. They are valid provided the planar dielectric beam splitters are lossless. In addition to the relationships given by Stokes equations, the coefficients must also satisfy the conservation of energy. In other words,

$$t^*t + r^*r = 1 \tag{9.1-5}$$

$$|t'|^2 + |r'|^2 = 1 \tag{9.1-6}$$

Using Eqs. (9.1-1) and (9.1-5), we obtain

$$t' = t \tag{9.1-7}$$

which is known as left-and-right incident theorem [1]. The above result, Eq. (9.1-7), holds even when the absorption is present in the beam splitter. According to the theorem, the transmittance and the phase shift when light is incident from the left side are identical to those when light is incident from the right side. There is no similar relationship for the reflection coefficients. However, for lossless beam splitter, it can be shown that

$$|r'|^2 = |r|^2 \tag{9.1-8}$$

which is a result of Eq. (9.1-7) plus the conservation of energy. In summary, it is important to remember that Stokes equations are results of time reversal symmetry. They can also be derived directly by using matrix theory of multilayers [1].

9.2 PHASE CONJUGATE FABRY–PEROT INTERFEROMETERS

In classical optics, Fabry–Perot interferometers consist of two parallel partially reflecting dielectric mirrors. Referring to Figure 9.2, we consider a phase conjugate Fabry–Perot interferometer which consists of a partially reflecting mirror and a phase conjugate mirror. Consider now the incidence of a beam of light. The beam is split into a reflected part and a transmitted part at the dielectric mirror. Let E_0 be the amplitude of the incident beam. The transmitted beam tE_0 is subsequently reflected back and forth between the ordinary mirror and the phase conjugate mirror. Let r, t, r', and t' be

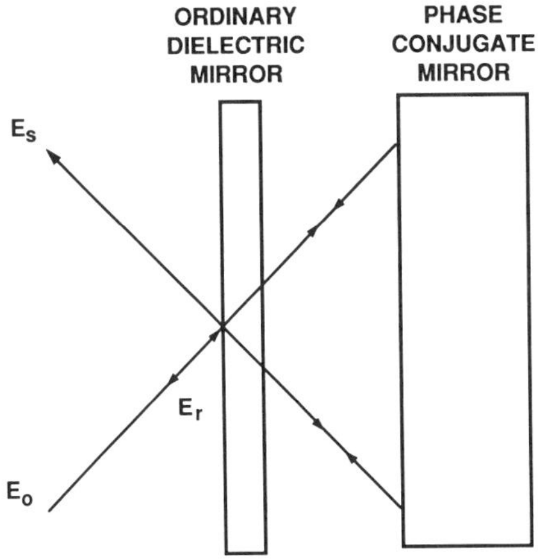

Figure 9.2. Phase conjugate Fabry–Perot interferometer.

the reflection and transmission coefficients of the ordinary mirror, and ρ be the phase conjugate reflection coefficient. It follows that the amplitude of the successively specular reflected beams are rE_0, $r'^*tt'|\rho|^2E_0$, $r'^*tt'|\rho|^4|r'|^2E_0, \ldots$. It can be easily shown that the phase difference between any two successive reflected beams is exactly zero by virtue of phase conjugation. Thus the total of the specularly reflected beams is

$$\begin{aligned} E_s &= rE_0 + r'^*tt'|\rho|^2E_0 + r'^*tt'|\rho|^4|r'|^2E_0 + \cdots \\ &= \frac{r(1-|\rho|^2)}{1-|\rho r|^2}\,E_0 \end{aligned} \tag{9.2-1}$$

where we have used the Stokes equations for $r'^*t = -t^*r$ and the conservation of energy for $|r'|^2 = |r|^2$. In a similar way, the total amplitude of the retroreflected beam is

$$E_r = \frac{\rho(1-|r|^2)}{1-|\rho r|^2}\,E_0^* \tag{9.2-2}$$

We notice that the retroreflected beam is a phase conjugate wave. Both E_s and E_r are independent of the distance between the ordinary mirror and the phase conjugate mirror. In addition, the specular reflection E_r is significantly reduced due to the presence of the phase conjugate mirror, and $E_s = 0$ when $\rho = 1$, according to Eq. (9.2-1). The reduction of specular reflection due to the onset of phase conjugation has been observed in photorefractive $BaTiO_3$ crystals which act as self-pumped phase conjugators [2].

If the phase conjugate reflectivity $|\rho|$ can be greater than unity, then self-oscillation is possible when

$$|\rho r| = 1 \tag{9.2-3}$$

According to Eqs. (9.2-1) and (9.2-2), $|\rho r| = 1$ would lead to finite E_s and E_r with an infinitesimal E_0. Thus Eq. (9.2-3) is an oscillation condition for a phase conjugate resonator.

In arriving at the results of Eqs. (9.2-1) and (9.2-2), we assume that the medium between the beam splitter and phase conjugate mirror is reciprocal. In other words, the phase shift is independent of the direction of propagation. If the medium contains nonreciprocal elements, then there will be a phase shift between the successive terms in Eq. (9.2-1). The nonreciprocal phase shift can be due to Faraday rotation, flow of liquids, etc. Let ϕ be the net nonreciprocal phase shift per roundtrip, the total of the specularly reflected beams becomes

$$E_s = \frac{r - r|\rho|^2e^{-2i\phi}}{1-|\rho r|^2e^{-2i\phi}}\,E_0 \tag{9.2-4}$$

and the total amplitude of the retroreflected beam is

$$E_r = \frac{\rho(1-|r|^2)e^{-i\phi}}{1-|\rho r|^2 e^{-2i\phi}} E_0^* \tag{9.2-5}$$

The dependence on ϕ is very similar to that of a regular Fabry–Perot interferometer and can be employed to measure nonreciprocal phase shift. The intensity of phase conjugate beam is given by, according to Eq. (9.2-5)

$$I_r = \frac{|\rho|^2(1-|r|^2)^2}{(1-|\rho r|^2)^2 + 4|\rho r|^2 \sin^2\phi} I_0 \tag{9.2-6}$$

where I_0 is the incident intensity. If the phase conjugate mirror has a unit reflectivity (i.e., $|\rho|^2 = 1$), then the intensity of the retroreflected beam becomes

$$I_r = \frac{(1-R)^2}{(1-R)^2 + 4R\sin^2\phi} I_0 \tag{9.2-7}$$

where $R = |r|^2$ is the mirror reflectivity. The result is formally identical to that of the transmittance of a Fabry–Perot etalon (Problem 9.3).

9.3 PHASE CONJUGATE MICHELSON INTERFEROMETER

In classical optics, a Michelson interferometer consists of a beam splitter and two planar mirrors. An incident beam of light is split into two by the beam splitter. These two parts of the beam are reflected back and recombined at the beam splitter. An interference pattern is observed at the output port. The interference fringes depend on the path difference between the two partial beams. Although the mirrors reverse the velocity of beams by retroreflection, the phases are not reversed. If these two mirrors are replaced with phase conjugate mirrors, then both the velocity and the phase of the waves are reversed.

Referring to Figure 9.3, we consider a phase conjugate interferometer which consists of a beam splitter, a mirror and a phase conjugator mirror (PCM). An incident beam of light is split into two by the beam splitter. Let E_0 be the amplitude of the incident light, the reflected and transmitted part are rE_0 and tE_0, respectively. These two parts are reflected by the phase conjugate mirror. Let ρ be the phase conjugate reflection coefficient. The amplitude of the electric field at the output port is given by

$$E = (r^*t + t^*r')\rho E_0^* = 0 \tag{9.3-1}$$

which is exactly zero according to Stokes equation (9.1-2). On the other hand, the amplitude of the retroreflected wave is

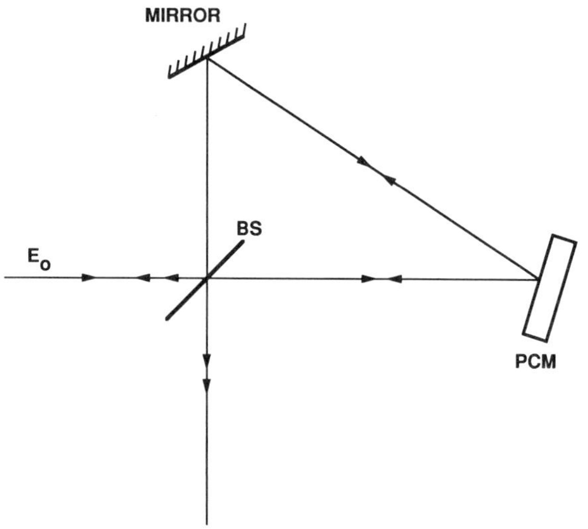

Figure 9.3. Phase conjugate Michelson interferometer.

$$E_r = (r^*r + t^*t')\rho E_0^* \tag{9.3-2}$$

or equivalently, according to Eq. (9.1-1)

$$E_r = \rho E_0^* \tag{9.3-3}$$

The above results agree with time reversal. Initially, only an incident beam enters the system. When time is reversed, the incident beam is reconstructed by the retroreflected beam. Thus the time reversal symmetry leads to the zero field at the output port. We also note that the zero field at the output port is independent of the path difference between the two ports of the beam. Optical time reversal using phase conjugate interferometers has been demonstrated using a self-pumped photorefractive phase conjugator as the PCM [3].

The zero field at the output port indicates that the two contributions are of the same magnitude and exactly 180° out of phase. This unique property is useful in parallel image subtraction. The application will be discussed further in Chapter 10.

9.4 PHASE CONJUGATE SAGNAC INTERFEROMETER

Sagnac interferometers were originally designed to measure the rotation rate of the earth by using light beams. A beam of light is split into two by a beam splitter. The two beams are caused to traverse opposing paths around a circuit formed by two or three mirrors. The beams recombine at the beam splitter and the interference fringes are observed. The whole apparatus can be mounted on a rigid platform that can be rotated about an axis. As a result of the rotation, the optical path lengths to traverse the circuit for the two beams are different. This leads to a phase difference which is proportional to the angular velocity of the platform. It can be shown that the path difference for the beams is given by [4, 5]

$$\Delta L = \frac{4}{c}\mathbf{A}\cdot\Omega \tag{9.4-1}$$

where **A** is the area of the circuit and Ω is the angular velocity.

In a phase conjugate Sagnac interferometer, one of the two mirrors is replaced by a mutually pumped phase conjugator (MPPC) (see Figure 9.4). Again, an incident light beam is split into two by the beam splitter. These two beams are directed into opposite faces of a photorefractive crystal. Under the appropriate conditions, mutually pumped phase conjugation occurs. It is important to note that the beams at the crystal need not be mutually coherent. The path length difference from the beam splitter can be greater than the coherent length of the light beam. The conjugate waves recombine at the beam splitter and the interference is obtained at the output port.

We now evaluate the amplitude of the field E at the output port (see Figure 9.4). Let A be the amplitude of the incident wave, and r, r', t be the

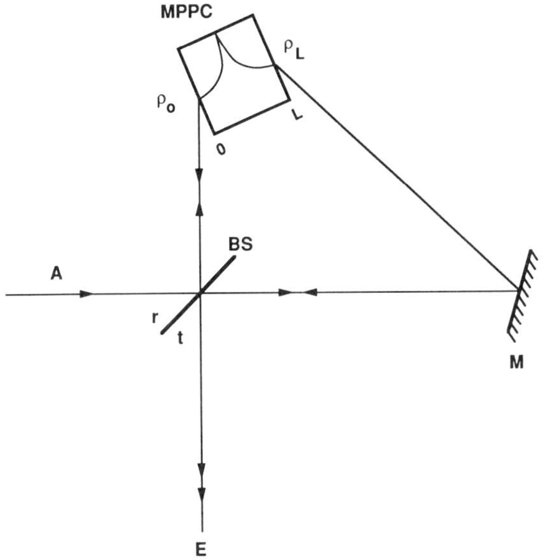

Figure 9.4. Phase conjugate Sagnac interferometer using MPPC.

reflection and transmission coefficients of the beam splitter, respectively. By following the propagation of the beams, we obtain

$$E = t\rho_0(rA)^* + r'\rho_L(tA)^* \tag{9.4-2}$$

where ρ_0 and ρ_L are the phase conjugate reflection coefficients of the MPPC. These two reflection coefficients are related, according to Eq. (6.4-14), by

$$\frac{\rho_0}{\rho_L} = \frac{|t|^2}{|r|^2} \tag{9.4-3}$$

Using $t^* r' = -tr^*$ according to Eq. (9.1-4) in conjunction with Eq. (9.4-3), we can rewrite Eq. (9.4-2) as

$$E = \rho_0 r^* t\left(1 - \frac{|r|^2}{|t|^2}\right)A^* \tag{9.4-4}$$

It follows that $E = 0$ for a beam splitter with $|r|^2 = |t|^2 = 1/2$. According to Eqs. (9.4-1) and (9.4-4), the two contributions are exactly π out of phase. Complete cancellation would require a 50–50 beam splitter.

We note that the output field Eq. (9.4-4) is similar to that of a Michelson interferometer discussed earlier in the chapter. The result is independent of the path difference between the two contributions. This unique property can also be employed for the parallel subtraction of optical images, as well as the detection of nonreciprocal phase shift due to rotation.

9.5 MACH–ZEHNDER HOLOGRAPHIC INTERFEROMETER

In classical optics, a straightforward method to obtain two-beam interference is to use a Mach–Zehnder interferometer which consists of a beam splitter and two mirrors (see Figure 9.5). An incident beam is split into two by the beam splitter. The mirrors are used to redirect the two beams so that they intersect and produce interference fringes. If a photorefractive crystal or a holographic medium is placed in the region of intersection, the interference pattern can be recorded as a volume grating or hologram.

Consider now the incidence of beams from both sides of the beam splitter such that the beam paths inside the interferometer are identical. The electric field due to each of the sources at the intersection region can be written

$$E_A = rA\exp(-i\mathbf{k}_1\cdot\mathbf{r}) + tA\exp(-i\mathbf{k}_2\cdot\mathbf{r}) \tag{9.5-1}$$

$$E_B = t'B\exp(-i\mathbf{k}_1\cdot\mathbf{r}) + r'B\exp(-i\mathbf{k}_2\cdot\mathbf{r}) \tag{9.5-2}$$

where $\mathbf{k}_1$, $\mathbf{k}_2$ are the wave vectors, r, r', t, t' are the reflection and

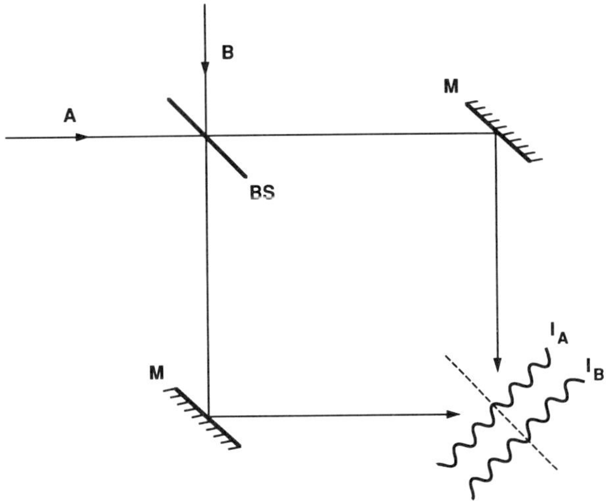

Figure 9.5. Mach–Zehnder holographic interferometer.

transmission coefficients of the beam splitter, respectively. The intensity, within a factor of proportionality, of these two fields can be written

$$I_A = |E_A|^2 = [|r|^2 + |t|^2 + rt^* e^{i\mathbf{K}\cdot\mathbf{r}} + r^* te^{-i\mathbf{K}\cdot\mathbf{r}}]|A|^2 \tag{9.5-3}$$

$$I_B = |E_B|^2 = [|r'|^2 + |t'|^2 + r't'^* e^{-i\mathbf{K}\cdot\mathbf{r}} + r'^* t' e^{i\mathbf{K}\cdot\mathbf{r}}]|B|^2 \tag{9.5-4}$$

where $\mathbf{K} = \mathbf{k}_2 - \mathbf{k}_1$.

Using $t' = t$ and $|r'|^2 = |r|^2$, we obtain

$$I_A = [|r|^2 + |t|^2 + rt^* e^{i\mathbf{K}\cdot\mathbf{r}} + r^* te^{-i\mathbf{K}\cdot\mathbf{r}}]|A|^2 \tag{9.5-5}$$

$$I_B = [|r|^2 + |t|^2 + r't^* e^{-i\mathbf{K}\cdot\mathbf{r}} + r'^* te^{i\mathbf{K}\cdot\mathbf{r}}]|B|^2 \tag{9.5-6}$$

The intensities I_A and I_B consist of a DC term and a sinusoidal term. As a result of the relationship $r't^* = -r^*t$, these two sinusoidal terms are spatially shifted by π (or 1/2 of a period). Physically, this means that the interference pattern due to source A is spatially shifted by 1/2 of the fringe relative to that of source B. In other words, these two interference patterns are exactly π out of phase. Thus if these two sources are of equal intensity, the simultaneous incidence of these two beams will lead to a complete cancellation of the interferences fringes. Using Eqs. (9.5-5) and (9.5-6) and $|A|^2 = |B|^2$, we obtain

$$I = I_A + I_B = 2|A|^2 \tag{9.5-7}$$

which indeed reflects the absence of interference fringes.

The unique property of π phase shift can be employed to implement parallel image subtraction. Such applications are discussed in the next chapter.

PROBLEMS

9.1 **Complex Conjugation and Time Reversal.** Let $\mathbf{E} = \mathbf{E}_0(\mathbf{r}) \exp(i\omega t)$ and $\mathbf{H} = \mathbf{H}_0(\mathbf{r}) \exp(i\omega t)$ be solutions to Maxwell's equations.

a. Show that $\mathbf{E}^*$ and $\mathbf{H}^*$ also satisfy Maxwell's equations. Note that $\mathbf{E}^*$, $\mathbf{H}^*$ and $\mathbf{E}$, $\mathbf{H}$ are, in fact, the same field since only the real parts have physical meaning.

b. Show that the conjugate waves

$$\mathbf{E}^c = \mathbf{E}_0^*(\mathbf{r}) \exp(i\omega t) \quad \text{and} \quad \mathbf{H}^c = \mathbf{H}_0^*(\mathbf{r}) \exp(i\omega t)$$

also satisfy the wave equation, provided the medium is lossless (i.e., ϵ and μ are the real tensors).

c. Show that $\mathbf{E}^c$, $\mathbf{H}^c$ are time-reversed replicas of $\mathbf{E}$, $\mathbf{H}$, respectively.

9.2 Derive Eqs. (9.2-4) and (9.2-5).

9.3 Derive Eqs. (9.2-6) and (9.2-7), and show that the structure is equivalent to a Fabry–Perot etalon with two mirrors each with a reflectivity of $R = |r|^2$, when $|\rho|^2 = 1$.

9.4 Let η be the grating diffraction coefficient of the MPPC in a phase conjugate Sagnac interferometer (see Figure 9.4), show that Eq. (9.4-4) can be written

$$E = \eta t^2 \left(1 - \frac{|r|^2}{|t|^2}\right) A$$

Also show that

$$\eta = \rho_0 \frac{r^*}{t} \frac{A^*}{A} = \rho_L \frac{t^*}{r} \frac{A^*}{A}$$

9.5 Show that, strictly speaking, Eq. (9.5-7) is valid only when the two sources are mutually incoherent but of nearly the wave wavelength. Find out how far part can the wavelength difference be?

REFERENCES

1. See, for example, P. Yeh, Optical Waves in Layered Media (Wiley, New York, 1988) p. 113.

2. D. M. Pepper, "Observation of Diminished Specular Reflectivity in Self-Pumped Photorefractive Phase Conjugators," Technical Digest, OSA Annual Meeting Seattle, WA, 1986, p. 51.
3. M. D. Ewbank, P. Yeh, M. Khoshnevisan, and J. Feinberg, "Time Reversal by an Interferometer with Coupled Phase Conjugate Reflectors," Opt. Lett. **10**, 282–284 (1985).
4. See, for example, J. Killpatrick, "The Laser Gyro," IEEE Spectrum (October, 1967), pp. 44–55.
5. E. J. Post, "Sagnac Effect," Rev. Mod. Phys. **39**, 475–493 (1967).

10

Optical Computing

The field of optics, specifically nonlinear optics, is emerging as an area of increasing importance for applications in optical computing because of its inherent parallelism and its extreme bandwidth of communication [1]. In addition, the nonlinear optical (NLO) response of materials allows for the possibility of manipulating the propagation of an optical beam by another beam of light inside a nonlinear medium. These manipulations include changes of phase, polarization, intensity, direction of propagation, and even the frequency of the light beam [2, 3]. In the area of optical computing, these NLO manipulations offer unique capabilities in real-time image processing, holographic interconnection, parallel image subtraction, phase conjugation, optical data storage, spatial light modulations and bistable optical devices [4].

In this chapter, we will describe various applications of photorefractive nonlinear optics in optical computing. These include real-time image processing, image subtraction and amplification, logic and matrix operations, optical interconnection and neural networks.

10.1 FOURIER OPTICS WITH PHOTOREFRACTIVE HOLOGRAMS

One of the most convincing examples of the power of parallel optical computing is the transformation property of a thin lens. It has been known for some time that a thin lens can perform a Fourier transform [5]. Referring to Figure 10.1, we consider a two-dimensional input object $f(x, y)$ located at the front focal plane of the lens. The object is illuminated with a plane wave of coherent radiation. As a result of the transformation property of the lens, the Fourier transform $F(u, v)$ is obtained at the rear focal plane. Within a factor of proportionality, the amplitude of the electric field at the rear focal plane can be written

$$F(u, v) = \int\int f(x, y) \exp\left[i \frac{2\pi}{\lambda f} (ux + vy) \right] dx\, dy \qquad (10.1\text{-}1)$$

where λ is the wavelength of light and f is the focal length.

Although Eq. (10.1-1) can be derived by using scalar diffraction theory [5], the basic principles can be explained in terms of the properties of thin

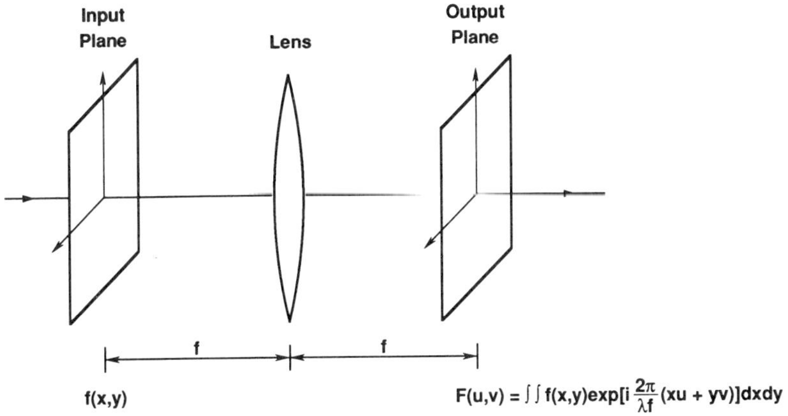

Figure 10.1. Fourier transformation using a thin lens. A transparency representing an arbitrary function $f(x, y)$, illuminated by a plane wave along the optical axis, is transformed into a Fourier transform $F(u, v)$ by the thin lens.

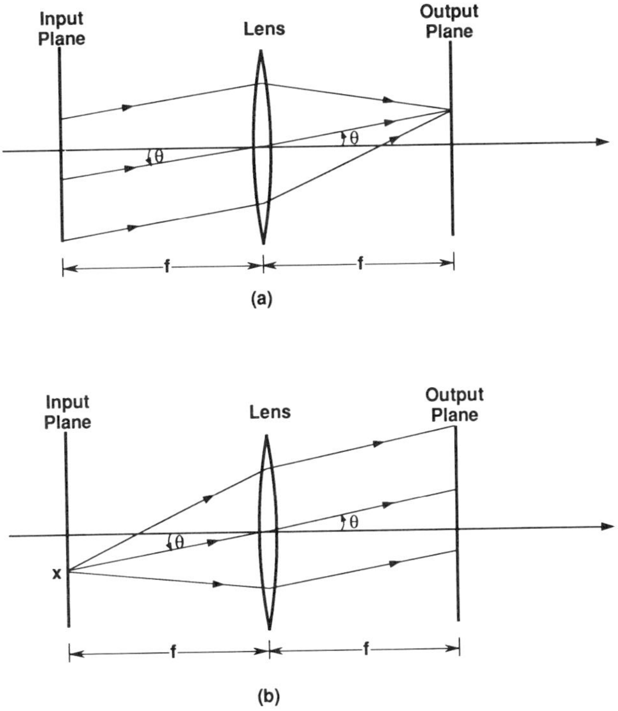

Figure 10.2. (a) A plane wave is transformed by a thin lens into a point at the rear focal plane. (b) A point at the front focal plane is transformed by a thin lens into a plane wave.

lenses. Referring to Figure 10.2, we consider the incidence of a plane wave onto the lens. The plane wave is focused into a spot in the focal plane. The location of the spot is given by

$$u = f\theta \tag{10.1-2}$$

where θ is the angle between the wave vector of the plane wave and the optic axis. The intensity of the spot will be proportional to the intensity of the incident plane wave. Since each spot corresponds uniquely to a spatial component of the object at the front focal plane, the distribution of spots at the rear focal plane represents the Fourier transform of the object.

If we assume paraxial ray approximation (i.e., $\theta \ll 1$), then the incident plane wave can be written

$$E = A \exp(-ik_x x - ikz) \tag{10.1-3}$$

where A is the amplitude, k is the wave number, and k_x is the x-component of the wave vector which is given by

$$k_x = k\theta \tag{10.1-4}$$

Using Eqs. (10.1-2) and (10.1-4), we obtain from Eq. (10.1-3)

$$E = A \exp(-ikz) \exp\left(-i \frac{kx}{f} u\right) \tag{10.1-5a}$$

or equivalently

$$E = A \exp(-ikz) \exp\left(-i \frac{2\pi}{\lambda f} ux\right) \tag{10.1-5b}$$

This is the electric field of the plane wave which produces a spot at u in the rear focal plane. Similarly, a spot in the front plane at x will be transformed into a plane wave by the thin lens. The electric field of the plane wave can be written

$$E = A \exp(-ikz) \exp\left(i \frac{2\pi}{\lambda f} ux\right) \tag{10.1-6}$$

We note that u is the variable and x is a parameter. Thus if the spot distribution is described by $f(x, y)$, then the electric field of the waves at the rear focal plane can be obtained by linear superposition. This leads to Eq. (10.1-1).

To appreciate the speed of computation, consider the Fourier transform of an input picture $f(x, y)$ which consists of 1000×1000 pixels. There are 10^6 algebraic operations for each point in (u, v) plane, according to Eq.

(10.1-1). Thus a Fourier transform would require 10^{12} algebraic operations. This would take an electronic computer a very very long time to process. Using a thin lens, the whole computation is done in less than 10^{-9} s. This is the power of parallelism.

One of the most useful applications of Fourier transformation is optical convolution and correlation for pattern recognition. Referring to Figure 10.3, we consider the basic principle of the Vander Lugt filter. An input pattern $f(x, y)$ is first transformed by a thin lens. Its Fourier transform $F(u, v)$ is then multiplied by a complex function $H(u, v)$. The product $F(u, v)H(u, v)$ is then transformed by another thin lens. If $H(u, v)$ is the Fourier transform of a function $h(x, y)$, then we obtain at the output plane

$$g(x, y) = \int\int f(\xi, \eta) h(x - \xi, y - \eta)\, d\xi\, d\eta \tag{10.1-7}$$

which is the convolution function of f and h. If $h(x, y)$ is given by

$$h(x, y) = s^*(-x, -y) \tag{10.1-8}$$

then

$$g(x, y) = \int\int f(\xi, \eta) s^*(\xi - x, \eta - y)\, d\xi\, d\eta \tag{10.1-9}$$

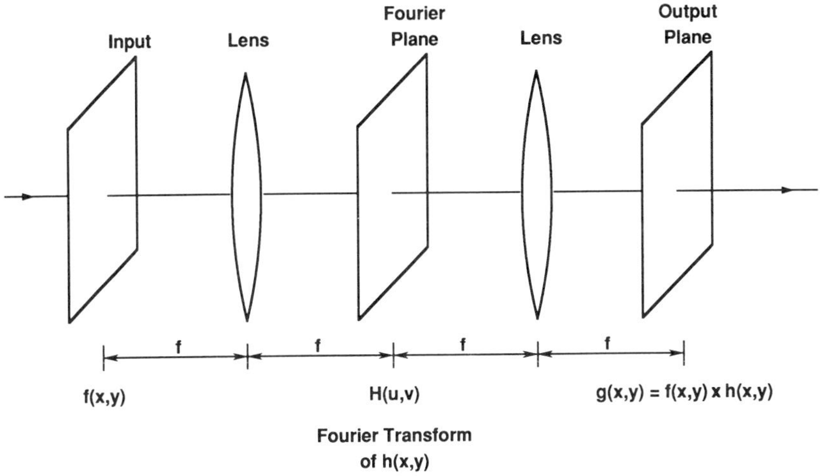

Figure 10.3. Vander Lugt filter. An input function $f(x, y)$ at the front focal plane is Fourier transformed by the first lens. The Fourier transform $F(u, v)$ is multiplied with $H(u, v)$ at the Fourier plane which is a Fourier transform of another function $h(x, y)$. The product $F(u, v) \times H(u, v)$ is then transformed by the second lens into $g(x, y)$ at the output plane which is a convolution function of f and h.

which is recognized as the cross correlation function of f and s. We note that the operation of the Vander Lugt filter requires the recording of the spatial spectrum (Fourier spectrum) of $h(x, y)$ or $s^*(-x, -y)$. Photorefractive crystals can be employed as the recording media. This opens the possibility of real-time optical correlations in which the recorded matched filter $H(u, v)$ can be changed dynamically.

Optical four-wave mixing in photorefractive media is suitable for real-time Fourier optics. Figure 10.4 illustrates a schematic drawing of such an optical correlator. We assume that the photorefractive crystal is thin so that it is approximately at the common Fourier plane of the two lenses. These lenses are used to transform the input patterns into their Fourier transforms. Optical four-wave mixing happens now in the Fourier domain.

Let $A_1(x, y)$, $A_2(x, y)$, and $A_3(x, y)$ be the three input patterns entering the system and let $a_1(u, v)$, $a_2(u, v)$, and $a_3(u, v)$ be their Fourier transforms, respectively. As a result of four-wave mixing, the phase conjugate wave is given by

$$a_4(u, v) = c a_1^* a_2 a_3 \tag{10.1-10}$$

where c is a constant. This wave, $a_4(u, v)$, will undergo Fourier transform and yield $A_4(x, y)$. In other words

$$A_4(x, y) = F\{c a_1^* a_2 a_3\} \tag{10.1-11}$$

where F represents the operation of Fourier transform.

Consider now some special cases of interest.

a. $A_3(x, y) = \delta(x, y)$. This corresponds to a pinhole and the Fourier transform $a_3(u, v)$ is a constant. The output in this case becomes

$$A_4(x, y) = c' F\{a_1^* a_2\} \tag{10.1-12}$$

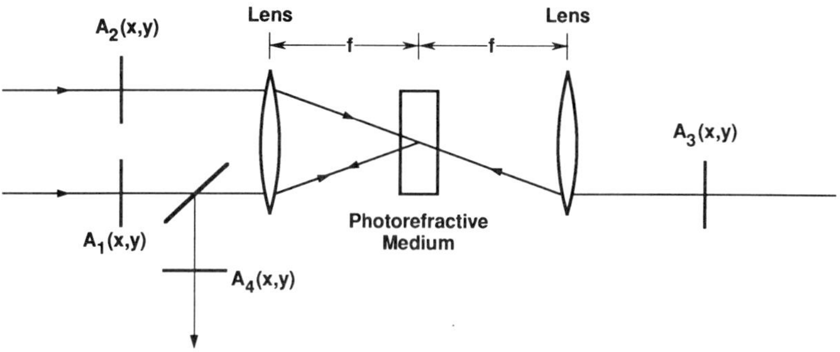

Figure 10.4. Real time optical correlation and convolution using four-wave mixing in a photorefractive medium. The output $A_4(x, y)$ is a correlation function of $A_1(x, y)$ with the convolution function of $A_2(x, y)$ and $A_3(x, y)$.

where c' is a constant and the result is the cross correlation function of A_1 and A_2. In other words,

$$A_4(x, y) = \int\int A_1^*(\xi - x, \eta - y) A_2(\xi, \eta)\, d\xi\, d\eta \qquad (10.1\text{-}13)$$

b. $A_1(x, y) = \delta(x, y)$. In this case a_1 is a constant, and the output becomes

$$A_4(x, y) = c' F\{a_2 a_3\} \qquad (10.1\text{-}14)$$

where c' is a constant and the result is exactly the convolution function of A_2 and A_3. In other words,

$$A_4(x, y) = \int\int A_2(x - \xi, y - \eta) A_3(\xi, \eta)\, d\xi\, d\eta \qquad (10.1\text{-}15)$$

Real time optical correlation and convolution by using Fourier optical four-wave mixing in photorefractive media has been demonstrated experimentally [6].

In general, when neither A_1 nor A_3 are delta-functions, the output is given by Eq. (10.1-11). Using the convolution theorem (see Problem 10.1), the output $A_4(x, y)$ is a correlation function of $A_1(x, y)$ and $F\{a_2 a_3\}$, which is a convolution function of $A_2(x, y)$ and $A_3(x, y)$. Thus, Eq. (10.1-11) can be written symbolically as

$$A_4 = cA_1 * A_2 \times A_3 \qquad (10.1\text{-}16)$$

where $*$ denotes correlation operation and $\times$ denotes convolution operation. Notice that these two operations are associative.

10.2 PARALLEL IMAGE SUBTRACTION

We have shown in the last chapter that exact cancellation is obtained at the output port of a phase conjugate Michelson interferometer. This means that the two contributions to the total electric field are exactly π out of phase. Such a relationship is ideal for subtraction.

Image subtraction has been a subject of considerable interest in signal processing. Electronic digital processing of images is slow because of its serial nature. Optical techniques offer the capability of parallel processing over the entire images. The parallel processing is versatile and inherently faster. The technique of optical image synthesis by the addition and subtraction of the complex amplitude of light was first described by Gabor et al. [7]. The basic principle consists of spatially modulating the two images by periodic waves that are mutually shifted by a phase of 180° [8, 9].

Although there are other techniques of image subtraction [10–13], interferometers such as the Mach–Zehnder or Michelson offer convenient ways for the addition and subtraction of the complex amplitude of images [14–16]. In the interferometric methods, the subtraction is obtained by introducing the two images symmetrically in the two arms of the interferometer where a path difference corresponding to π phase shift exists between them. It is known that the interferometers are extremely difficult to adjust and that they cannot easily maintain the fixed path difference. In addition, only the center fringe is useful for image subtraction or addition. In many cases, the center fringe is not large enough to cover all the images. The method of parallel image subtraction by phase-conjugate interferometry described in the following eliminates these two problems.

Referring to Figure 10.5, we consider the insertion of one transparency at each arm of a phase conjugate Michelson interferometer. Each transparency represents an image. Let $T_1(x, y)$ and $T_2(x, y)$ be the intensity transmittance of the transparencies. Using proper optics for reimaging, the electric field at the output port is given by

$$E_S = \rho A^*[r^* t T_1(x, y) + r' t^* T_2(x, y)] \tag{10.2-1}$$

where A is the amplitude of the incident wave, ρ is the reflection coefficient of the self-pumped phase conjugator (SPPC), r, r', t are reflection and transmission coefficients, respectively. It is important to note that the use of SPPC is only a matter of convenience. Externally pumped phase conjugators (EPPCs) work as well. In Eq. (10.2-1) we neglect the effect of beam splitter BS2 which only affects the overall intensity of the results. Using the Stokes relation, Eq. (10.2-1) becomes

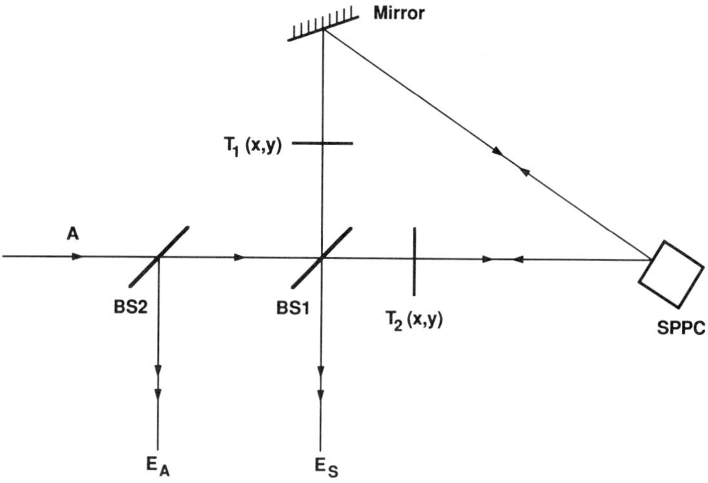

Figure 10.5. Parallel image subtraction using a phase conjugate Michelson interferometer.

$$E_S = \rho A^* r^* t[T_1(x, y) - T_2(x, y)] \tag{10.2-2}$$

which is exactly the difference between the two images. At the other output port created by BS2, the amplitude of the electric field is

$$E_A = \rho A^*[|r|^2 T_1(x, y) + |t|^2 T_2(x, y)] \tag{10.2-3}$$

For a beam splitter with $|r|^2 = |t|^2 = 1/2$, the result becomes

$$E_A = \rho A^* |r|^2 [T_1(x, y) + T_2(x, y)] \tag{10.2-4}$$

which is exactly the addition of the two images. Parallel subtraction and addition of two images have been demonstrated experimentally using $BaTiO_3$ SPPC [17–19].

Figure 10.5 shows the parallel image subtractor using a phase conjugate Michelson interferometer. The image subtractor can also perform logic operations. Consider the case where both transparencies are either 1 or 0. According to Eq. (10.2-2), a complete cancellation would require that these two transparencies be identical. An output intensity of 1 will appear at port E_S when only one of these two transparencies transmits. Thus such an image subtractor can act as an XOR gate. The truth table of such a logic operation is given in Table 10.1. In the case when the transparencies are encoded with a matrix of binary data, such an image subtractor acts as a two-dimensional array of XOR gates.

A special case of image subtraction is intensity inversion, which is obtained by removing one of the transparencies so that the transmittance $T_1(x, y)$ becomes unity in one arm. The result becomes

$$E_s = \rho A^* r^* t[1 - T_2(x, y)] \tag{10.2-5}$$

which represents an inverted image of $T_2(x, y)$. The subtraction technique described above is good for intensity subtraction only. This is a result of double passages through the transparency. Amplitude subtraction and single passage may be useful in situations when successive (or cascaded) operations are required. The earlier technique, proposed by Gabor et al., requires a

Table 10.1 The Logic Operations Represented by the Intensity I_s

T_1	T_2	$I \propto \|T_1 - T_2\|^2$
1	1	0
1	0	1
0	1	1
0	0	0

successive recording of the two images on a holographic plate with a 180° phase shift and the subsequent readout of the composite hologram. This 180° phase shift must remain uniform across the entire images during the second recording.

In what follows, we describe a new method that utilizes the simultaneous recording and readout of the two images in photorefractive medium (or other nonlinear medium) by using a double Mach–Zehnder interferometric arrangement. The 180° phase shift is automatically present between the two sets of beams writing the image-bearing hologram by virtue of the Stokes principle [8]. When such a composite hologram is read out by a reference beam, parallel subtraction of the two images is obtained.

Figure 10.6 shows the subtraction of two images on transparencies with amplitudes T_1 and T_2. These two transparencies are illuminated with two mutually incoherent light sources. For transparencies, the illumination can be from behind the images, as opposed to frontal illumination in the case of reflective objects. The two image-bearing beams are directed toward the two input ports of the Mach–Zehnder interferometer. They are mixed into two beams, which are then recombined at a photorefractive crystal (or other nonlinear medium). Let E_1 and E_2 be the field amplitudes of the mixed beams. At the photorefractive crystal, these two amplitudes can be written

$$E_1 = (tT_1A + r'T_2B)\exp(-i\mathbf{k}_1 \cdot \mathbf{r}) \tag{10.2-6}$$

$$E_2 = (rT_1A + t'T_2B)\exp(-i\mathbf{k}_2 \cdot \mathbf{r}) \tag{10.2-7}$$

where $\mathbf{k}_1$, $\mathbf{k}_2$ are the wave vectors, A and B are the amplitudes of the two

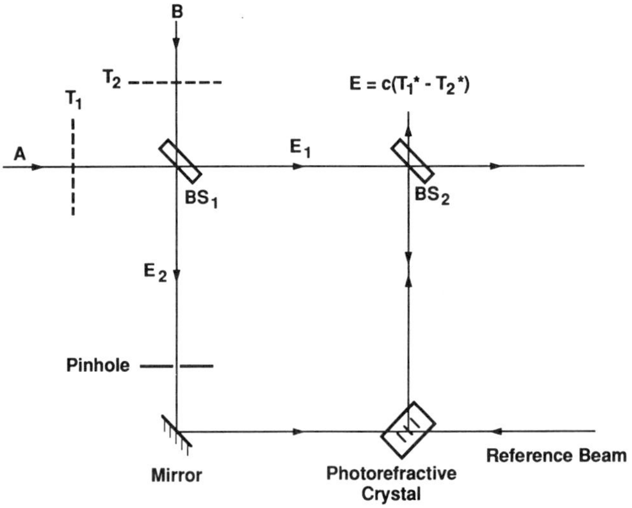

Figure 10.6. Parallel image subtraction using Mach–Zehnder holographic interferometer.

incoherent illuminating beams, t, r, t', and r' are the Fresnel transmission and reflection coefficients of beam splitter BS_1 (the prime indicates incidence from the back side of the beam splitter). These two field amplitudes form an interference pattern in the photorefractive crystal. The fundamental component of the intensity pattern can be written

$$I = (rt^*|T_1|^2|A|^2 + t'r'^*|T_2|^2|B|^2)\exp(-i\mathbf{K}\cdot\mathbf{r}) + \text{c.c.} \qquad (10.2\text{-}8)$$

where $\mathbf{K} = \mathbf{k}_2 - \mathbf{k}_1$ and we have neglected the cross terms A^*B and B^*A because A and B are the amplitudes of two mutually incoherent beams. According to the Stokes principle, the Fresnel reflection and transmission coefficients for a lossless beam splitter satisfy the relation

$$rt^* + tr'^* = 0 \qquad (10.2\text{-}9)$$

where we have used $t' = t$. Substituting Eq. (10.2-9) into Eq. (10.2-8), we obtain

$$I = rt^*(|T_1|^2|A|^2 - |T_2|^2|B|^2)\exp(-i\mathbf{K}\cdot\mathbf{r}) + \text{c.c.} \qquad (10.2\text{-}10)$$

Thus the interference pattern (and recorded hologram) that contains image information T_1 is 180° out of phase with respect to that containing image information T_2. This 180° phase shift is independent of the path length of the two arms of the Mach–Zehnder interferometer. Thus we have a simultaneous recording with an inherent phase shift of 180°. When such a hologram is read out by a reference beam, the diffracted beam that is sampled through beam splitter BS_2 contains information about the difference between the intensities of the two images. To obtain high-contrast image subtraction, it is desirable that we have $|A|^2 = |B|^2$; note that the amplitude reflection and transmission coefficients (r, r', t) are arbitrary.

We now modify the above analysis in order to perform algebraic subtraction of the two complex amplitudes. If we place a pinhole in one arm of the interferometer, the pinhole will act as a spatial filter and eliminate spatial information about the images. Thus, for this case, the field amplitude E_2 becomes

$$E_2 = (rA + t'B)\exp(-i\mathbf{k}_2\cdot\mathbf{r}) \qquad (10.2\text{-}11)$$

provided that the dc (spatial) components of images T_1 and T_2 are of the same magnitude. The intensity pattern of the interference becomes

$$I = (rt^*T_1^*|A|^2 + t'r'^*T_2^*|B|^2)\exp(-i\mathbf{K}\cdot\mathbf{r}) + \text{c.c.} \qquad (10.2\text{-}12)$$

and when we substitute in Eq. (10.2-9) this becomes

$$I = rt^*(T_1^*|A|^2 - T_2^*|B|^2)\exp(-i\mathbf{K}\cdot\mathbf{r}) + \text{c.c.} \qquad (10.2\text{-}13)$$

We note that, according to Eq. (10.2-13), the holograms now contain the complex field amplitudes of the images instead of the intensities (absolute square of the amplitudes) of the images. Again, because of the Stokes principle, the holographic gratings formed by beams from these two input ports are shifted relative to each other by exactly 180°. In other words, the holographic grating that contains complex image T_1 is shifted 180° relative to that of complex image T_2 regardless of the path difference. When such a superposition of holograms is read out by the reference beam, the diffracted beam consists of the algebraic difference of the two complex fields. Strictly speaking, the amplitude of the diffracted beam when observed at the difference plane is proportional to the complex conjugate of $(T_1 - T_2)$. The reference beam must be counterpropagating with respect to E_2 in order to read out the hologram efficiently (i.e., select only the fundamental component of the intensity pattern). The reference beam can be either coherent or incoherent with respect to the two illumination beams, as it simply reads the gratings written in the crystal.

This new technique can also be used for the subtraction of two three-dimensional objects. When used for this purpose, the difference must be observed in a volume instead of a plane. In practice, this could be achieved by examining the beam in a scattering medium such as liquid-nitrogen vapor. Since this new method does not require that there be a round-trip propagation through the transparencies, it can be used for multiple operations, cascaded operations, or even do-loop-type operations. Parallel subtraction of complex images using Mach–Zehnder holographic interferometers has been demonstrated experimentally [20].

In summary, we have described techniques for parallel subtraction of two images in real-time by using photorefractive phase conjugation or holographic interference. The latter method provides an automatic phase shift of 180° between the complex images and may be viewed as a real-time implementation of Gabor's image-subtraction technique. The major advantages of this technique over previous image subtraction techniques are the amplitude subtraction capability (instead of intensity subtraction) and the ability to process three-dimensional images and perform multiple operations, since there is only a single pass through the transparencies. In addition, the phase information contained in the transparencies can be determined by this technique, since the output from the interferometer is proportional to the difference in the field amplitudes. For example, if $T_1 = 1$ and $T_2 = e^{i\Psi}$ ($\sim 1 + i\Psi$, for small Ψ), the output will be proportional to Ψ.

10.3 OPTICAL INTERCONNECTION

The VLSI technology for high speed computing, has the potential of creating systems with 10^{10} interconnected logic gates on a single wafer. Each of these devices can perform a logic operation in a time period less than

10^{-10} s. The awesome functional capability of such an integration is not matched by a corresponding communication bandwidth. Even at present complexity, designs are severely restricted by wiring congestion and delays due to wiring capacitance [21]. As the desired throughput rate continues to increase, the combination of decreasing gate sizes and increasing chip area can lead to a communication crisis [22, 23].

There are many advantages of optical interconnection. The inherent parallism and the extremely wide bandwidth for communication can be exploited for demanding computations. Optical processors are inherently two-dimensional for the processing of images. Optical signals can propagate through each other in separate channels with essentially no interaction. Optical signals can also propagate in parallel channels without any interference and crosstalk. In the VLSI systems, optical interconnection can be used to solve the problem of communication as well as the clock distribution.

It is generally believed that a hybrid technology organized to take advantage of the strengths of both electronics and optics could have a revolutionary impact on future computing systems. In the opto-electronic systems, electronics is employed for the high speed switching (logic operations) and optics is employed for communication. Thus, in opto-electronic integrated circuits (OEIC) lasers are interfaced with electronic circuits so that signals can be carried by the emitted light beams. In addition, photo detectors are also interfaced with the systems so that signals can be received and then converted to electronic signals for further processing.

Optical elements such as lenses, mirrors, gratings, holograms, fibers, and waveguides are essential for efficient optical interconnection. With the aid of these optical elements, signal-bearing optical beams can be efficiently collected, optically processed and then efficiently redirected into photodetectors for further electronic processing. Holograms (or computer generated holograms CGH) are especially useful for interconnection between elements of different spatial geometry, for example, coupling light from an edge emitter to a spatial light modulator with square pixels. Without spatial mode conversion, there is a severe loss of energy. Holograms can be tailored made to perform highly efficient fixed interconnection between elements with different spatial geometry. Figure 10.7 shows a schematic drawing of a holographic optical interconnection. The hologram serves as the routing element. Light emitting from a laser will be diffracted by the hologram and will be redirected toward one or many of the photodetectors. Such a fixed interconnection pattern is adequate for many applications, including the point operation, matrix operation, the clock distribution and a globally synchronous systolic processor. However, there are many situations in which a dynamic optical interconnect is required.

Optical interconnection linking laser arrays and detector arrays plays a key role in parallel computing [23, 24]. Conceptually, such an interconnection can be achieved by using an optical matrix-vector multiplication,

$$\mathbf{v}' = M\mathbf{v} \tag{10.3-1}$$

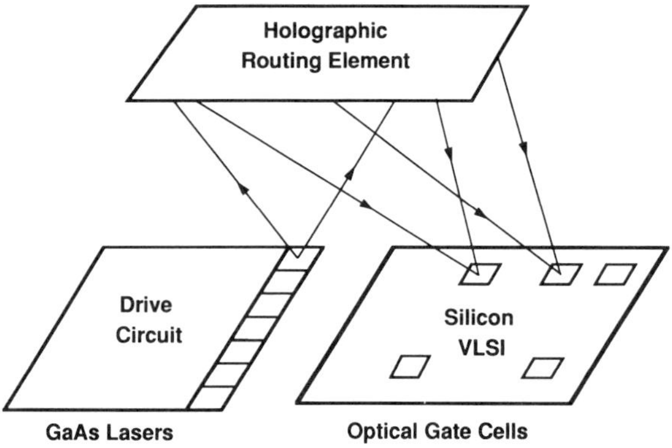

Figure 10.7. Holographic optical interconnection.

where **v** is the input vector representing the signals carried by an array of N lasers and $\mathbf{v}'$ is the output vector representing the signals received by the array of N detectors. M is an $N \times N$ matrix representing the interconnection pattern.

When a transparency or a spatial light modulator (SLM) is used as the interconnection mask, parallel matrix-vector multiplication can be achieved. Let the interconnection mask consist of an $N \times N$ array of transmission windows. The input vector (a row of N elements) is fanned out by using, for example, a cylindrical lens pair, so that the light from each element is broadcast over a corresponding column of windows. After transmitting through the interconnection mask, similar optics are used to collect the light from each row of the windows and to sum the output into a column vector of N elements. Such an architecture provides the N^2-parallelism. However, a large fraction of energy is absorbed by the transparency or SLM. When used as a crossbar switch, such an architecture has an energy efficiency of only $1/N$, where N is the dimension of the array [25]. This occurs because $(N-1)/N$ of the light energy from each element of the input vector does not pass through the crossbar mask. The energy loss increases as the dimension of the array increases. This is sometimes referred to as the fanout energy loss. For a 1000×1000 crossbar switch, the loss due to fanout can be as big as 99.9%. This loss is not acceptable in high speed computing because high gains need to be provided to overcome the loss. In addition to the inherent fanout energy loss, all SLM have finite insertion loss due to imperfect transmissions and scatterings. If we include such insertion losses, the energy efficiency for a crossbar switch would become t/N, where t is the transmittance ($t < 1$) of each of the windows.

It has been known for some time that holographic optical element (HOE) can be employed for free space optical interconnection. Light from each laser source of the input array is Bragg scattered and redirected to one or more detectors of the output array. There are several specific requirements that need to be met for an HOE to be used for interconnection of VLSI circuits. These include alignment, diffraction efficiency, etc. In addition, a new hologram is needed for each new interconnection pattern. In view of these issues, it is desirable to use dynamic holographic media such as photorefractive crystals. Photorefractive crystals such as $BaTiO_3$, SBN are by far the most efficient holographic media. The recording of a volume hologram inside these crystals requires a time of the order of 1 ms using light energy of the order of 1 W/cm^2.

In this section, we describe a scheme of forming a dynamic hologram inside a photorefractive crystal such that the hologram is capable of performing reconfigurable optical interconnection with a very high energy efficiency. Such a scheme utilizes the nonreciprocal energy transfer in two-wave mixing in order to achieve an extremely high diffraction efficiency.

Referring to Figures 10.8 and 10.9, we describe this new method of reconfigurable optical interconnection which utilizes the nonreciprocal energy transfer in photorefractive two-wave mixing to improve the energy efficiency. Figure 10.8 describes a one-dimensional case for the sake of clarity in explaining the concept. A small fraction of this incident laser beam is coupled out of the beam by using a beam splitter (BS). This small fraction (called probe beam) is then expanded by using a cylindrical lens and passes through the spatial light modulator (SLM). In the example shown, the laser beam is to be connected to detectors *b* and *d* as prescribed by the SLM. The transmitted probe team is then recombined with the main beam inside a photorefractive crystal. As a result of nonreciprocal energy coupling, almost all the energy in the main beam is transferred to the probe beam which carries the interconnection pattern. The result is an optical interconnection with a very high energy efficiency.

Figure 10.9 describes the reconfigurable interconnection for laser arrays and detector arrays. In the example (a 4 × 4 interconnection) shown, laser 1

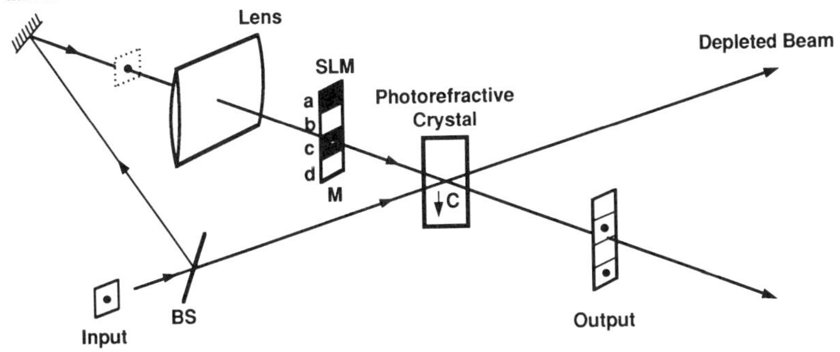

Figure 10.8. A schematic drawing of $1 \times N$ optical interconnection using dynamic photorefractive holograms with $N = 4$.

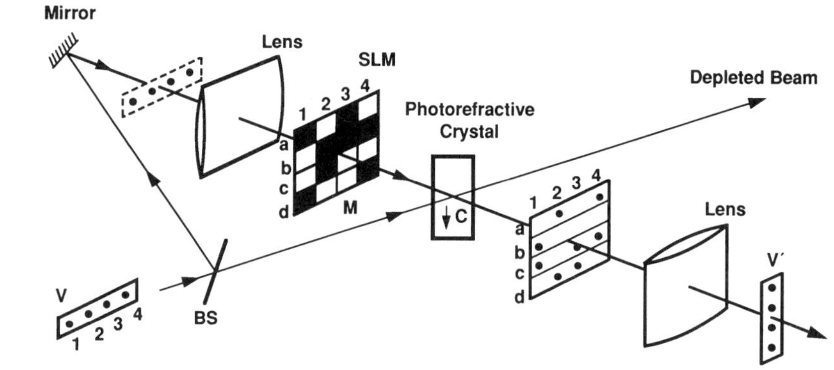

Figure 10.9. A schematic drawing of $N \times N$ optical interconnection using dynamic photorefractive holograms with $N = 4$.

is to be connected to detectors *b* and *c*, laser 2 is to be connected to detectors *a* and *d*, laser 3 is to be connected to detectors *c* and *d*, laser 4 is to be connected to detectors *a* and *c*. In terms of the matrix-vector multiplication, such an interconnection can be written as

$$\mathbf{v}' = \begin{bmatrix} 0 & 1 & 0 & 1 \\ 1 & 0 & 0 & 0 \\ 1 & 0 & 1 & 1 \\ 0 & 1 & 1 & 0 \end{bmatrix} \begin{bmatrix} v_1 \\ v_2 \\ v_3 \\ v_4 \end{bmatrix} = \begin{bmatrix} v_2 + v_4 \\ v_1 \\ v_1 + v_3 + v_4 \\ v_2 + v_3 \end{bmatrix} \qquad (10.3\text{-}2)$$

where v_1, v_2, and v_4 are signals carried by the lasers.

A cylindrical lens is used to focus the 2D array of beams into a vector (1D array). As a result, detector *a* receives signals from lasers 2 and 4, detector *b* receives signals from laser 1, detector *c* receives signals from lasers 1, 3 and 4, and detector *d* receives signals from lasers 2 and 3. Such a concept can be extended to interconnect *N* lasers with *M* detectors where *N* and *M* are two large numbers.

The two-wave mixing described in Figures 10.8 and 10.9 may be viewed as a real-time holography in which the recording and readout occur simultaneously inside the photorefractive crystal. The beam splitter and the SLM are used to record a volume hologram which represents the interconnection pattern as prescribed by the SLM. The energy coupling involved in the two-wave mixing ensures that the diffraction efficiency during the readout is almost 100%. This, of course, requires a proper orientation of the crystal so that the energy of the readout beam is greatly depleted. The high energy efficiency comes from the fact that most of the energy is carried by the readout beam which does not pass through the SLM but is diffracted into the interconnection pattern by the hologram.

The energy efficiency of such an interconnection pattern can be estimated for a crossbar switch as follows. Let *R* be the reflectance of the beam splitter. It is legitimate to assume that the beam splitter is practically lossless. We may also assume that the surface of the photorefractive crystal is antireflection-coated so that the Fresnel reflection loss can be neglected. Under these conditions, the two beams that arrive at the photorefractive crystal have energies $(1 - R)$ and Rt/N, respectively. Inside the crystal, these two beams undergo photorefractive coupling. As a result, most of the energy of the pump beam is transferred to the probe beam which contains the interconnection pattern. The energy efficiency can be easily derived and is given by

$$\eta = \frac{tR}{N} \frac{1+m}{1+me^{-\gamma L}} e^{-\alpha L} \tag{10.3-3}$$

where we recall that t is the transmittance of the SLM, m is the beam intensity ratio,

$$m = \frac{(1-R)N}{tR} \tag{10.3-4}$$

and L is the interaction length, γ is the coupling constant, and α is the bulk absorption coefficient. For photorefractive crystals such as $BaTiO_3$, SBN, the coupling constant is very large (i.e., $\gamma L \gg 1$). The efficiency can be written approximately

$$\eta = \left[\frac{tR}{N} + (1-R)\right] e^{-\alpha L} \tag{10.3-5}$$

We notice that for large N, the ultimate energy efficiency is $\exp(-\alpha L)$ by using beam splitter with a a very small reflectance R (i.e., $R \sim 0$).

In summary, we have described a method of reconfigurable optical interconnection using dynamic holograms in photorefractive crystals. Such a method minimizes or eliminates the fan-out energy loss and provides a very high energy efficiency. The interconnection can be reconfigured by using a different SLM pattern. The hologram formation time will limit the reconfiguration time. Once the hologram which contains the interconnection pattern is formed inside the photorefractive crystal, such a scheme is capable of providing the interconnection for high data rate transmission. In such an interconnection, the output of each laser can be input to any one or all of the detectors. Optical phase conjugation can also be used in conjunction with the two-wave mixing to correct for any phase aberration that may be caused by the crystal imperfection [26]. Reconfigurable optical interconnection using photorefractive holograms has been demonstrated experimentally in $BaTiO_3$ crystals [27–29].

10.4 OPTICAL NEURAL NETWORKS

All neural network models follow the same basic architecture, simulating that of the biological neural system. A neural network consists of many identical elements (neurons) linked by interconnections (synapses). The differences between different models lie in the way neurons are connected and the way interconnections are formed. Neurons are usually depicted as points and interconnections as lines. Figure 10.10 shows the basic architecture.

Each neuron receives signals from and sends signals to many other neurons. The incoming signals to a neuron are first amplified or attenuated by the interconnections and then summed by the neuron. Suppose neuron i receives a total input signal x_i^{in} from c different neurons, and the output signal sent by neuron j is x_j^{out}. The relationship between the received and transmitted signals is

$$x_i^{\text{in}} = \sum_{j=1}^{c} w_{ij} x_j^{\text{out}} \tag{10.4-1}$$

where w_{ij} is the interconnection weight between neurons i and j. The total received signal is then processed by neuron i to send out an output signal [30]

$$x_i^{\text{out}} = g(x_i^{\text{in}}) \tag{10.4-2}$$

where g is a thresholding function. The commonly used function, $g(x)$, has a lower bound and an upper bound with continuous values in between, for example, a sigmoid function, as in Figure 10.11. When the input is very small, the output is the minimum value. When the input is some moderate value, the output is something in between. When the input is very high, the output reaches its maximum value. An extreme case of the function is

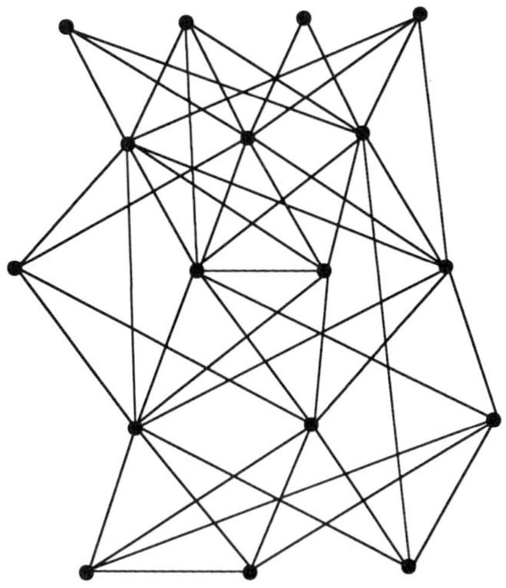

Figure 10.10. Architecture of neural networks. The neurons (dots) are interconnected by synapses (lines). Acting as logic gates, each neuron fires signals when the total received signal is above certain threshold. The strength of the synapses represents information stored in the network.

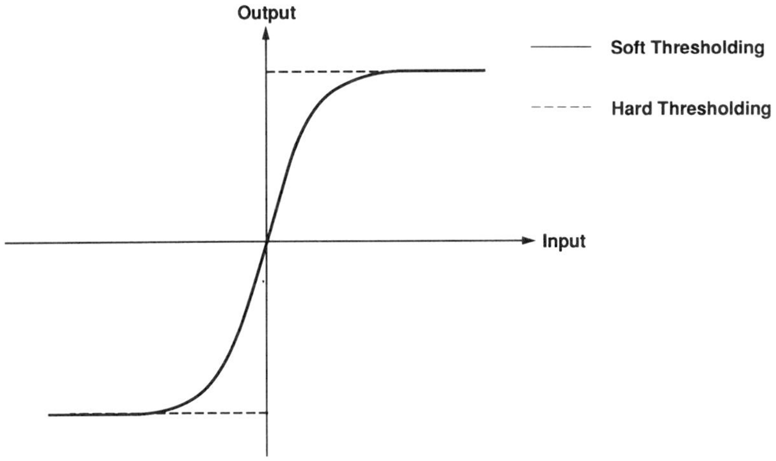

Figure 10.11. Thresholding functions.

obtained when the output jumps from its minimum to its maximum at certain threshold input value, also shown in Fig. 10.11. In this case, the output has only two possibilities, low, if the input is below the threshold value and high, otherwise. The processing performed by the neuron in this case is a threshold-logic function. This step function is called a hard thresholding function. In analogy, a continuous function $f(x)$ is called a soft thresholding function.

Many models assume that neurons are divided into different layers, as shown in Figure 10.12. A neuron receives signals only from neurons in the previous layer and sends a signal only to neurons in the next layer. The simplest network is a single layer network. It contains one layer of inter-

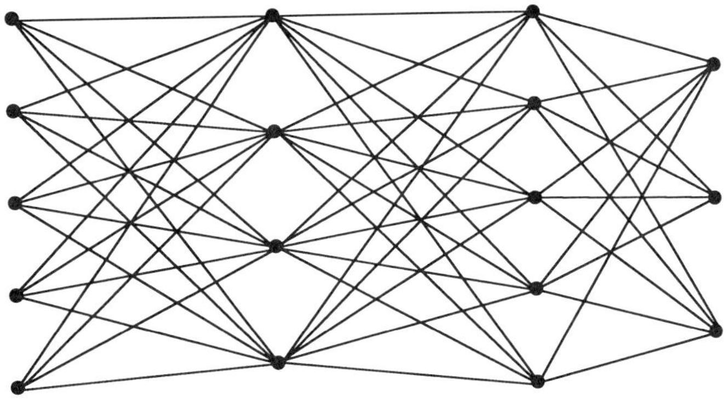

Figure 10.12. Multilayer neural network. Only neurons in the neighboring layers are interconnected.

connections and two layers of neurons. The neurons before and after the interconnections are called the input and output neurons, respectively. Multilayer networks are needed to achieve complicated functions.

One simple neural network is a two-state classifier. It contains two layers of neurons, N input neurons and only one output neuron. The task is to classify many objects into two different classes. An object can be specified by a given set of values for all N neurons in the first layer. The class which the given object belongs to corresponds to the output value of the neuron in the second layer. This neuron uses a hard thresholding function to give an output value of either high or low, usually 1 or -1. All objects corresponding to a final output value 1 are classified into class 1, otherwise class 2. By selecting the proper weight matrix $\mathbf{w}$ and the threshold w_0, a number of objects can be classified into two desired classes. As N becomes very large, the maximum number of objects that can be correctly classified into any possible dichotomy is defined as the capacity of this neural network. The capacity of a linear classifier is $2(N + 1)$ [31].

More complicated neural networks can be constructed from a number of simple classifiers. A general single layer network can be generated by adding more output neurons, each of those being a classifier. Multilayer networks can be created by cascading single layer networks. In general, the set of functions that are implementable by a neural network depends on its structure and interconnections.

Once the structure of a neural network is given, the interconnections need to be specified. The structure determines the capacity of the neural network. The interconnections depend on the particular task. For example, let us assume that a neural network is designed to recognize several faces. How many faces can be recognized depends on the structure. Whose face can be recognized depends on the interconnections, that is, the weight matrix. The procedure to specify the interconnections according to given samples is called training.

A number of training methods have been found. Some treat neural networks as dynamical systems, and they are known as dynamical training algorithms. Some calculate the weight matrix directly from the given samples, known as direct calculations.

Dynamical training typically specifies an energy function in weight space. Different training methods define different energy functions. The global minimum of the energy function is the point corresponding to the correct weight matrix. A training procedure starts from any point in the weight space, where the energy function is usually not at a minimum value. Then the first point is moved along the surface according to the steepest descent until the global minimum point is reached.

An example of dynamical training is the *Perceptron*. Consider only a single output neuron with a hard thresholding function. The weight matrix becomes a weight vector $\mathbf{w}$. The goal is to find a weight vector such that

$$\begin{cases} \mathbf{w} \cdot \mathbf{x}^i > 0, & \mathbf{x}^i \in \text{Class } 1 \\ \mathbf{w} \cdot \mathbf{x}^i < 0, & \mathbf{x}^i \in \text{Class } 2 \end{cases} \tag{10.4-3}$$

where $\mathbf{x}^i$ is a sample vector whose elements specify the values of the first-layer neurons. (Here the threshold value w_0 was absorbed into the weight vector by adding an element $x_0^i = 1$ for all sample vectors.) The total number of arbitrary vectors to be classified, M, must not exceed the capacity of the neural network. The perceptron algorithm starts from an arbitrary weight vector $\mathbf{w}_1$. We then try to classify the sample vectors with $\mathbf{w}_1$. If a sample vector is correctly classified, we go to the next sample. If a sample vector is misclassified, we change the weight vector according to

$$\mathbf{w}_{i+1} = \mathbf{w}_i + \sigma^i \mathbf{x}^i \tag{10.4-4}$$

where $\mathbf{w}_i$ is the weight vector at time i and

$$\sigma^i = \begin{cases} 0, & \text{if } \mathbf{x}^i \text{ is correctly classified} \\ 1, & \text{if } \mathbf{x}^i \in \text{Class 1 and } \mathbf{w} \cdot \mathbf{x}^i < 0 \\ -1, & \text{if } \mathbf{x}^i \in \text{Class 2 and } \mathbf{w} \cdot \mathbf{x}^i > 0 \end{cases} \tag{10.4-5}$$

After the weight change, we continue the process to classify the sample vectors with $\mathbf{w}_{i+1}$. We change the weight vector whenever there is a misclassified sample vector $\mathbf{x}^i$, until all the sample vectors are correctly classified. The final weight vector is not unique but can always be found as a result of this training, provided a solution exists. The convergence proof can be found in Appendix A.

In some cases, the weight matrix can be easily calculated and direct calculation methods can be used. An example of direct calculation is the *Simple Sum*. Given the same set of sample vectors as before, the Simple Sum algorithm calculates the weight vector as

$$\mathbf{w} = \sum_{i=1}^{M} \alpha^i \mathbf{x}^i \tag{10.4-6}$$

where $\alpha^i = 1$ if $\mathbf{x}^i$ is in class 1 and $\alpha^i = -1$ otherwise. To verify that this weight vector can correctly classify a sample vector $\mathbf{x}^j$, consider the product $\mathbf{w} \cdot \mathbf{x}^j$.

$$\begin{aligned} \mathbf{w} \cdot \mathbf{x}^j &= \sum_{i=1}^{M} \alpha^i \mathbf{x}^i \cdot \mathbf{x}^j \\ &= \alpha^j \mathbf{x}^j \cdot \mathbf{x}^j + \sum_{i \neq j} \alpha^i \mathbf{x}^i \cdot \mathbf{x}^j \end{aligned} \tag{10.4-7}$$

When all the sample vectors are statistically independent, the first term in the last expression will dominate. The product will be greater than 0 if α^j is

1 and less than 0 if α^j is -1. Therefore the weight vector so calculated will correctly classify the sample vectors.

The above methods and many others used in training a classifier also apply to the training of neural networks with more than one output neuron. Each neuron can be trained as a classifier, and the weight matrix can be decomposed into weight vectors with respect to each output neuron. For each input vector, the neural network will give a series of classifications, resulting in an output vector.

Direct calculation and dynamical training are used in different situations. For multilayer networks, dynamical training can obtain interconnections without knowing the outputs of the middle-layer neurons. The most commonly used method is backward error propagation (BEP). For single layer networks, direct calculation can reach the optimal solution without changing the weights iteratively. The interconnections are stored in the network as memories, once they are established. When the input to the network is one of the sample vectors, the network will recall the related output stored during the training procedure. This is known as associative memory.

The outer product scheme is one of the models for associative memory. Suppose the input vectors $\{\mathbf{x}^k\}$ are to be associated with the output vectors $\{\mathbf{y}^k\}$ $(k = 1, 2, \ldots, M)$, respectively. The weight matrix, according to the outer product scheme, is a sum of the outer product of the input and output vectors

$$\mathbf{w} = \sum_{k=1}^{M} |y^k\rangle\langle x^k| \tag{10.4-8}$$

where the element w_{ij} is

$$w_{ij} = \sum_{k=1}^{M} y_i^k x_j^k \tag{10.4-9}$$

The outer product scheme can be employed for implementing both auto-associative memory $(\mathbf{y}^k = \mathbf{x}^k)$ and hetero-associative memory $(\mathbf{y}^k \neq \mathbf{x}^k)$. It works in the same way as the Simple Sum algorithm, as discussed earlier in the section. The Simple Sum algorithm in Eq. (10.4-6) is a special case when y_i^k is either 1 or -1.

When an input vector is used to recall the corresponding output vector, the network performs an inner product of the weight matrix and the vector. Suppose the first input vector, $\mathbf{x}^1$, is presented to the network; the inner product

$$\mathbf{w} \cdot \mathbf{x}^1 = |y^1\rangle\langle x^1|x^1\rangle + \sum_{k=2}^{M} |y^k\rangle\langle x^k|x^1\rangle \tag{10.4-10}$$

consists of two terms. The first term is the desired output $\mathbf{y}^1$ multiplied by $\langle x^1|x^1\rangle$. The second term represents the noise resulting from the cross correlation between different input vectors. When the number of memories, M, does not exceed the capacity, the thresholding function of the output neurons will eliminate the crosstalk noise and give the correct answer.

The capacity is the maximum number of associative memories that can be stored and correctly recalled by the network. It can be calculated under the assumption that the input and output vectors are independent random vectors and that each component of a random vector has equal probability of being 1 or -1. In this case, each component in the vector equation, Eq. (10.4-10), is a random variable with Gaussian probability distribution, according to the Central Limit Theorem. The threshold value is set at 0. The probability of correctly classifying each output element is required to approach 1 when the number of input neurons approaches infinity. It can be shown that the maximum number of associate pairs of random vectors that can be stored and recalled is (see Appendix B)

$$M_{\max} = \frac{N}{2(\log N + \log N_1)} \tag{10.4-11}$$

where we recall that N is the dimension of the input vectors and N_1 is the number of neurons at the output layer.

The outer product scheme can tolerate errors in the input vectors, that is, if the input vector used in the recall process is not exactly the same as that used in the training process, the output vector will still be the desired one. For example, an input vector used in the recall process is $\mathbf{x}^1_{\text{error}}$, which is close to the input vector $\mathbf{x}^1$ used in the training process but different in ρN bits. Here ρ is the fraction of error bits. The inner product is written

$$\begin{aligned} \mathbf{w} \cdot \mathbf{x}^1_{\text{error}} &= |y^1\rangle\langle x^1|x^1_{\text{error}}\rangle + \sum_{k=2}^{M} |y^k\rangle\langle x^k|x^1_{\text{error}}\rangle \\ &= |y^1\rangle(1-2\rho)N + \sum_{k=2}^{M} |y^k\rangle\langle x^k|x^1_{\text{error}}\rangle \end{aligned} \tag{10.4-12}$$

We note that the signal term is decreased by a factor $(1-2\rho)$. The noise term remains near zero due to the random nature of the binary vectors. As long as the signal term still dominates, the output after thresholding will be $\mathbf{y}^1$, the desired output vector.

However, tolerating errors in the input vectors decreases the maximum number of vectors that can be correctly recalled, that is, the capacity. It can be shown that the capacity of a network capable of tolerating any input vector with ρN bits errors is

$$M_{\max} = \frac{(1-2\rho)^2 N}{2(\log N + \log N_1)} \tag{10.4-13}$$

The Hopfield model [30, 32] modifies the outer product scheme and introduces feedback to realize an auto-associative memory. The weight matrix, in this model, is a sum of outer products of the training vectors without the diagonal terms,

$$w_{ij} = \begin{cases} \sum_{k=1}^{M} x_j^k x_i^k , & \text{if } i \neq j \\ 0 , & \text{if } i = j \end{cases} \qquad (10.4\text{-}14)$$

In this network, the output at the second layer is redirected to the first layer as a new input. The iterations result in a series of output vectors. As the feedback continues, the output vectors will converge to a stable state corresponding to the stored vector which is nearest to the input vector.

10.5 OPTICAL IMPLEMENTATION

The interconnections play an essential role in the processing of information in neural networks. The interconnections serve the purpose of communication pathways. In addition their weight w_{ij} (strength of connection) provides all the memory of the system. It has been recognized that interconnection-intensive computation can be achieved by using optical techniques. By virtue of their potentially high capacity, holographic techniques have been recognized as some of the most promising approaches [33–35]. The volume hologram in particular offers the most compact means of storing interconnection patterns [34, 35]. Although the interconnection pattern can be computed and fixed in such holograms for prescribed tasks in which the problem parameters do not change, the idea of a neural network that can be adapted on-line to solve problems is especially appealing.

We discuss here an implementation of a perceptron using photorefractive crystals as the holographic medium. By examining the perceptron learning algorithm, we find that three mathematical operations are needed. According to Eqs. (10.4-4), (10.4-5), inner product, addition, and subtraction are essential for the implementation of a perceptron. If the interconnection $\mathbf{w}$ is represented as volume holograms, then the inner product can be easily achieved by the readout process and some optics. Both addition and subtraction weight modification can be achieved by using a double Mach–Zehnder interferometer in conjunction with photorefractive hologram recording. Shown in Figure 10.13 is a diagram depicting the most basic one-layer network with N input elements and one output. The weighted sum of the input pattern elements is thresholded to yield the output

$$y = g\left(\sum_{i=1}^{N} w_i x_i\right), \qquad g(z) = \begin{cases} 1 , & \text{if } z > 0 \\ 0 , & \text{if otherwise} \end{cases} \qquad (10.5\text{-}1)$$

where $g(z)$ is the threshold nonlinearity, w_i is the ith weight, and x_i is the ith

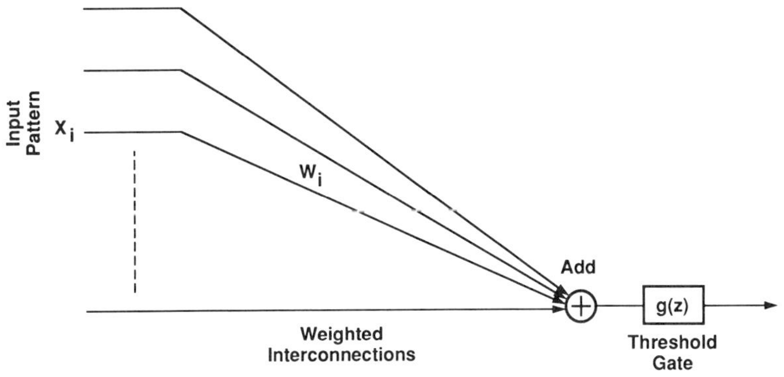

Figure 10.13. One layer neural network.

element of the input pattern. Such a system can be used to dichotomize a set of patterns into two prescribed classes, and more complex, multiple layered networks can be built up using this as the basic building block. Also, extension to multi-category pattern classification can be achieved by having a matrix of weights and a multiplicity of output units as is discussed later in this section.

Simple learning algorithms can be characterized by the update equation

$$\mathbf{w}(p+1) = \mathbf{w}(p) + \sigma(p)\mathbf{x}(p) \tag{10.5-2}$$

where $\mathbf{w}(p)$ is the weight vector at time p, $\mathbf{x}(p)$ is the input vector of the pattern shown at time p, and $\sigma(p)$ is a multiplier that depends on the particular learning algorithm. This includes most of deterministic algorithms where some kind of descent procedure is involved and incremental changes are made to the set of weights during each iteration. For perceptron learning we have, according to Eq. (10.4-5)

$$\sigma(p) = \begin{cases} 0, & \text{if output } y(p) \text{ was correct} \\ 1, & \text{if } y(p) = 0 \text{ but should have been } 1 \\ -1, & \text{if } y(p) = 1 \text{ but should have been } 0 \end{cases} \tag{10.5-3}$$

The threshold bias can be absorbed into the patterns by choosing one element of each pattern to be always equal to a nonzero constant [31]. Note that both additive and subtractive changes to the weight $\mathbf{w}$ must be made to implement the algorithm directly. Although multiple layer networks require more complex routines such as Back Propagation, most algorithms share the common requirement of bipolar weight changes.

The basic components to implement the network described above are an input device to convert the patterns into the appropriate format (e.g., electrical to optical, incoherent to coherent optical), and interconnection device, and a thresholding device for the output unit. The function of the interconnections in this context is to simply compute the inner product between the input pattern **x** and the weight **w**. Volume holograms can be used to implement such functions in a way that is extendable to the multiple category case (i.e., multiple inner products). Consider the arrangement shown in Figure 10.14 where a holographic medium is positioned at the Fourier plane of lens L1. The input pattern is displayed in the spatial light modulator (SLM) which is positioned at the front focal plane of the same lens. A hologram is exposed with a pattern $f(x, y)$ in the SLM and a reference plane wave as shown. After development, another pattern $g(x, y)$ is loaded into the SLM. The light passing through the SLM is diffracted by the hologram and the diffracted amplitude can be shown to be the inner product between the two patterns f and g (see Problem 10.7). Clearly, this is an overkill since the same function could have been achieved with a planar hologram. However, in the multiple category case where a number of different inner products need to be computed simultaneously, the added dimension afforded by the volume hologram is necessary unless one resorts to spatial multiplexing of the planar hologram.

By virtue of their dynamic nature, photorefractive crystals are ideal candidates for the holographic medium. In addition, crystals such as $LiNbO_3$, $BaTiO_3$, and SBN are by far the most efficient holographic media requiring relatively low optical intensity levels (e.g., $1\,W/cm^2$); the most efficient photorefractive crystals exhibit photosensitivities approaching that of photographic films. The holographic process records both amplitude and phase of patterns and thus holographic interconnection techniques can be

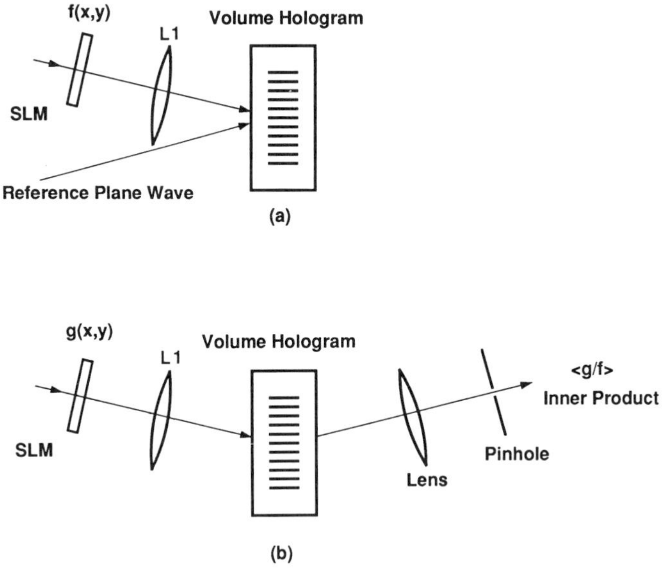

Figure 10.14. Inner product computation using volume hologram.

used to store bipolar valued weights and adaptable changes, which are either subtractive or additive, are possible.

The Perceptron implementation reported by previous workers [37] employed photorefractive crystal for adaptable interconnections but used incoherent erasure to achieve subtractive weight changes. In their system, the photorefractive crystal was placed at the image plane of the SLM and a device (movable piezoelectric mirror) was used to provide either a coherent reference or an incoherent one. For subtractive changes to the hologram, a coherent reference beam was made to be incoherent with respect to the object beam so that nonuniform incoherent erasure resulted. This is roughly equivalent to providing a weight bias in which erasure will result in subtraction. Such methods involving incoherent erasure do not fully take advantage of the phase sensitive nature of holography.

To accurately implement the learning algorithm described by Eq. (10.5-2) and to exploit the hologram's coherent capability just described, a method by which both subtractive as well as additive changes to the weights can be made must be devised. Physically, such a method would amount to a multiple exposure hologram in which each exposure results in a grating (index grating in the case of photorefractive crystals) whose phase is either 0 or π (relative phase difference is π). An obvious method of achieving this is shown in Figure 10.15 in which two coherent waves intersect in a medium to create a holographic grating. One of the beams passes through a phase shifting device (e.g., an electro-optic phase modulator, liquid crystal phase modulating device, piezoelectric movable mirror) to acquire a phase of either 0 or π relative to the other writing beam. After writing a hologram with one phase setting, holographic subtraction can be performed by using the other phase setting in the subsequent exposure, and addition can be performed by using the original phase setting. Such phase shifting devices, however, may suffer from inaccuracies and also performance may degrade with time depending on various device characteristics (e.g., backlash in

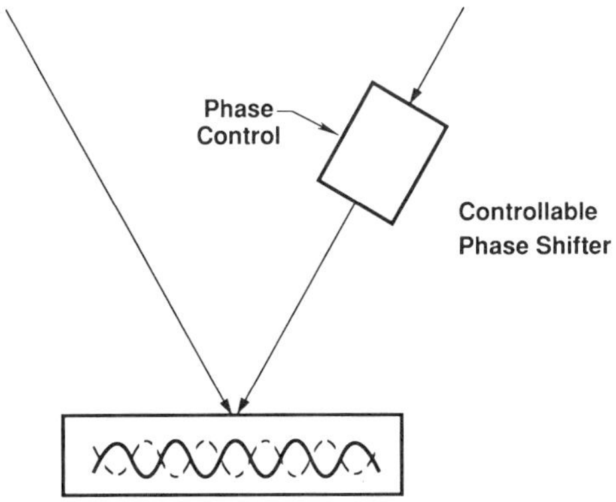

Figure 10.15. Hologram phase control using phase shifting device.

piezo mirrors, voltage inaccuracies in phase modulation devices (electro-optic, liquid crystal). Also the continuous phase variability of these devices is an overkill in the present application in which only two phase settings, 0 and π, are required. For this purpose, the Mach–Zehnder holographic interferometer is ideal.

The basic principle of operation of the phase control system has been described in Chapter 9 (see Figure 10.16). This configuration is also known as a double Mach–Zehnder interferometer because both input ports are used. The beam from laser source 1 (marked by the dotted line) is split into two paths by the dielectric beamsplitter to intersect within the holographic medium to create a grating. Another laser source (which can be derived from laser source 1 or a separate laser with the same nominal wavelength) is positioned so that the directions of the beams transmitted and reflected by the beamsplitter traverse precisely the same paths as those resulting from source 1.

Photorefractive gratings can be formed by either source 1 or source 2. These two gratings are mutually π out of phase. Of course, the accuracy of this result is dependent on how well the beams can be aligned. Such a phase shift of π has been employed for the parallel subtraction of images as discussed earlier in this chapter.

We now describe the overall optical system as shown in Figure 10.17, in which an arrangement to allow for multiple exposures of a photorefractive hologram is shown. This setup uses the holographic phase control method just described. If the SLM contains a picture whose amplitude distribution is given by $a(x', y')$, then the grating amplitude written in the crystal due to source 1 can be described by

$$\Gamma_1(x, y) = K[1 - \exp(-t_1/\tau)] \frac{A(x, y)}{I_0 + |A(x, y)|^2} \qquad (10.5\text{-}4)$$

where τ is the time constant of the medium (assuming intensity is kept constant for all exposures), t_1 is the exposure time, $A(x, y)$ is the Fourier

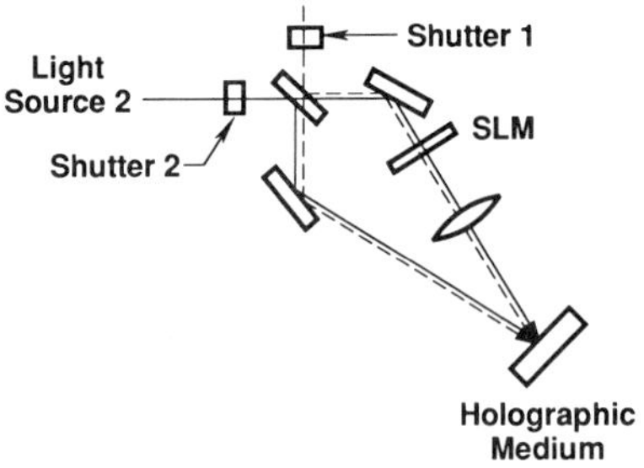

Figure 10.16. Double Mach–Zehnder interferometer for the recording of holograms with a phase of either 0 or π by controlling the shutters.

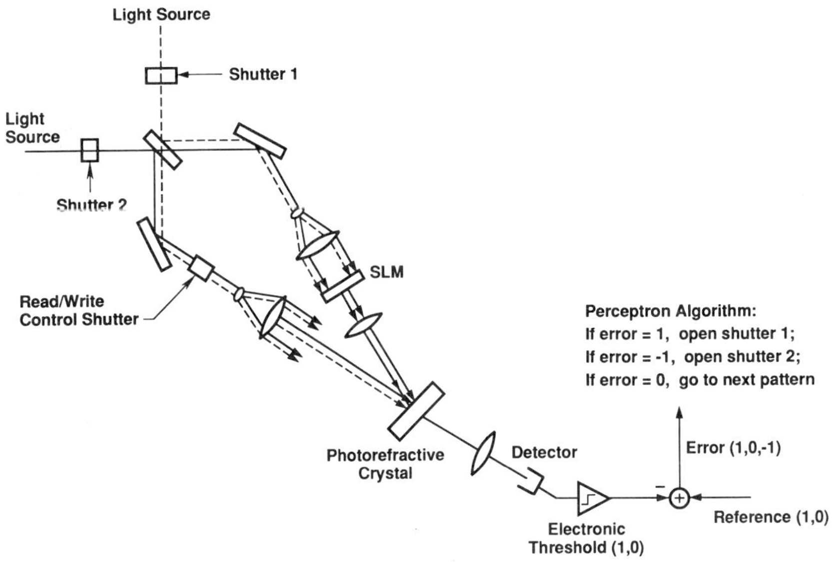

Figure 10.17. A learning machine for optical pattern classification.

transform of $a(x', y')$, I_0 is the reference beam intensity, and K is a constant determined by the characteristics of the particular crystal. If, without changing the picture, source 1 is turned off and source 2 is turned on, the new grating can be shown to be proportional to the first with the opposite sign. In particular, if the second exposure time duration is t_2, then

$$\Gamma(x, y) = K\{[1 - \exp(-t_1/\tau)]\exp(-t_2/\tau) - [1 - \exp(t_2/\tau)]\} \frac{A(x, y)}{I_0 + |A(x, y)|^2} \tag{10.5-5}$$

True subtractive weight changes are thus possible without the use of external phase shifters.

Note in Eq. (10.5-5) that as the new out of phase hologram is being written, the initial hologram is partially erased due to the presence of the writing beams (incoherent erasure). This will modify the actual learning algorithm that is used in the following way. The weight update equation in comparison with the ideal rule of Eq. (10.5-2) is given by

$$\mathbf{w}(p + 1) = \exp(-t_e/\tau)\mathbf{w}(p) + [1 - \exp(-t_e/\tau)]\sigma(p)\mathbf{x}(p) \tag{10.5-6}$$

where t_e is the exposure time, so that the old weight distribution is diminished slightly as new holograms are written. The "forgetting" effect actually may prove to be beneficial in the optical implementation where the amplitude of the hologram has a finite dynamic range since it helps to somewhat normalize the weight values during learning to keep the index gratings from saturating.

After the modification of the interconnection weight, the next picture is input to the SLM. The image-bearing beam is diffracted off the hologram in the photorefractive crystal. The intensity of the diffracted beam is measured by the photodetector and the output is given by

$$y(p) = \left| \sum_{i=1}^{N} w_i(p) x_i(p) \right|^2 \tag{10.5-7}$$

The detector output is then thresholded and the result is compared with the reference. The error signal is then employed to control the weight changes.

The extension of this concept to the multiple category case, which requires multiplicity of output lines as opposed to a single output sufficient for dichotomies, is shown in Figure 10.18 where the single detector and reference beam shutter have been replaced by a detector array/shutter combination and an SLM/shutter combination, respectively. Here, the reference for each pattern is no longer a single plane wave but a set of plane waves dictated by the collection of openings in the one-dimensional (1D) SLM. Initially, the crystal contains no holograms and interconnections are built up by simply exposing the hologram with light source 1 with the pattern in the 2D SLM and its associated reference pattern in the 1D SLM.

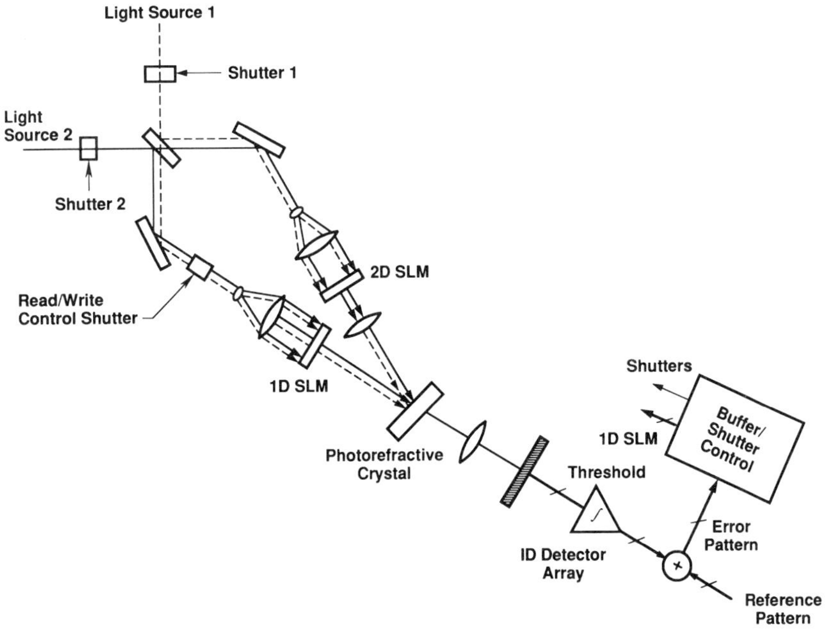

Figure 10.18. A learning machine for optical pattern classification into multiple categories.

This process is repeated for each pattern in the training set. After the initialization, the first pattern is loaded into the 2D SLM and the reference beam shutter is closed to interrogate the system. The reconstructed output pattern is then compared against the desired output pattern to yield the error vector. The algorithm must now be performed in two steps; first, only those portions of the 1D SLM corresponding to the positive portions of the error vector are opened and light source 1 is turned on to strengthen certain interconnections following the Perceptron recipe. Then, only those portions corresponding to negative elements of the error vector are loaded into the 1D SLM and light source 2 is turned on to weaken the appropriate weights. The buffer needed to store the error result needed for this function is especially simple and can be integrated into the detector array/amplifier assembly. The subtractive capability allows for a more exact implementation of the learning algorithms even for the multiple category case.

In summary, we have described an optical implementation of a learning machine which implements the Perceptron algorithm for pattern dichotomy. The optical system implements the weight storage and update functions using coherent means and, in particular, makes novel use of the Mach–Zehnder holographic interferometry to achieve truly subtractive as well as additive weight changes, precluding the need for biases which are typically used in incoherent implementations as discussed earlier.

Such an implementation scheme has been demonstrated experimentally using $LiNbO_3$ as the holographic medium [38–40]. Figure 10.19 shows the experimental results on learning. In this example, the system has an error of 50% at the beginning. This is natural because the crystal contains no hologram and the diffraction efficiency is zero. The system is, however, trained at the end of the 10th cycle when all the classifications are correct. As a result of the erasure, the interconnection weight **w** continues to decay during the interrogation. The system starts to make mistakes at the 14th cycle. The errors are corrected by the training algorithm. This will happen repeatedly as a result of the grating erasure.

Hologram fixing is desirable in the optical implementation of perceptron learning. It is desirable to fix the interconnection weight at the end of the

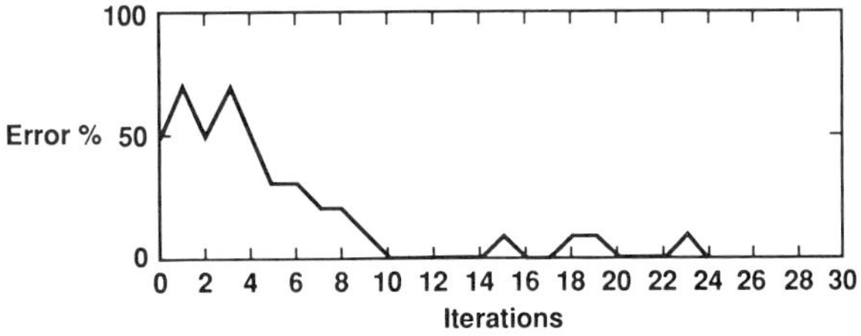

Figure 10.19. Learning curve for a two-state photorefractive perceptron using eight images. A $LiNbO_3$ crystal is used as the holographic recording medium.

10th cycle so that no erasure occurs. This will ensure the pattern classification without errors after the training.

PROBLEMS

10.1 Shift Invariance. Let $G(u, v)$ be the Fourier transform of $g(x, y)$.

a. Show that the Fourier transform of $g(x-a, y-b)$ is $G(u, v) \times \exp[i2\pi(ua + vb)]$.

b. Show that the power spectrum of $g(x, y)$ and $g(x-a, y-b)$ are identical. This is the shift invariance of Fourier transform.

10.2 Convolution Theorem

a. If $F\{g(x, y)\} = G(u, v)$ and $F\{h(x, y)\} = H(u, v)$, show that

$$F\left\{\int\int g(\xi, n) h(x-\xi, y-\eta)\, d\xi\, d\eta\right\} = G(u, v) H(u, v)$$

b. Derive Eq. (10.1-16).

10.3 Fourier Transform of Gaussian Beams

a. Show that the Fourier transform of a Hermite–Gaussian beam is still a Hermite–Gaussian beam. For the expression of Hermite–Gaussian, see, for example, A. Yariv and P. Yeh, Optical Waves in Crystals (Wiley, New York, 1984). The following integral may be useful:

$$\int_{-\infty}^{\infty} \exp(ixy) \exp\left(-\frac{x^2}{2}\right) H_n(x)\, dx = (2\pi)^{1/2} \exp\left(-\frac{y^2}{2}\right) H_n(y) i^n$$

b. Show that the Fourier transform property of a thin lens can be derived by using the ABCD-Law of Gaussian beams. Hint: Consider the illumination of a transparency $f(x, y)$ by a plane wave $\exp[-ikz]$ at the front focal plane of the lens. The transmitted beam can be expressed in terms of a complete set of Hermite–Gaussian beams.

10.4 Fundamental Fan-in Loss. Show that there is a fundamental fan-in loss of $(N-1)/N$ when single mode fibers are used in an $N \times N$ crossbar switch.

10.5 Parallel Encoding and Decoding. Parallel image subtraction using a phase conjugate Michelson interferometer can be employed for encryption and decryption. Referring to Figure 10.5, let $T_1(x, y)$ be the original image to be encoded. The image is assumed to be an array of pixels with transmittance 0 or 1.

a. Let $T_2(x, y)$ be a mask consisting of an array of 0 or 1 pixels with random distribution. Show that the output is encoded with the random mask.

b. Show that the encoded image can be decoded by using a similar subtraction with the same random mask.

10.6 **a.** Derive Eq. (10.3-3).
b. Derive Eq. (10.3-5).

10.7 **Holographic Inner Product.** Referring to Figure 10.14, we consider the recording of a pattern $f(x, y)$ and then the readout with a pattern $g(x, y)$.
a. Show that the output is simply a cross correlation of $f(x, y)$ and $g(x, y)$ provided that the hologram is not too thick.
b. Show that using a pinhole at the rear focal plane, we can obtain the inner product.

REFERENCES

1. J. A. Neff, "The Role of Optics in Future Computational Systems," Topical Meeting on Optical Computing, Incline Village, NV, 1985.
2. See, for example, A. Yariv and P. Yeh, Optical Waves in Crystals (Wiley, New York, 1984).
3. See, for example, Y. R. Shen, Principles of Nonlinear Optics (Wiley, New York, 1985).
4. See, for example, P. Yeh et al., "Photorefractive Nonlinear Optics and Optical Computing," Opt. Eng. **28**, 328–343 (1989).
5. J. W. Goodman, Introduction to Fourier Optics (McGraw-Hill, New York, 1968).
6. J. O. White and A. Yariv, "Real-Time Image Processing via Four-Wave Mixing in a Photorefractive Medium," Appl. Phys. **37**, 5–7 (1980).
7. D. Gabor, G. W. Stroke, R. Restrict, A. Funkhouser, and D. Brumm, "Optical Image Synthesis (Complex Amplitude Addition and Subtraction) by Holographic Fourier Transformation," Phys. Lett. **18**, 116 (1965).
8. See, for example, J. F. Ebersole, "Optical Image Subtraction," Opt. Eng. **14**, 436 (1975).
9. See, for example, G. Idebetouw, L. Bernardo, and M. Miller, "Real-Time Incoherent Subtraction of Irradiance," Appl. Opt. **19**, 1218 (1980).
10. Y. H. Ja, "Real-Time Image Subtraction in Four-Wave Mixing with Photorefractive $Bi_{12}GeO_{20}$ Crystals," Opt. Comm. **42**, 377 (1982).
11. C. P. Grover and R. Trembley, "Real-Time Image Subtraction Using Complementary Photographic Diffusers," Appl. Opt. **21**, 2666 (1982).
12. C. Warde and J. I. Thackara, "Oblique-Cut $LiNbO_3$ Microchannel Spatial Light Modulators," Opt. Lett. **7**, 344 (1982).
13. G. G. Mu, C. K. Chiang, and H. K. Liu, "White-Light Image Addition and Subtraction by Colorimetric Measurement," Opt. Lett. **6**, 389 (1981).
14. K. Matsuda, N. Takeya, T. Tsujiuchi, and M. Shimoda, "An Experiment of Image Subtraction Using Holographic Beam Splitter," Opt. Comm. **2**, 425 (1971).

15. K. Patroski, S. Yokozeki, and T. Suzuki, "Image Subtraction by Using Fourier-Imaging Phenomenon," Nouv. Rev. Opt. **6**, 25 (1975).
16. W. T. Cathey, Jr. and J. G. Dodge, "Image Comparison by Interference," J. Opt. Soc. Am. **56**, 1139 (1966).
17. A. E. T. Chiou, P. Yeh, and M. Khoshnevisan, "Coherent Image Subtraction Using Phase-Conjugate Interferometry," in Nonlinear Optics and Applications, Pochi Yeh (Ed.), Proc. SPIE **613**, 201–206 (1986).
18. A. E. T. Chiou and P. Yeh, "Parallel Image Subtraction Using Phase-Conjugate Michelson Interferometer," Opt. Lett. **11**, 306–308 (1986).
19. S. K. Kwong, G. A. Rakuljic, and A. Yariv, "Real-Time Image Subtraction and "XOR" Operation Using a Self-Pumped Phase Conjugate Mirror," Appl. Phys. Lett. **48**, 201–203 (1986).
20. P. Yeh, T. Y. Chang, and P. H. Beckwith, "Real-Time Optical Image Subtraction Using Dynamic Holographic Interference in Photorefractive Media," Opt. Lett. **13**, 586–588 (1988).
21. C. Mead, "Potential and Limitations of VLSI," Topical Meeting on Optical Computing, Incline Village, NV, 1985.
22. J. A. Neff, "The Role of Optics in Future Computational Systems," Topical Meeting on Optical Computing, Incline Village, NV, 1985.
23. See, for example, J. W. Goodman, F. J. Leonberger, S.-Y. Kung, and R. A. Athale, "Optical Interconnections for VLSI Systems," Proc. IEEE **72**, 850–866 (1984).
24. See, for example, "Optical Computing," John A. Neff (Ed.), Proc. SPIE **625**, 109–172 (1986).
25. A. A. Sawchuk and B. K. Jenkins, "Dynamic Optical Interconnections for Parallel Processors," in Optical Computing, J. A. Neff (Ed.), Proc. SPIE **625**, 143–153 (1986).
26. A. E. T. Chiou and P. Yeh, Opt. Lett. **11**, 461 (1986).
27. P. Yeh, A. E. T. Chiou, and J. Hong, "Optical Interconnection Using Photorefractive Dynamic Holograms," Appl. Opt. **27**, 2093–2096 (1988).
28. A. E. T. Chiou, P. Yeh and J. Hong, "Reconfigurable Interconnection Using Photorefractive Holograms," Proc. SPIE **1151**, 24–32 (1989).
29. A. E. T. Chiou and P. Yeh, "Energy Efficiency of Optical Interconnection Using Photorefractive Holograms," Appl. Opt. **29**, 1111 (1990).
30. J. J. Hopfield, "Neural Networks and Physical Systems with Emergent Collective Computational Abilities," Proc. Natl. Acad. Sci. USA **79**, 2554 (1982).
31. R. O. Duda and P. E. Hart, Pattern Classification and Scene Analysis (Wiley, New York, 1973).
32. J. J. Hopfield, "Neurons with Graded Response Have Collective Computational Properties Like Those of Two-State Neurons," Proc. Natl. Acad. Sci. USA **81**, 3088 (1984).
33. D. Psaltis, X. Gu, and D. Brady, "Holographic Implementations of Neural Networks," in An Introduction to Neural and Electronic Networks, S. F. Zornetzer, J. L. Davis, and Clifford Lau (Eds.) (Academic Press, New York, 1989).

34. D. Brady, X. Gu, and D. Psaltis, "Photorefractive Crystals in Optical Neural Computers," Proc. SPIE **882**, paper No. 20 (1988).
35. D. Psaltis, X. Gu, and D. Brady, "Fractal Sampling Grids for Holographic Interconnections," Proc. SPIE **963**, paper No. 70 (1988).
36. J. J. Caulfield, "Parallel N^4 Weighted Optical Interconnections," Appl. Opt. **26**, 4039 (1987).
37. D. Psaltis, D. Brady, and K. Wagner, "Adaptive Optical Neural Computers," Appl. Opt. **27**, 1752 (1988).
38. J. Hong, S. Campbell, and P. Yeh, "Optical Pattern Classifier with Perceptron Learning," Appl. Opt. **29**, 3019–3025 (1990).
39. J. Hong, S. Campbell, and P. Yeh, "Adaptive Optical Pattern Classifier," Proc. Int. Topical Meeting on Optical Computing, 10F4, 1990, p. 266.
40. J. Hong, S. Campbell, and P. Yeh, "Optical Implementation of Perceptrons Using Photorefractive Media," Tech. Digest, II Topical Meeting on Photorefractive Materials, 1990, p. 265.

11

Other Applications

In addition to their applications in optical computing, photorefractive materials can also be used for many other applications. These include image amplification, beam cleanup, spatial light modulators, thresholding and power limiting devices, etc. Most of these applications utilize the unique properties of photo-induced gratings and energy coupling in these materials. In this chapter, we briefly describe some of the important applications.

11.1 IMAGE AMPLIFICATION

The energy transfer in photorefractive two-wave mixing discussed in Section 4.1 can be used for image amplification [1]. Specifically, the interaction of a strong pump beam with a weak image-bearing beam inside a photorefractive medium results in an amplification of the image information (see Figure 11.1). To preserve the contrast of a grey scale transparency, it is crucial that the amplification factor be uniform spatially and throughout the intensity dynamic range of the image-carrying beam. According to Eq. (4.1-28), the intensity gain for two-wave mixing is given by

$$g = \frac{1+m}{1+me^{-\gamma L}}\, e^{-\alpha L} \tag{11.1-1}$$

where γ is the coupling constant, α is the absorption coefficient, L is the interaction length and m is the input intensity ratio

$$m = \frac{I_1(0)}{I_2(0)} \tag{11.1-2}$$

Here, we recall that I_2 is the intensity of the weak signal beam and I_1 is intensity of the strong pump beam.

We note that the signal gain depends on the beam intensity ratio m. A high fidelity image amplifier requires that the gain be independent of m. This is possible only when

$$me^{-\gamma L} \gg 1 \tag{11.1-3}$$

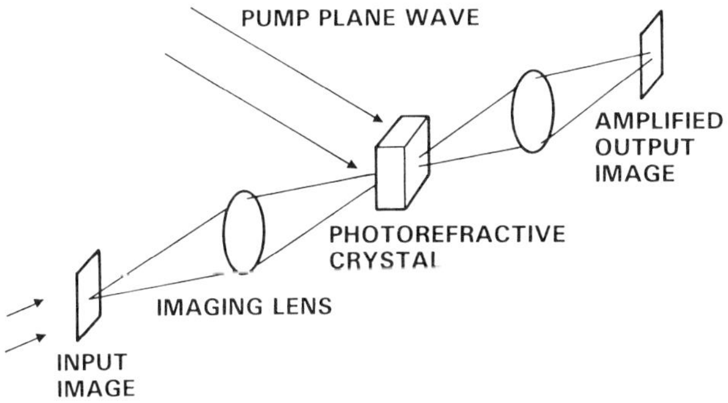

Figure 11.1. Optical image amplification using two-wave mixing in photorefractive crystals.

or equivalently

$$e^{\gamma L} \ll m \tag{11.1-4}$$

Under these conditions, the depletion of the pump beam is not significant. In practice, m can be of the order of 10^6. In other words, th pump beam is 10^6 stronger than the signal beam. At the intensity ratio, an amplification of 10^4 for the signal beam will only deplete 1% of the pump beam.

Strictly speaking, Eq. (11.1-1) is derived for the beam coupling of two plane waves. An image-bearing beam is not a plane wave. It actually consists of many plane wave components. The simple plane wave model must be modified to be more accurately applicable. In what follows, we present a simple modification of the theory of two-wave mixing involving image bearing beams and discuss in particular the amplification process when the crystal is placed at the Fourier transform plane.

Shown in Figure 11.2 is the image amplification geometry in which the amplitudes of the image bearing waves are represented by $\{E_j\}$, $j = 1, 2, \ldots, N$, and the pump plane wave is denoted by E_p. In photorefractive crystals, the index modulation depends on the intensity interference pattern so that the present analysis begins with the intensity expression which is given by

$$I(\mathbf{r}) = I_0 + \mathrm{Re}\left\{ E_p \sum_{m=1}^{N} E_m^* \exp[-i(\mathbf{k}_p - \mathbf{k}_m)\cdot\mathbf{r}] + \sum_{q\neq m}^{N} \sum_{m=1}^{N} E_q E_m^* \exp[-i(\mathbf{k}_q - \mathbf{k}_m)\cdot\mathbf{r}] \right\} \tag{11.1-5}$$

where $\mathbf{k}_p$ and $\mathbf{k}_m$ are the wave vectors of the pump and the mth probe beams. Let xz-plane be the plane of incidence.

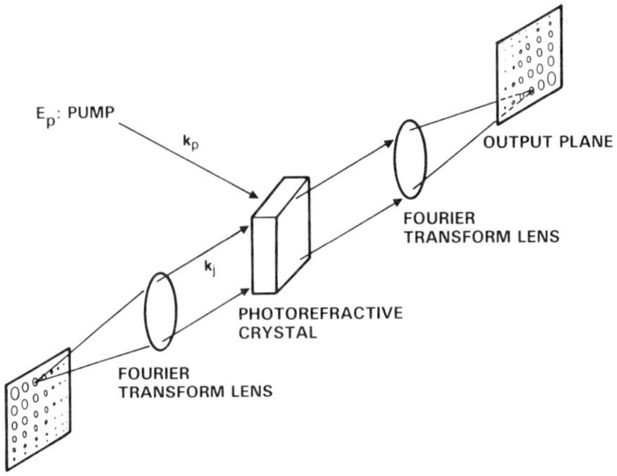

Figure 11.2. Two-wave mixing between a plane wave pump E_p and an image-bearing beam which consists of a set of point sources. The image bearing beam is transformed by using a thin lens into a set of plane waves at the rear focal plane where the photorefractive crystal is located.

These wave vectors can be written,

$$\mathbf{k}_p = \frac{2\pi}{\lambda}\left(-\hat{x}\sin\theta + \hat{z}\cos\theta\right), \qquad \mathbf{k}_m = \frac{2\pi}{\lambda}(\hat{x}\sin\theta_m + \hat{z}\cos\theta_m) \quad (11.1\text{-}6)$$

where $\hat{x}$, $\hat{z}$ are unit vectors, θ, θ_m are beam angles and the total intensity is $I_0 = |E_p|^2 + \Sigma_{m=1}^{N} |E_m|^2$. If the probe waves span a sufficiently small angular spectrum (much smaller than the angle between the pump and the average probe direction such that $|\theta_m - \theta_q| \ll |\theta - \bar{\theta}|$, where $\bar{\theta}$ is the average probe direction, then the interference between the probe waves will induce index gratings that are much weaker than those written by the interference between the probe waves and the pump wave. This is the scenario of interest, and the index of refraction can then be approximated, according to Eq. (4.1-5), as

$$n = n_0 + \frac{1}{2} e^{i\phi} \frac{n_1}{I_0} \sum_{m=1}^{N} E_p^* E_m \exp[-i(\mathbf{k}_m - \mathbf{k}_p)\cdot\mathbf{r}] \qquad (11.1\text{-}7)$$

where n_1 is the index modulation coefficient which is assumed to be the same for all of the gratings. With this and the assumption of slowly varying amplitudes for each plane wave involved in the interaction, Maxwell's equations yield the following coupled system:

$$\begin{aligned} \frac{dE_p}{dz} &= -\frac{\gamma_p}{I_0}\sum_{m=1}^{N}|E_m|^2 E_p\,, & \gamma_p &= \frac{n_1\pi}{\lambda\cos\theta} \\ \frac{dE_m}{dz} &= \frac{\gamma_m}{I_0}|E_p|^2 E_m\,, \quad m = 1, 2, \ldots, N\,, & \gamma_m &= \frac{n_1\pi}{\lambda\cos\theta_m} \end{aligned} \qquad (11.1\text{-}8)$$

where we assumed that $\phi = \pi/2$. We now separate the intensity and phase by employing the same technique used in Chapter 4. In particular, the intensities of the pump and probes, $I_p = |E_p|^2$ and $I_m = |E_m|^2$, are governed by

$$\begin{aligned} \frac{dI_p}{dz} &= -\frac{\Gamma_p}{I_0}\sum_{m=1}^{N} I_m I_p\,, && \Gamma_p = 2\gamma_p \\ \frac{dI_m}{dz} &= \frac{\Gamma_m}{I_0} I_p I_m\,, \quad m = 1, 2, \ldots, N\,, && \Gamma_m = 2\gamma_m \end{aligned} \tag{11.1-9}$$

In the assumed diffusion driven scenario ($\phi = \pi/2$), the phases of the beams are decoupled so that the intensity equations describe the two wave-mixing process completely. To obtain the final result, we make one further approximation which is consistent with the assumptions made thus far. We have assumed that $\cos\theta_m \approx \cos\theta$, so that it is justified to approximate $\Gamma_p = \Gamma_m = \Gamma$. With this, the equations become

$$\begin{aligned} \frac{dI_p}{dz} &= -\frac{\Gamma}{I_0}\sum_{m=1}^{N} I_m I_p \\ \frac{dI_m}{dz} &= \frac{\Gamma}{I_0} I_p I_m\,, \quad m = 1, 2, \ldots, N \end{aligned} \tag{11.1-10}$$

These equations are easily integrated to yield the following solutions:

$$\begin{aligned} I_p(z) &= \frac{mI_0 \exp(-\Gamma z)}{1 + m\exp(-\Gamma z)} \\ \frac{I_m(z)}{I_m(0)} &= \frac{1+m}{1 + m\exp(-\Gamma z)} \\ m &= \frac{I_p(0)}{\sum_{m=1}^{N} I_m(0)} \end{aligned} \tag{11.1-11}$$

The solutions indicate that the probe waves are amplified uniformly with a gain factor which is a function of the ratio of the pump beam to the total intensity of the probes. Small deviations from this are expected as the coupling coefficient is a function of the actual probe beam angle. The uniformity of the amplification has been verified experimentally [2, 3].

11.2 SPATIAL AMPLIFICATION

In image amplification, we assume that the pump beam is uniform over the entire image beam. This is essential to ensure uniform amplification. Spatial amplification is a technique in which only a set of selected spatial frequency components of an image are amplified.

Spatial filtering, the technique of enhancing image features by filtering out appropriate spatial frequency components, has been established as a powerful coherent image-processing tool [4]. It is frequently used for applications such as edge enhancement, character recognition, and image correlation [5]. The technique consists of Fourier transforming an input (coherent) image by a lens, masking out certain spatial frequencies by placing a spatial filter at the Fourier plane, and then re-imaging it using a second lens. For example, if the spatial filter blocks out the low spatial frequencies, then the processed image will show edge enhancement.

Despite its utility, spatial filtering has some practical limitations. One immediate limitation derives from the fact that some spatial frequency components are eliminated in the blockout procedure. The consequence is that spatial filtering systems typically cannot be cascaded in stages. The problem is due to information loss at the filters. There are situations that the stages following the first one may require some (or all) of the components that are removed in the first stage. Another limitation is that spatial filtering is inherently energy inefficient. In a typical image the dc component contains a major portion of the total energy and is usually eliminated in many spatial-filtering applications for various reasons (e.g., for minimizing the cross correlation between the images in pattern recognition).

In what follows, we describe a technique, called *spatial amplification*, that overcomes the limitations of spatial filtering mentioned above. Like spatial filtering, the technique consists of manipulating an input image at its Fourier plane (FP) as shown in Figure 11.3. Rather than blocking out the appropriate set of spatial frequency components for a desired effect, however, the complementary set of spatial frequency components is amplified through photorefractive two-beam coupling with a nonuniform pump beam. For example, by selectively amplifying only the spatial frequencies corresponding to horizontal lines, the processed image will show more-intense horizontal lines, as depicted in Figure 11.3. The intensity of the enhanced features will depend on the gain of the amplifier and can easily be a factor of 1000 greater than that of the unamplified components by using photorefractive two-beam coupling in $BaTiO_3$ [6]. Therefore, certain features of the image can be enhanced with respect to the overall image intensity by using the appropriate pump-beam distribution. Also, spatial amplification makes it possible to cascade several stages, with each stage corresponding to a specific type of processing since no part of the original image information is lost at any stage. Another consequence of not having to eliminate any spatial frequency component is the high energy efficiency of the process since there is a net energy gain in spatial amplification (as opposed to loss in spatial filtering due to blocking), which would be useful for low-intensity applications.

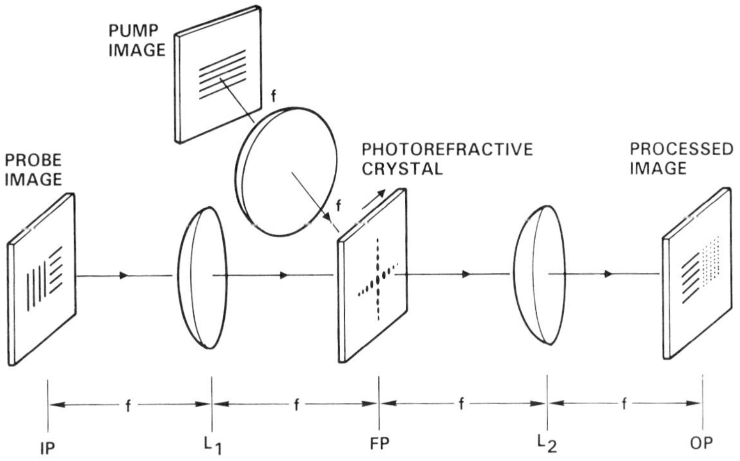

Figure 11.3. Concept of spatial amplification: an input image is Fourier transformed into an amplifying medium, where spatial frequencies are selectively amplified, then it is re-imaged by another Fourier transformation. Since no spatial frequency component is removed, this process can be cascaded for further processing. IP, input plane; OP, output plane.

The amplification required in spatial amplification is provided by photorefractive two-beam coupling. It is important that the phases of the various spatial frequency components remain constant through the amplification process to preserve the integrity of the original image. Two-beam coupling in a photorefractive crystal in the diffusion regime provides precisely this function and is used as the gain medium in the experiments. The interaction between a pump beam and a probe beam with their amplitudes and phases given by I_1 and ψ_1 and I_2 and ψ_2, respectively, can be described by the following coupled equations:

$$\frac{d}{dz} I_1 = -\Gamma \frac{I_1 I_2}{I_1 + I_2} - \alpha I_1 \tag{11.2-1}$$

$$\frac{d}{dz} I_2 = \Gamma \frac{I_1 I_2}{I_1 + I_2} - \alpha I_2 \tag{11.2-2}$$

$$\frac{d}{dz} \psi_1 = \beta \frac{I_2}{I_1 + I_2}, \qquad \frac{d}{dz} \psi_2 = \beta \frac{I_1}{I_1 + I_2} \tag{11.2-3}$$

where α is the (linear) bulk absorption coefficient, Γ is the two-beam coupling constant, and β is the phase-coupling constant. The coupling coefficient Γ is a function of the light-induced space-charge electric field, the linear electro-optic coefficient, and other material parameters. Equations (11.2-1) and (11.2-2) can be solved to show that the probe beam can gain energy at the expense of the pump beam as they propagate along the crystal in the z-direction, which is the amplification process. More importantly, however, it can be shown that the phase-coupling constant β is zero for a photorefractive material whose charge-transport mechanism is predominantly diffusion. In such a case, it is apparent from Eqs. (11.2-3) that there is no phase coupling between the two beams. Therefore, the probe beam can undergo significant amplification in amplitude, but its phase will remain unchanged regardless of the phase of the pump beam. The two-beam coupling process in $BaTiO_3$ without the application of an external electric field satisfies this requirement of amplification without the phase cross talk. Spatial amplification has been demonstrated experimentally using a photorefractive $BaTiO_3$ crystal [7].

11.3 LASER BEAM CLEANUP

In photorefractive materials that operate by pure diffusion ($\phi = \pi/2$), the phase coupling constant β is zero, according to Eq. (4.1-20). Thus the phases ψ_1 and ψ_2 of the beams remain constant regardless of the energy transfer. This unique property is also useful for laser beam cleanup.

The lack of phase cross talk can be understood also in terms of the diffraction from the self-induced index grating in the photorefractive crystal. Normally, if a beam that contains phase information $\psi(\mathbf{r}, t)$ is diffracted from a fixed grating, the same phase information also appears in the diffracted beam. In the self-induced index grating, the same phase information $\psi(\mathbf{r}, t)$ is impressed onto the grating in such a way that diffraction from such a grating will be accompanied by a phase shift $-\psi(\mathbf{r}, t)$. Such a dynamic hologram makes self-cancellation of phase information possible when the incident beam is diffracted from the grating produced by the incident and the reference beams. Such a self-cancellation of phase information is actually equivalent to the reconstruction of the reference beam when the hologram is read out by the object beam.

Energy transfer without phase cross talk can be employed to compress both the spatial and the temporal spectra of a light beam [8, 9]. In other words, the energy transfer without phase cross talk can be utilized to clean up the spatial wave front aberrations.

In the cleanup of spatial aberration, a spatial mode filter (e.g., a pinhole mirror) is used to select a clean part of the aberrated beam. The rest of the beam consists of several spatial-frequency components. After the separation, these two portions of the beam are brought together at a photorefractive crystal (Figure 11.4). Because of the energy transfer without phase cross talk, the signal beam can be amplified without bearing any phase information from the aberrated part of the beam.

Laser beam cleanup using photorefractive two-wave mixing has been demonstrated experimentally using SBN crystals and argon ion lasers [8, 9].

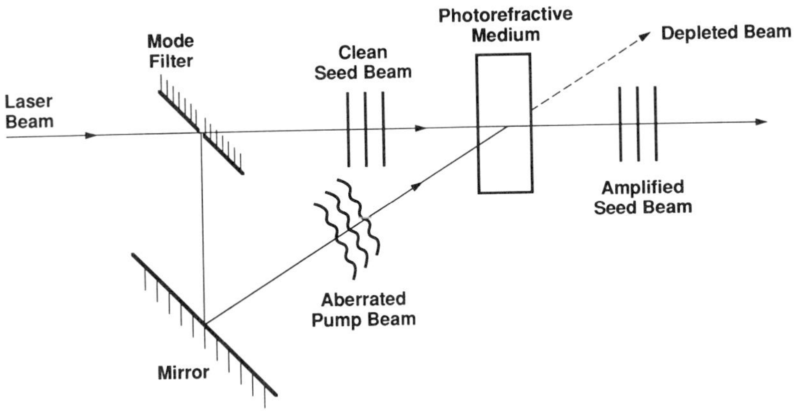

Figure 11.4. Laser beam clean-up using two-wave mixing in photorefractive crystals.

Phase conjugation can also be employed to eliminate phase aberration due to crystal imperfection. To achieve aberration correction, the seed beam is first transmitted through the photorefractive crystal. The transmitted seed beam (now aberrated) is then reflected by a phase conjugator. The phase conjugated seed beam will retrace its earlier path and return to the photorefractive crystal. A pump beam is now recombined with the phase conjugated seed beam. As a result of two-wave mixing and phase conjugation, the seed beam will be amplified and its aberration eliminated upon transmission through the crystal. This technique has also been demonstrated experimentally using a $BaTiO_3$ crystal as the phase conjugator and an SBN crystal as the two-wave mixing medium [9].

11.4 PHOTOREFRACTIVE SPATIAL LIGHT MODULATORS

Spatial light modulators (SLM) are devices that can be used to impose a given set of information on an optical wave by spatially modulating its intensity and/or phase distribution. They can be used to serve the purpose of interfacing in many optical information processing systems to provide input to and to extract information from the system. Two classes of SLMs are distinguished by the way information is provided to the device. First, the electrically addressed SLM is a device that changes the local intensity/phase transmittance property of the device via electrical control by using the electro-optical or magneto-optical effects found in many materials. The magneto-optic SLM [10] and the liquid crystal television [11] are two popular examples of this class of devices. The optically addressed SLM, on the other hand, is a device whose optical transmittance properties are controlled by another optical beam. Often, such devices are used to convert an incoherent optical beam into a coherent beam with the same spatial modulation. The liquid crystal light valve, PROM, and PRIZ are good examples of this class of SLMs. By virtue of their optical response, photorefractive materials are ideal media for optically addressed SLM. In what follows, we describe several schemes of photorefractive SLM.

Referring to Figure 11.5, we consider a photorefractive SLM that can be used to convert an incoherent image into a coherent one. An image described by a transparency $T(x, y)$ is imaged onto a thin plate of photorefractive crystal by using an incoherent source S. At the same time, the crystal plate is used in a four-wave mixing process with coherent light. Specifically, A_2 and A_3 are counterpropagating pump beams and A_1 is a probe beam. In the photorefractive medium, A_1 and A_2 form an interference pattern which induces an index grating. The amplitude of the index grating can be written

$$\Delta n = n_1 e^{i\phi} \frac{A_1^* A_2}{I_0 + I_s} e^{-i\mathbf{K}\cdot\mathbf{r}} + \text{c.c.} \tag{11.4-1}$$

where ϕ is a spatial phase shift, $\mathbf{K}$ is the grating wave vector, n_1 is a constant and I_0 is the sum of the intensities of the coherent beams and I_s is the intensity of the incoherent image inside the crystal.

We assume that all the coherent beams are uniform in intensity so that I_0 is a constant. The incoherent image I_s, however, is a function of (x, y) reflecting the image information. The incoherent image is now impressed into the index grating as an intensity denominator. When the grating is read by the readout beam A_3, the diffracted beam is a phase conjugate of A_1 except that the beam is now spatially modulated. The amplitude of the diffracted beam can be written

$$A_4 = cn_1 \frac{A_1^* A_2 A_3}{I_0 + I_s(x, y)} \tag{11.4-2}$$

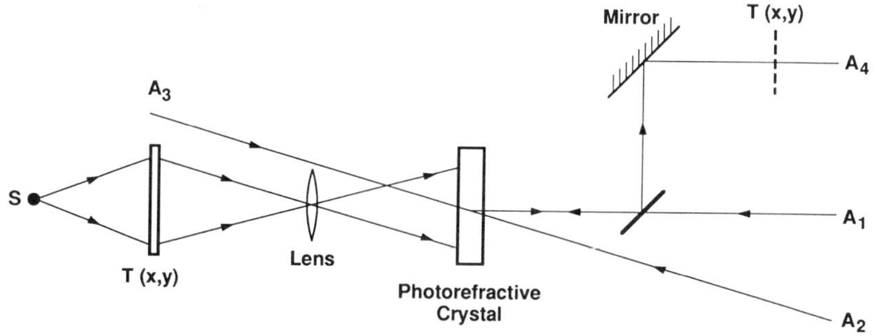

Figure 11.5. A photorefractive spatial light modulator (SLM) using four-wave mixing. The device is also known as photorefractive incoherent–coherent optical converter (PICOC) [12].

where c is a constant. If we assume that $I_s(x, y) \ll I_0$, then Eq. (11.4-2) can be written

$$A_4 = A_4(I_s = 0)\left[1 - \frac{I_s(x, y)}{I_0} + O\left(\frac{I_s}{I_0}\right)^2\right] \tag{11.4-3}$$

By examining the above equation, we note that the diffracted beam is spatially modulated with $I_s(x, y)$. This is the principle of operation of the photorefractive SLM described in Figure 11.5.

The spatial light modulator described above is often referred to as the photorefractive incoherent–coherent optical converter (PICOC) [12]. The operation of the device is based on local erasure produced by the input image $I_s(x, y)$. According to Eq. (11.4-3), a high fidelity spatial light modulation requires that the image intensity $I_s(x, y)$ be small compared to beams writing and reading the grating (i.e., $I_s(x, y) \ll I_0$). Such a weak input image unfortunately is represented as a weak modulation imposed upon a strong uniform bias signal. The result is a device with an inherently low dynamic range which may be unsuitable for use in some image processing applications.

We note that the poor contrast of spatial modulation in PICOC is a result of the fact that the image bearing beam does not write the photorefractive grating but only erases it. This gives rise to the background bias due to the writing and reading beams described in the preceding paragraph. A high contrast SLM would require that the image bearing beam be used to write the grating. This can be achieved by using an amplitude grating mask. Figure 11.6 depicts a schematic drawing of such an SLM. Here an image mask is placed in contact with a one-dimensional sinusoidal amplitude grating. This composite image target is then illuminated by an incoherent source and imaged on a photorefractive crystal. The photorefractive material then responds to the incident periodic intensity pattern exactly as it would

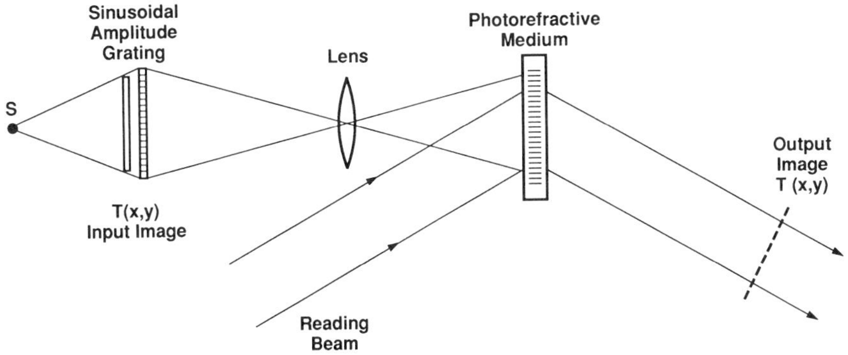

Figure 11.6. A high contrast photorefractive SLM.

if this pattern had resulted from the interference of two coherent beams, and thus a phase grating is written in the material. This grating is simultaneously read out by a separate coherent beam in the usual real-time holographic fashion.

Let $T(x, y)$ be the input image and the transmittance of the sinusoidal amplitude grating be written

$$t = 1/2(\cos \mathbf{K} \cdot \mathbf{r} + 1) \tag{11.4-4}$$

where $\mathbf{K}$ is the grating wave vector. The intensity pattern at the photorefractive crystal can be written

$$I = I_0 T(x, y) t + I_R \tag{11.4-5}$$

where I_0 is the intensity of the incoherent source and I_R is the intensity of the read beam. We rewrite the intensity as

$$I = (1/2) I_0 T(x, y)(1 + \cos \mathbf{K} \cdot \mathbf{r}) + I_R \tag{11.4-6}$$

Such an intensity pattern will induce an index grating inside the photorefractive medium. According to the discussion in Chapter 3, the index grating can be written

$$\Delta n = n_1 e^{i\phi} \frac{I_0 T(x, y)}{I_0 T(x, y) + 2I_R} e^{-i\mathbf{K} \cdot \mathbf{r}} + \text{c.c.} \tag{11.4-7}$$

where ϕ is a spatial phase shift. If we further assume that $I_0 T(x, y) \ll 2I_R$, then

$$\Delta n = n_1 e^{i\phi} \left(\frac{I_0}{2I_R} \right) T(x, y) e^{-i\mathbf{K} \cdot \mathbf{r}} + \text{c.c.} \tag{11.4-8}$$

Thus, for sufficiently weak input image, the induced index grating is proportional to the input image. Unlike the response of the PICOC, we see that there is no bias term present in this expression. When such an index grating is read by a plane wave, the amplitude of the diffracted beam can be written

$$A = c \frac{I_0}{\sqrt{I_R}} T(x, y) \tag{11.4-9}$$

where c is a constant.

The photorefractive SLMs described earlier are holographic in nature. Optical alignment is critical in the read out process. Information (i.e., images) can be imprinted in photorefractive crystals without resorting to holographic techniques as shown in Figure 11.7. The crystal is located at the

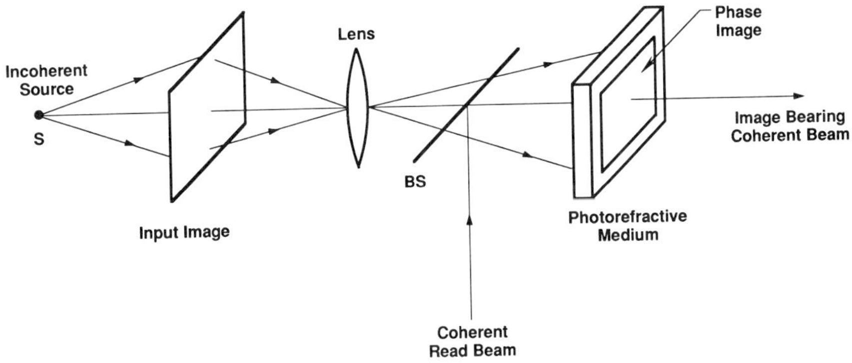

Figure 11.7. A nonholographic photorefractive SLM.

image plane of the object (the image bearing beam can be incoherent or coherent) and the crystal responds to the nonuniform beam with spatial modulations of its refractive index via the same mechanisms described earlier. The illumination of input images leads to a spatial modulation of the index of refraction. When a coherent beam transmits through the crystal, the image information is impressed onto the beam via spatial phase modulation. For details about these non-holographic photorefractive SLMs, interested readers are referred to Refs. [12, 13].

11.5 PHOTOREFRACTIVE TIME INTEGRATORS

The formation of photorefractive gratings requires the absorption of photons and the transport of photo-induced charge carriers. The process is very similar to the charge buildup in a capacitor. If the medium is illuminated with an intensity

$$I = I_0 + \mathrm{Re}\{I_1 e^{-i\mathbf{K}\cdot\mathbf{r}}\} \tag{11.5-1}$$

where I_0, I_1 are constant and $\mathbf{K}$ is the wave vector of the interference pattern, the photorefractive index grating can be written

$$\Delta n = C \frac{I_1}{I_0} e^{-i\mathbf{K}\cdot\mathbf{r}} \tag{11.5-2}$$

where C is a complex constant. The above result is a steady state solution derived in Chapter 3. We now assume that $I_1(x, t)$ is a slowly varying function of x compared to $\cos(\mathbf{K}\cdot\mathbf{r})$. The index grating can be written (see Problem 3.3)

$$\Delta n = \frac{C}{\tau} \int_{-\infty}^{t} e^{(t'-t)/\tau} \frac{I_1(x, t')}{I_0}\, dt' e^{-i\mathbf{K}\cdot\mathbf{r}} \tag{11.5-3}$$

where τ is the decay time constant of the index grating. Physically, the above equation indicates that the index grating is an accumulation of all the index gratings produced by earlier illumination weighted by an exponential decaying factor $\exp[(t' - t)/\tau]$. The integration can be approximated as

$$\Delta n = \frac{C}{\tau I_0} \int_{t-\tau}^{t} I_1(x, t')\, dt' e^{-i\mathbf{K}\cdot\mathbf{r}} \tag{11.5-4}$$

We note that the amplitude of the index grating is proportional to the running integration of the signal intensity with integration time τ. When the grating is read out by an optical beam, the diffracted intensity is proportional to the square of the integration of the signal intensity I_1 without any bias. In other words, I_0 does not contribute to the diffraction.

The time-integration property is useful in the correlation of RF signals [14]. In conventional acousto-optical signal correlations, multiplication of two RF signals is achieved by using two Bragg cells in series [15]. When an optical beam is diffracted from the first Bragg cell, the first RF signal is impressed onto the beam. A subsequent diffraction from the second Bragg cell yields the multiplication. In a photorefractive time-integrating correlator, the multiplication is achieved in the formation of index grating in the medium. Referring to Figure 11.8, we consider the incidence of a collimated light beam onto a diffraction grating. The beam is split into two. These two beams are then directed toward the Bragg cells. The RF signals are fed to the Bragg cells so that the diffracted beams can be written

$$E_1 = S_1(t - x/v) \exp[i(\omega' t - \mathbf{k}_1 \cdot \mathbf{r})] \tag{11.5-5}$$

$$E_2 = S_2(t + x/v) \exp[i(\omega' t - \mathbf{k}_2 \cdot \mathbf{r})] \tag{11.5-6}$$

where ω' is the frequency of the diffracted beam, $\mathbf{k}_1$, $\mathbf{k}_2$ are the wave

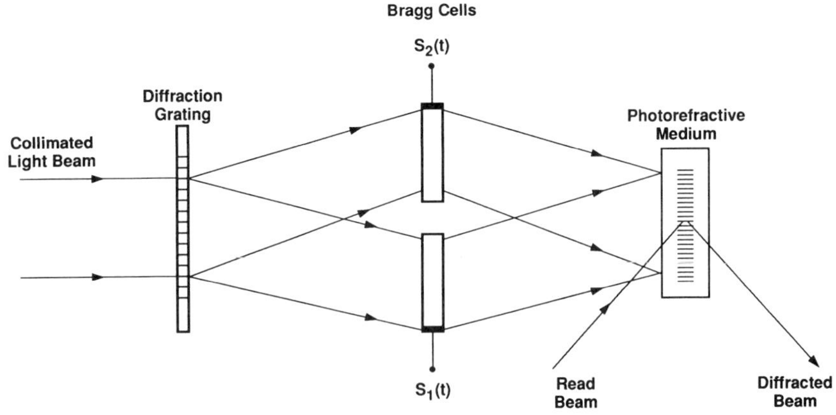

Figure 11.8. Photorefractive time-integrator.

vectors, v is the acoustic velocity and S_1, S_2 are RF signals. These diffracted beams are then recombined at the photorefractive medium. The intensity inside the photorefractive medium can be written

$$I = |S_1(t - x/v)|^2 + |S_2(t + x/v)|^2 + 2\,\mathrm{Re}[S_1^*(t - x/v)S_2(t + x/v)e^{-i\mathbf{K}\cdot\mathbf{r}}] \tag{11.5-7}$$

where $\mathbf{K}$ is the grating wave vector. We will treat the case where $|S_1(t)|^2$ and $|S_2(t)|^2$ can both be approximated as constants, as in the case of FM signals. Then the intensity can be separated into a dc term I_0 and a signal term $I_1(x, t)$ modulating a spatial carrier $\exp(-i\mathbf{K}\cdot\mathbf{r})$. This signal intensity will result in the formation of an index grating in the photorefractive medium described by Eq. (11.5-4). The grating can then be read out by an auxiliary beam and can be imaged onto a charge coupled device (CCD) detector for readout. The intensity of the diffracted beam can be written

$$I_{\mathrm{out}}(x) = C'\left|\int_{t-\tau}^{t} S_1^*(t' - x/v)S_2(t' + x/v)\,dt'\right|^2 \tag{11.5-8}$$

where C' is a constant. By defining a new variable t' as $t' - x/v$, the above integral can be written

$$I_{\mathrm{out}} = C'\left|\int_{t-x/v-\tau}^{t-x/v} S_1^*(t')S_2(t' + 2x/v)\,dt'\right|^2 \tag{11.5-9}$$

The output is proportional to the magnitude square of the correlation between the signals $S_1(t)$ and $S_2(t)$ integrated over a finite time interval τ. We note that the dc term does not appear in the output. This greatly reduces the noise in the detection.

11.6 NONRECIPROCAL OPTICAL TRANSMISSION

The subject of wave propagation in layered media has been treated by many workers. It is known that the transmissivity as well as the phase shift of beams traversing through an arbitrary layered dielectric structure (including absorbing media) is independent of the side of incidence. In other words, beams incident from the right side will experience a transmittance that is exactly the same as that experienced by the beams that are incident from the left side. This is the so-called left-and-right-incidence theorem [16] and is always true, provided that the materials are linear (i.e., that ϵ and μ are independent of the field strength). In photorefractive materials such as $BaTiO_3$, $LiNbO_3$, $Bi_{12}GeO_{20}$, and $Bi_{12}SiO_{20}$, the dielectric response is nonlinear and even nonlocal. Therefore this theorem may no longer be true. Thus a photorefractive crystal plate may be used as a nonreciprocal optical element.

Consider the transmission and reflection of electromagnetic radiation from a slab of photorefractive material. The incident and reflected waves inside the material form an index grating by means of the photorefractive effect. This refractive-index modulation provides the basis for the coupling of these two waves. As a result of this coupling, the amplitudes of these two waves are no longer constant. This leads to an interesting nonreciprocal optical phenomenon.

Referring to Figure 11.9, we consider a simple example in which a slab of photorefractive crystal plate is placed in front of a partial reflecting dielectric mirror. The crystal plate is cut such that its *c*-axis is perpendicular to the partial reflecting mirror. Let R be the intensity reflectance of the mirror, and $T = 1 - R$ is the intensity transmittance. The surfaces of the photorefractive crystal are assumed to be antireflection coated. When a beam of light is incident from the left side, the transmittance of the composite structure is T, neglecting the material absorption of the photorefractive crystal. If, however, the beam of light is incident from the right side as shown in Figure 11.9, the transmittance is no longer T, as a result of the presence of the index grating in the photorefractive crystal. In terms of two-wave mixing, the reflected light due to the mirror will undergo beam coupling with the incident beam inside the photorefractive medium.

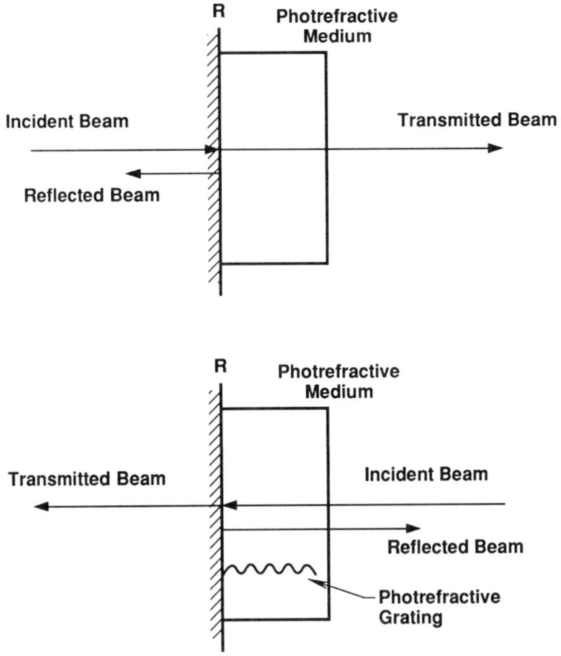

Figure 11.9. Transmission of coherent light through a photorefractive crystal plate with coated surfaces.

Let the intensity of the incident beam be I_2 and that of the reflected beam be I_1. The beam incident at $z = L$ has an intensity of $I_2(L)$. At the mirror surface, the intensities are related by

$$I_1(0) = RI_2(0) \tag{11.6-1}$$

and the transmitted intensity is $TI_2(0)$. Using the results obtained in Chapter 4, we have

$$I_2(L) = C + [C^2 + B\exp(-\gamma L)]^{1/2} \tag{11.6-2}$$

where

$$C = (1 - R)I_2(0)/2 \tag{11.6-3}$$

$$B = RI_2^2(0) \tag{11.6-4}$$

The intensity transmittance of the whole structure is given by

$$T' = \frac{TI_2(0)}{I_2(L)} \tag{11.6-5}$$

Using Eqs. (11.6-2)–(11.6-4), we obtain

$$T' = \frac{2T}{T + [T^2 + 4R\exp(-\gamma L)]^{1/2}} \tag{11.6-6}$$

which reduces to

$$T' = T \quad \text{at } \gamma = 0 \tag{11.6-7}$$

We note that T' can be greater than or less than T depending on the sign of γL. For $\gamma L > 0$, the reflected wave $I_1(z)$ is a decreasing function of z, indicating the donation of energy back to beam I_2. This leads to an increase in transmittance. For $\gamma L < 0$, the reflected wave $I_1(z)$ is an increasing function of z, indicating the amplification of the reflected wave due to two-wave mixing. This leads to a decrease in the transmittance. Figure 11.10 plots T' as a function of γL for various values of R.

We consider now the case of a photorefractive crystal plate whose surfaces both exhibit finite reflectance (see Figure 11.11). Let the electric field of the waves be written

$$E = \begin{cases} A_1\exp(-ik_1 z) + B_1\exp(ik_1 z), & z < 0 \\ A_2(z)\exp(-ik_2 z) + B_2(z)\exp(ik_2 z), & 0 < z < L \\ A_3\exp[-ik_3(z - L)], & L < z \end{cases} \tag{11.6-8}$$

where A_1 is the amplitude of the incident wave, B_1 is the amplitude of the

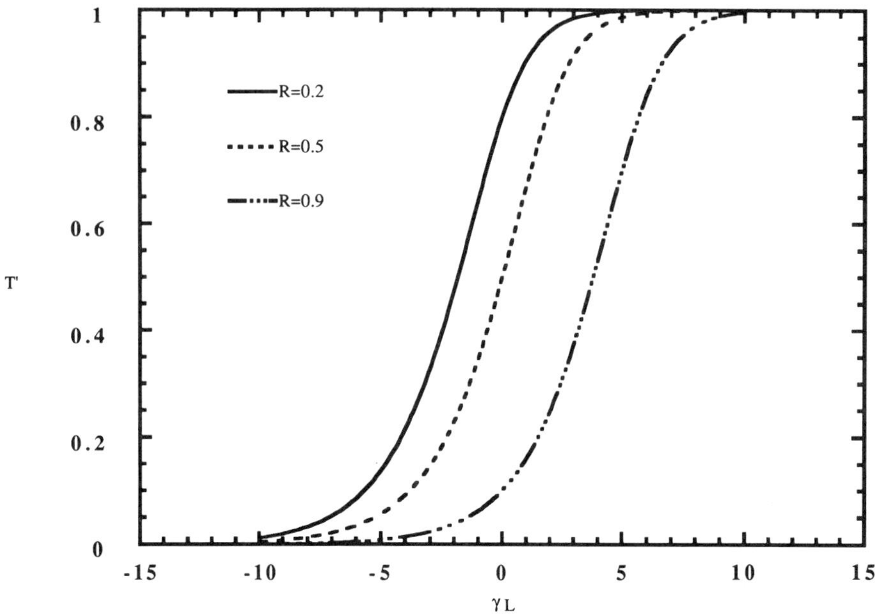

Figure 11.10. Transmittance of the coated photorefractive crystal plate as a function of the coupling constant γL for $R = 0.2$, 0.5, 0.9.

reflected wave, and A_3 is the amplitude of the transmitted wave. $A_2(z)$ and $B_2(z)$ are the amplitudes of the wave inside the photorefractive material. k_1, k_2, and k_3 are the wave numbers.

In the photorefractive medium, the two waves $A_2(z)\exp(-ik_2z)$ and $B_2(z)\exp(ik_2z)$ generate an index grating and undergo two-wave mixing. The energy coupling is determined by the beam ratio and the coupling constant γL.

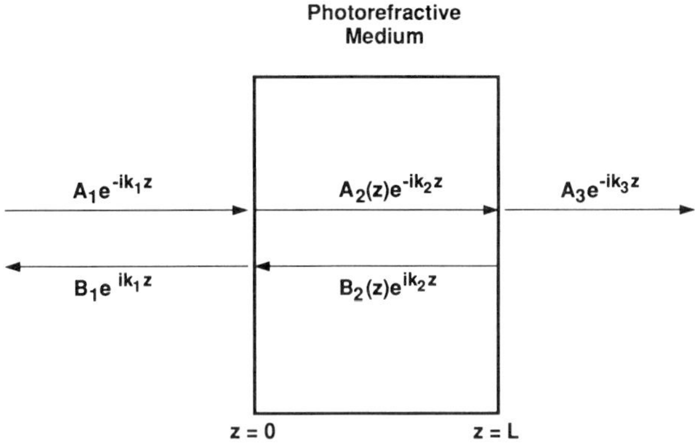

Figure 11.11. Transmission of coherent light through an uncoated photorefractive crystal plate.

We assume that the photorefractive medium operates by diffusion only so that there is no phase coupling. Using the results obtained in Chapter 4, the intensities of the beams inside the photorefractive medium can be written

$$|A_2|^2 = C + [C^2 + B\exp(-\gamma z)]^{1/2}$$
$$|B_2|^2 = -C + [C^2 + B\exp(-\gamma z)]^{1/2} \tag{11.6-9}$$

where

$$C = \tfrac{1}{2}[|A_2(L)|^2 - |B_2(L)|^2] \tag{11.6-10}$$

$$B = |A_2(L)B_2(L)|^2 \exp(\gamma L) \tag{11.6-11}$$

The boundary condition at $z = L$ requires that

$$|B_2(L)|^2 = R_{23}|A_2(L)|^2 \tag{11.6-12}$$

where R_{23} is the intensity reflectance of the the surface at $z = L$. Since there is no phase coupling, we may define

$$A_2(L) = g_+ A_2(0) \tag{11.6-13}$$

$$B_2(0) = g_- B_2(L) \tag{11.6-14}$$

where g_+ and g_- are real amplification factors. Using Eqs. (11.6-9)–(11.6-12), we obtain

$$|g_+|^{-2} = \frac{1-R_{23}}{2} + \left[\left(\frac{1-R_{23}}{2}\right)^2 + R_{23}\exp(\gamma L)\right]^{1/2} \tag{11.6-15}$$

$$|g_-|^2 = \frac{1-R_{23}^{-1}}{2} + \left[\left(\frac{1-R_{23}^{-1}}{2}\right)^2 + R_{23}^{-1}\exp(\gamma L)\right]^{1/2} \tag{11.6-16}$$

In addition to the relationship given by Eqs. (11.6-13) and (11.6-14), the amplitudes are also related by

$$A_2(0) = A_1 t_{12} + B_2(0) r_{21} \tag{11.6-17}$$

$$B_1 = B_2(0) t_{21} + A_1 r_{12} \tag{11.6-18}$$

$$A_3 = A_2(L) t_{23} \exp(-ik_2 L) \tag{11.6-19}$$

$$B_2(L) = A_2(L) r_{23} \exp(-2ik_2 L) \tag{11.6-20}$$

where t_{12}, t_{21}, t_{23}, r_{12}, r_{21}, and r_{23} are transmission and reflection coefficients. The indices reflects the incidence condition. For example, t_{12} is the transmission coefficient for incident beam from $z < 0$ and entering the medium between $0 < z < L$.

By eliminating $A_2(0)$, $A_2(L)$, $B_2(0)$, and $B_2(L)$, we obtain the following expressions:

$$r = \frac{B_1}{A_1} = \frac{r_{12} + r_{23} g_+ g_- \exp(-2ik_2L)}{1 + g_+ g_- r_{12} r_{23} \exp(-2ik_2L)} \tag{11.6-21}$$

$$t = \frac{A_3}{A_1} = \frac{t_{12} t_{23} g_+ \exp(-ik_2L)}{1 + g_+ g_- r_{12} r_{23} \exp(-2ik_2L)} \tag{11.6-22}$$

As a result of the photorefractive coupling, the parameters g_+ and g_- depend on the direction of the c-axis. Reversing the c-axis of the crystal leads to a change of sign of γ. This results in a different transmittance, according to Eq. (11.6-22). The nonreciprocal behavior is a result of the finite phase shift ϕ, which leads to a finite coupling constant γ. For very weak coupling (i.e., $\gamma L \ll 1$), g_+ and g_- tend to unity, and the nonreciprocity disappears. For moderate or strong coupling (i.e., $|\gamma L| > 1$), the transmission or reflection depends strongly on the sign γL. The phase factor $k_2 L$ in Eqs. (11.6-21) and (11.6-22) is a fast-varying function of frequency ω as well as of length L. Such a fast-varying dependence on frequency when the light transmits may not be observable through a thick plate, even though there are multiple reflections inside the plate. There are two reasons for this. First, the plate is so thick that the spectral fringes that are due to Fabry–Perot interference are so finely spaced that they are beyond the resolution of many spectrometers. Second, the plate surfaces may not be exactly flat and parallel. This smears the interference. Therefore the transmission of light through a thick plate is obtained by averaging over $k_2 L$, that is,

$$\langle T \rangle = \frac{1}{\pi} \int_0^{\pi} \frac{k_3}{k_1} |t|^2 \, d\theta \tag{11.6-23}$$

where $\theta = k_2 L$. Using Eq. (11.6-22) and carrying out the integration of Eq. (11.6-23), we obtain

$$\langle T \rangle = \frac{k_3}{k_1} \frac{|t_{12} t_{23} g_+|^2}{1 - |g_+ g_- r_{12} r_{23}|^2} \tag{11.6-24}$$

If we define

$$T_{12} = \frac{k_2}{k_1} |t_{12}|^2 \tag{11.6-25}$$

$$T_{23} = \frac{k_3}{k_2} |t_{23}|^2 \tag{11.6-26}$$

$$R_{12} = |r_{12}|^2 \tag{11.6-27}$$

$$R_{23} = |r_{23}|^2 \tag{11.6-28}$$

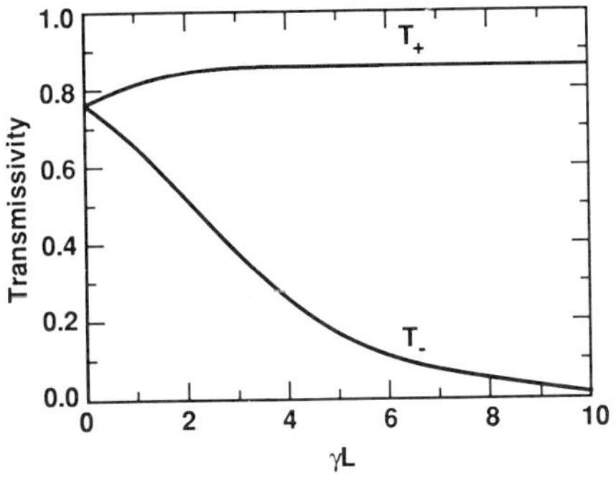

Figure 11.12. Transmissivities from right side and left side (T_+, T_-) as functions of the photorefractive coupling constant γL.

$$G_+ = |g_+|^2 \tag{11.6-29}$$

$$G_- = |g_-|^2 \tag{11.6-30}$$

then Eq. (11.6-24) can be written as

$$\langle T \rangle = \frac{G_+ T_{12} T_{23}}{1 - G_+ G_- R_{12} R_{23}} \tag{11.6-31}$$

This expression can also be obtained by adding, instead of the amplitudes, the intensities of the multiple reflections. Note that, even with this spectral averaging, the transmissivity $\langle T \rangle$ is still nonreciprocal. Figure 11.2 shows the transmissivity $\langle T \rangle$ for both right-side and left-side incidence. Depending on the angle of incidence, the coupling constant γ may vary from 0.1 to 40 cm^{-1} for many of the photorefractive crystals. Taking $\gamma L = 1$, this asymmetry in transmission should be experimentally observable according to Figure 11.12. Although we assume that the c-axis is perpendicular to the surface, optimized nonreciprocity may require some other orientations.

In summary, we have considered the propagation of electromagnetic radiation in a photorefractive layered medium. The reflectance and the transmittance are derived. It is found that the photorefractive coupling of the reflected wave with the incident wave in the medium leads to a nonreciprocal optical phenomenon. This property makes it possible to construct a one-way glass that favors transmission from one side only.

PROBLEMS

11.1 Energy Transfer in TWM. Consider the nonreciprocal transfer of energy in TWM. Using the results obtained in Chapter 4, the beam intensities after the coupling can be written

$$I_1(L) = I_1(0)\,\frac{I + m^{-1}}{1 + m^{-1}e^{\gamma L}}\,e^{-\alpha L}$$

$$I_2(L) = I_2(0)\,\frac{I + m}{1 + me^{-\gamma L}}\,e^{-\gamma L}$$

where $m = I_1(0)/I_2(0)$.

a. Let $I_1(0)$ and $I_2(0)$ be derived from a single beam with intensity I_0 and a beam splitter with transmittance T and reflectance R. Let $I_1(0) = TI_0$ and $I_2(0) = RI_0$ with $R + T = 1$. Show that

$$I_2(L) = I_0\,\frac{e^{-\alpha L}}{1 + (T/R)e^{-\gamma L}}$$

$$I_1(L) = I_0\,\frac{e^{-\alpha L}}{1 + (R/T)e^{\gamma L}}$$

b. Define the energy transfer efficiency as

$$\eta = \frac{I_1(0) - I_1(L)}{I_1(0)}$$

show that

$$\eta = \frac{T + Re^{\alpha L} - e^{-\alpha L}}{T + Re^{\alpha L}}$$

c. For lossless media, $\alpha = 0$, the efficiency becomes

$$\eta = \frac{R(e^{\alpha L} - 1)}{1 + R(e^{\alpha L} - 1)}$$

Find the reflectance R that optimizes the energy transfer efficiency.

11.2 Image Amplification

a. Show that Eq. (11.1-10) can be written

$$\frac{d}{dz}\,I_p = -\frac{\Gamma}{I_0}\,I_sI_p\,, \qquad \frac{d}{dz}\,I_s = \frac{\Gamma}{I_0}\,I_sI_p$$

where

$$I_s = \sum_{m=1}^{N} I_m = \sum_{m=1}^{N} |E_m|^2$$

b. Show that the space-average signal intensity can be written

$$\langle |E_s|^2 \rangle = \sum_{m=1}^{N} |E_m|^2$$

11.3 **Photorefractive Gratings via Imaging.** Consider the presence of a sinusoidal amplitude grating in front of a photorefractive medium. Show that a volume index grating can be induced by the illumination of an incoherent beam.

11.4 An image bearing beam is split into two by using a beam splitter. These two beams are recombined in a photorefractive medium using a Mach–Zehnder interferometry. Show that, at steady state, no image information is recorded.

11.5 **Photorefractive Polarizers.** Consider the possibility of converting an unpolarized beam into a polarized one. Using a polarizing beam splitter, an unpolarized beam can be split into two. One of the beams can be transformed by using a half-wave plate so that both beams have the same polarization state. These two beams can then be combined by using photorefractive two-wave mixing. Discuss the energy coupling.

REFERENCES

1. J. P. Huignard and A. Marrakchi, "Two-Wave Mixing and Energy Transfer in $Bi_{12}SiO_{20}$ Crystal: Application to Image Amplification and Vibration Analysis," Opt. Lett. **6**, 622–624 (1981).
2. Y. Fainman and S. H. Lee, "Applications of Photorefractive Crystals to Optical Signal Processing," in Optical and Hybrid Computing, H.H. Szu (Ed.), Proc. SPIE **634**, 380–392 (1986).
3. J. Hong, A. E. Chiou, and P. Yeh, "Image Amplification by Two-Wave Mixing in Photorefractive Crystals," Appl. Opt. **39**, 3026–3029 (1990).
4. J. W. Goodman, Introduction to Fourier Optics (McGraw-Hill, San Francisco, CA, 1968).
5. J. Tsujiuchi, in Progress Optics, E. Wolf (Ed.) (North-Holland, Amsterdam, 1963), Vol. II, Chap. IV.
6. Y. Fainman, E. Klancnik, and S. H. Lee, "Optimal Coherent Image Amplification by Two-Wave Coupling in Photorefractive $BaTiO_3$," Opt. Eng. **25**, 228 (1986).
7. T. Y. Chang, J. Hong, and P. Yeh, "Spatial Amplification," Opt. Lett. **15**, 743–745 (1990).
8. P. Yeh and A. E. T. Chiou, in Digest of Conference on Lasers and Electro-Optics (Optical Society of America, Washington, DC, 1985), p. 274.
9. A. E. T. Chiou and P. Yeh, "Beam Cleanup Using Photorefractive Two-Wave

Mixing," Opt. Lett. **10**, 621–623 (1985); "Laser-Beam Cleanup Using Photorefractive Two-Wave Mixing and Optical Phase Conjugation," Opt. Lett. **11**, 461–463 (1986).

10. C. Warde, A. M. Weiss, A. D. Fisher, and J. I. Thackara, "Optical Information Processing Characteristics of the Microchannel Spatial Light Modulator," Appl. Opt. **20**, 2066 (1981).
11. F. Blechman, "Radio Electronics" **57**, 39 (1986).
12. Y. Shi, D. Psaltis, A. Marrakchi, and A. R. Tanguay, Jr., "Photorefractive Incoherent-to-Coherent Optical Converter," Appl. Opt. **22**, 3665 (1983).
13. J. Hong, S. Campbell, and P. Yeh, "Photorefractive Spatial Light Modulator," Proc. SPIE **1151**, 476–484 (1989).
14. D. Psaltis, J. Yu, and J. Hong, "Bias Free Time-Integrating Optical Correlator Using a Photorefractive Crystal," Appl. Opt. **24**, 3860–3865 (1985).
15. See, for example, A. Yariv and P. Yeh, Optical Waves in Crystals (Wiley, New York, 1984).
16. See, for example, P. Yeh, Optical Waves in Layerd Media (Wiley, New York, 1988).

12

Higher Order Photo-Induced Gratings

Up to now, we have considered only gratings which are represented by a periodic variation of the index of refraction. In this chapter, we consider periodic media in which the nonlinear optical susceptibilities are a periodic function of position. In particular, we consider $\chi^{(2)}$-gratings in which a periodic variation of the second-order dielectric susceptibility $\chi^{(2)}$ is produced by photo-illumination.

12.1 INTRODUCTION

Ge-doped optical fibers are found to exhibit photosensitivity. The photosensitivity is manifested by photo-induced index changes as well as $\chi^{(2)}$-changes in the core of silica fibers. When the core of a single-mode optical fiber is exposed to a high intensity laser beam, a periodic modification of the refractive index as well as the nonlinear susceptibility $\chi^{(2)}$ result along the axis of the fiber. The index grating is known as the Hill grating [1]. Formation of the Hill grating is believed to be initiated by the existence of a standing wave caused by Fresnel reflection at the far end of the fiber. Nonlinear optical absorption (e.g., 2-photon absorption) may be responsible for the formation of the index grating with a period of $\Lambda = \lambda/(2n)$. The grating period exactly matches with the wavelength of the laser beam and the light is reflected via Bragg scattering.

The physical mechanism responsible for the formation of $\chi^{(2)}$-gratings in optical fibers is currently undergoing intense investigation. The presence of such a nonlinear grating leads to efficient frequency doubling of cw Q-switched and mode-locked Nd:YAG laser in single-mode optical glass fibers. When a cw mode-locked and Q-switched Nd:YAG laser at $\lambda =$ 1.06 μm was coupled into a single mode glass fiber, green lights at $\lambda =$ 0.53 μm were observed after steady illumination of few hours [2]. Second harmonic generation in glass was widely accepted, until now, as a forbidden process since glass is an amorphous medium possessing an inversion symmetry. All second-order processes should therefore be absent [3]. The long build-up time for the conversion to second harmonic suggests that an essential ingredient of the process is a "permanent" photo-induced change in the fiber analogous to the photorefractive Hill gratings reported in

Ge-doped single-mode fibers [1]. It is believed that the intense fields at the pump (ω) and second harmonic (2ω) frequencies mix to orient defects that destroy the local inversion symmetry [4]. Consider the photo-excitation of an electron from the ground state to an excited state. The transition can be achieved by either absorbing one photon at 2ω or absorbing two photons at ω. The transition amplitude for the two-photon nonlinear optical absorption is proportional to $\exp(-ik^{2\omega}z)$, whereas that of the other process is proportional to $\exp(-i2k^{\omega}z)$. The interference of these two quantum processes leads to a spatially periodic distribution of the photo-generated charge carriers. In a macroscopic picture, two pump fields and the second harmonic field mix and produce a periodic space-charge field. The period of the space-charge field is exactly the phase-matching periodicity for second harmonic generation and is given by (see Problem 12.1)

$$\Lambda = \frac{2\pi}{(k^{2\omega} - 2k^{\omega})} \tag{12.1-1}$$

where $k^{2\omega}$ is the wave number of the second harmonic field and k^{ω} is that of the fundamental field.

If the amplitude of the space-charge field is large enough (e.g., 10^4–10^6 V/cm), then a significant change in $\chi^{(2)}$ is induced, that is,

$$\Delta\chi^{(2)} = \chi^{(3)}E^{\text{dc}} \tag{12.1-2}$$

where $\chi^{(3)}$ is the third-order nonlinear susceptibility and E^{dc} is the space-charge field. We note that $\Delta\chi^{(2)}$ has exactly the same periodicity as the periodic space-charge field. In what follows, we consider the propagation of optical waves in periodic $\chi^{(2)}$ media.

12.2 THREE-WAVE MIXING IN PERIODIC $\chi^{(2)}$-MEDIA

Consider the mixing of three electromagnetic waves, $A_j \exp[i(\omega_j t - k_j z)]$ with $j = 1, 2, 3$, in a medium which exhibits second-order optical nonlinearity (i.e., $\chi^{(2)}$). The frequencies of the waves are ω_1, ω_2, and ω_3 such that $\omega_3 = \omega_1 + \omega_2$. As a result of the nonlinear response of the medium, nonlinear polarization at ω_1 is produced by the mixing of waves at ω_3 and ω_2. This nonlinear polarization is proportional to $\chi^{(2)}A_3A_2^* \exp[i\omega_1 t - i(k_3 - k_2)z]$ which, acting as a distributed dipole, will radiate and generate $A_1 \exp[i(\omega_1 t - k_1 z)]$. Similarly, waves at ω_2 or ω_3 can be generated by the mixing of waves at ω_3 and ω_1 or ω_1 and ω_2, respectively. The coupling of thc three waves is described by the following coupled equations [3]:

$$\frac{d}{dz} A_1 = -i\kappa A_2^* A_3 e^{-i\,\Delta k\,z}$$

$$\frac{d}{dz} A_2 = -i\kappa A_1^* A_3 e^{-i\,\Delta k\, z}$$

$$\frac{d}{dz} A_2 = -i\kappa A_1^* A_3 e^{i\,\Delta k\, z} \tag{12.2-1}$$

where

$$A_j = \left(\frac{n_j}{\omega_j}\right)^{1/2} E_j\,, \quad j = 1, 2, 3 \tag{12.2-2}$$

are normalized amplitudes of the waves, the E_j are electric field amplitudes of the waves, the n_j are indices of refraction of the medium at each of the frequencies ω_j, and

$$\Delta k = k_3 - (k_1 + k_2) \tag{12.2-3}$$

$$\kappa = \frac{1}{2c}\left(\frac{\omega_1 \omega_2 \omega_3}{n_1 n_2 n_3}\right)^{1/2} \chi^{(2)} \tag{12.2-4}$$

The coupled equations can be solved for sum frequency generation ($\omega_3 = \omega_1 + \omega_2$). In the case of no pump-depletion, a direct integration of the third equation in Eq. (12.2-1) leads to

$$A_3(L) = -ikA_1 A_2\left(\frac{e^{i\,\Delta k\, L} - 1}{i\,\Delta k}\right) \tag{12.2-5}$$

where L is the length of interaction and we assume a boundary condition of $A_3(0) = 0$. We note that the amplitude $A_3(L)$ is a periodic function of the interaction length. In other words, the conversion cannot be increased further by increasing the interaction length L. This is the result of the phase mismatch Δk. Thus, a prerequisite for efficient sum frequency generation is that

$$\Delta k = 0 \tag{12.2-6}$$

which is known as the phase-matching condition. According to Eq. (12.2-5), $A_3(L)$ is linearly proportional to L with $\Delta k = 0$.

In anisotropic media, the phase matching condition $\Delta k = 0$ can be achieved by taking advantage of the natural birefringence of the media. In isotropic media such as glass, the condition $\Delta k = 0$ cannot be satisfied in the transparent regime where $dn/d\lambda < 0$. It has been known for some time that efficient three-wave mixing is still possible provided the coupling constant κ is modulated periodically with a spatial period [5]

$$\Lambda = \frac{2\pi}{\Delta k} \tag{12.2-7}$$

Let the nonlinear susceptibility be written

$$\chi^{(2)} = \chi_o^{(2)} \cos Kz \tag{12.2-8}$$

where $\chi_o^{(2)}$ is a constant and K is related to the period by $K = 2\pi/\Lambda = \Delta k$. Using Eqs. (12.2-4), (12.2-8), and (12.2-1) and keeping only the terms without fast varying phases, we obtain

$$\begin{aligned} \frac{d}{dz} A_1 &= -i\kappa_o A_2^* A_3 \\ \frac{d}{dz} A_2 &= -i\kappa_o A_1^* A_3 \\ \frac{d}{dz} A_3 &= -i\kappa_o A_1 A_2 \end{aligned} \tag{12.2-9}$$

where κ_o is given by

$$\kappa_o = \frac{1}{4c}\left(\frac{\omega_1\omega_2\omega_3}{n_1 n_2 n_3}\right)^{1/2} \chi_o^{(2)} \tag{12.2-10}$$

We note that the phase mismatch factor $\exp(-i\,\Delta k\, z)$ disappears in Eq. (12.1-9). The coupled Eqs. (12.2-9) can be solved in closed form. Before we proceed with the solution, we note that there are conservation laws governing the amplitudes A_i. They are

$$|A_1|^2 + |A_3|^2 = c_1^2, \qquad |A_2|^2 + |A_3|^2 = c_2^2 \tag{12.2-11}$$

where c_1, c_2 are constants. Since $|A_i|^2$ is proportional to the photon flux at ω_i, Eqs. (12.2-11) are conservation laws for the photon flux. Physically, photon flux at ω_1 plus photon flux at ω_3 are conserved and similarly for photon flux at ω_2 and at ω_3. Since one photon at ω_3 is formed by one photon at ω_1 and one photon at ω_2, the photon-flux must obey the conservation laws (12.2-11).

Let

$$u_j = A_j \exp(-i\phi_j)\,, \quad j = 1, 2, 3 \tag{12.2-12}$$

where both u_j and ϕ_j are real. We obtain

$$\begin{aligned} u_1' &= \kappa_o u_2 u_3 \sin\phi \\ u_2' &= \kappa_o u_1 u_3 \sin\phi \\ u_3' &= -\kappa_o u_1 u_2 \sin\phi \end{aligned} \tag{12.2-13}$$

$$\begin{aligned} \phi_1' u_1 &= -\kappa_o u_2 u_3 \cos\phi \\ \phi_2' u_2 &= -\kappa_o u_1 u_3 \cos\phi \\ \phi_3' u_3 &= -\kappa_o u_1 u_2 \cos\phi \end{aligned} \tag{12.2-14}$$

with

$$\phi = \phi_3 - \phi_1 - \phi_2 \tag{12.2-15}$$

In the above, a prime (′) indicates differentiation with respect to z.

Using Eqs. (12.2-14) and (12.2-15), we obtain

$$u_1 u_2 u_3 \cos\phi = \Gamma \tag{12.2-16}$$

where Γ is a constant.

These equations can be integrated for various boundary conditions.

a. *Sum Frequency Generation* $\omega_3 = \omega_1 + \omega_2$. In the sum frequency generation (or up-conversion), input waves at ω_1 and ω_2 are mixed and a new wave at $\omega_1 + \omega_2$ is generated. If we set $u_3(0) = 0$, which corresponds to zero photon flux at ω_3, then $\Gamma = 0$, according to Eq. (12.2-16). This leads to

$$\phi = \pm\pi/2 \tag{12.2-17}$$

and ϕ_1, ϕ_2, ϕ_3 are all constants, according to Eq. (12.2-14). After an inspection of the coupled Eqs. (12.2-9), we find that $\phi = -\pi/2$ is a good solution. Thus, we have

$$\begin{aligned} u_1' &= -\kappa_o u_2 u_3 \\ u_2' &= -\kappa_o u_1 u_3 \\ u_3' &= \kappa_o u_1 u_2 \end{aligned} \tag{12.2-18}$$

Using Eqs. (12.2-11), (12.2-12), and (12.2-18), we obtain

$$u_3' = \kappa_o (c_1^2 - u_3^2)^{1/2} (c_2^2 - u_3^2)^{1/2} \tag{12.2-19}$$

The equation can be integrated and the solution can be expressed in terms of elliptic functions:

$$\begin{aligned} u_3^2(z) &= u_2^2(0)\,\mathrm{sn}^2(u_1(0)\kappa_o z, k) \\ u_2^2(z) &= u_2^2(0) - u_3^2(z) \\ u_1^2(z) &= u_1^2(0) - u_3^2(z) \end{aligned} \tag{12.2-20}$$

where

$$k = c_2/c_1 = u_2(0)/u_1(0) \tag{12.2-21}$$

We note that the beams are labeled such that $0 \leq k \leq 1$. According to the solution Eq. (12.2-20), the intensity of the sum-frequency ω_3, is a periodic function of interaction distance. The maximum is $u_2^2(0)$, provided $u_2(0) \leq u_1(0)$. Figure 12.1 illustrates the photon flux as functions of interaction

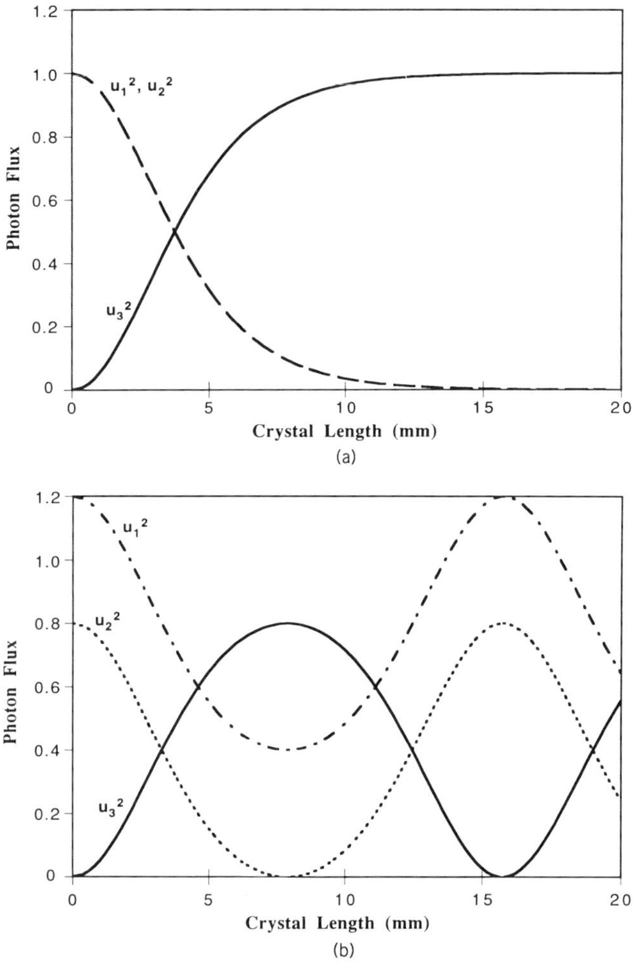

Figure 12.1. (a) Photon flux, in units of $2\pi c\epsilon_0/h$, of the three waves in the sum frequency generation ($\omega_1 + \omega_2 = \omega_3$) in a $KTiOPO_4$ crystal with initial conditions $u_1^2(0) = 1$, $u_2^2(0) = 1$, and $u_3^2(0) = 0$ ($\lambda_1 = 0.9\ \mu m$, $\lambda_2 = 1.3\ \mu m$, and $\lambda_3 = 0.53\ \mu m$). (b) Photon flux, in units of $2\pi c\epsilon_0/h$, of the three waves in the sum frequency generation ($\omega_1 + \omega_2 = \omega_3$) in a $KTiOPO_4$ crystal with initial conditions $u_1^2(0) = 1.2$, $u_2^2(0) = 0.8$, and $u_3^2(0) = 0$ ($\lambda_1 = 0.9\ \mu m$, $\lambda_2 = 1.3\ \mu m$, and $\lambda_3 = 0.53\ \mu m$).

length. In the sum-frequency generation, pump photons at ω_1 and ω_2 are combined to form the sum photon at $\omega_3 = \omega_1 + \omega_2$. Thus, for each pair of the pump photons (ω_1, ω_2) a sum-frequency photon is generated. The photon flux at ω_3 continue to increase as the beams propagate in the nonlinear medium until one of the pump beams is depleted. At this point, the photon flux u_2^2 is zero, and the photon flux u_3^2 reaches its maximum (see Figure 12-1). Beyond this point in the medium, the process reverses. In other words, the partially depleted photons at ω_1 and the sum-frequency photons at ω_3 mix to form pump photons at ω_2 leading to an increase of the photon flux u_2^2. This is the process of optical parametric amplification (OPA) which is discussed next. Thus the energies are coupled back and forth periodically between the sum-frequency beam and the pump beams.

In the case when $c_2 = c_1$, that is, the photon flux at ω_1 and ω_2 are identical, the elliptic function $\mathrm{sn}(u_1(0)\kappa_o z, k)$ becomes a hyperbolic tangent function. This is the situation when a complete conversion to ω_3 becomes possible. The energy exchange in this case is no longer periodic. This is illustrated in Figure 12.1(a).

b. *Difference Frequency Generation* $\omega_2 = \omega_3 - \omega_1$. In the difference frequency generation, an incident beam at ω_1 is mixed with a pump beam at ω_3. As a result of three-wave mixing, a third beam at frequency $\omega_3 - \omega_1$ is generated. This leads to an amplification of the incident beam at ω_1 and is known as optical parametric amplification (OPA). The boundary condition requires that $u_2(0) = 0$. This leads to $\Gamma = 0$ and, according to Eq. (12.2-16)

$$\phi = \pm \pi/2 \tag{12.2-22}$$

Again ϕ_1, ϕ_2, and ϕ_3 are constants according to Eq. (12.2-14). By an inspection of the coupled Eqs. (12.2-9), we find that $\phi = \pi/2$ is a good solution. Thus, we have

$$u_1' = \kappa_o u_2 u_3\,, \qquad u_2' = \kappa_o u_1 u_3\,, \qquad u_3' = -\kappa_o u_1 u_2 \tag{12.2-23}$$

Using Eqs. (12.2-11) and (12.2-12), the third equation in Eqs. (12.2-23) can be written

$$u_3' = -\kappa_o (c_1^2 - u_3^2)^{1/2} (c_2^2 - u_3^2)^{1/2} \tag{12.2-24}$$

where $c_1^2 = u_1^2(0) + u_3^2(0)$ and $c_2^2 = u_3^2(0)$. An integration of the above equation leads to

$$-\kappa_o z + c' = \frac{1}{c_1}\,\mathrm{sn}^{-1}\left(\frac{u_3}{c_2}, k\right) \tag{12.2-25}$$

where c' is a constant, $k = c_2/c_1 \leq 1$. The boundary condition at $z = 0$

requires that

$$c' = \frac{1}{c_1}\,\mathrm{sn}^{-1}(1, k) \equiv \frac{K}{c_1} \tag{12.2-26}$$

where K is the elliptic integral

$$K = \int_0^1 \frac{dz}{(1 - z^2)^{1/2}(1 - k^2 z^2)^{1/2}} \tag{12.2-27}$$

Thus, we have

$$u_3(z) = u_3(0)\,\mathrm{sn}(K - c_1 \kappa_o z, k) \tag{12.2-28}$$

The photon flux amplitude $u_1(z)$ and $u_2(z)$ are given by

$$u_1^2(z) = c_1^2 - u_3^2(z)\,, \qquad u_2^2(z) = u_3^2(0) - u_3^2(z) \tag{12.2-29}$$

Again, we note that all three photon fluxes are periodic functions of interaction distance. The maximum photon flux for the signal beam is $[u_1^2(0) + u_3^2(0)]$. Figure 12.2 illustrates the photon flux as functions of interaction length. In an optical parametric amplifier (OPA) u_1^2 is the photon flux of the weak incoming signal to be amplified, whereas u_3^2 is the photon flux of the pump beam. Each pump photon at ω_3 disintegrates into two photons at ω_1, and ω_2 such that $\omega_3 = \omega_1 + \omega_2$. This three-photon process leads to an increase of the signal photon flux and thus the amplification of the signal beam. The amplification continues as the beams propagate in the nonlinear medium until the pump beam is depleted. At this point, the photon flux u_1^2 reaches its maximum, whereas the photon flux u_3^2 is complete-

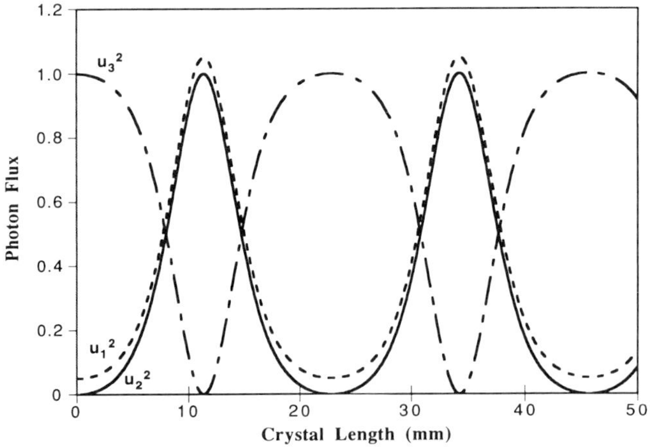

Figure 12.2. Photon flux, in units of $2\pi c\epsilon_0/h$, of the three waves in the difference frequency generation ($\omega_3 - \omega_1 = \omega_2$) in a $KTiOPO_4$ crystal with initial conditions $u_1^2(0) = 0.05$, $u_2^2(0) = 0$, and $u_3^2(0) = 1$ ($\lambda_1 = 0.9\ \mu m$, $\lambda_2 = 1.3\ \mu m$, and $\lambda_3 = 0.53\ \mu m$).

ly depleted. Beyond this point in space, the process reverses. In other words, the signal photons at ω_1 and the idler photons at ω_2 combine to form pump photons at ω_3 leading to an increase of the photon flux u_3^2. Thus the energies are coupled back and forth periodically between the signal beam and the pump beam.

c. *Second Harmonic Generation*. Second harmonic generation can be viewed as a special case of sum-frequency generation. If we set $\omega_1 - \omega_2 = \omega$, then $\omega_3 = 2\omega$. The results obtained in (**a**) can be employed to treat type II second harmonic generation [6]. Although the results can be modified to treat type I second harmonic generation, it is often easier to derive the coupled equations directly. In type I second harmonic generation, $E_1(\omega_1)$ and $E_2(\omega_2)$ are amplitudes of the same mode. We have to be very careful in treating the power of the fundamental wave. For this reason, we will treat the problem by rederiving the coupled equations.

In the case of perfect phase matching, the coupled equations for type I second harmonic generation are given by

$$\frac{d}{dz} E_1 = -i\kappa E_1^* E_2 , \qquad \frac{d}{dz} E_2 = -i\kappa E_1^2 \tag{12.2-30}$$

where E_1, E_2 are amplitudes of the fundamental and the second harmonic, respectively, and κ is the the coupling constant given by

$$\kappa = \frac{\omega_1}{2n_1 c} \chi^{(2)} \tag{12.2-31}$$

where n_1 is the index of refraction of the fundamental and the second harmonic. It can be shown that

$$\frac{d}{dz} \{|E_1|^2 + |E_2|^2\} = 0 \tag{12.2-32}$$

and the power is conserved.

Using a similar technique, the coupled equations can be integrated and the solutions are

$$\begin{aligned} |E_1|^2 &= c^2 \operatorname{sech}^2(c\kappa z + c') \\ |E_2|^2 &= c^2 \tanh^2(c\kappa z + c') \end{aligned} \tag{12.2-33}$$

where c and c' are constants with

$$c^2 = |E_1|^2 + |E_2|^2$$

and

$$\sinh^2 c' = \left| \frac{E_2}{E_1} \right|^2_{z=0} \tag{12.2-34}$$

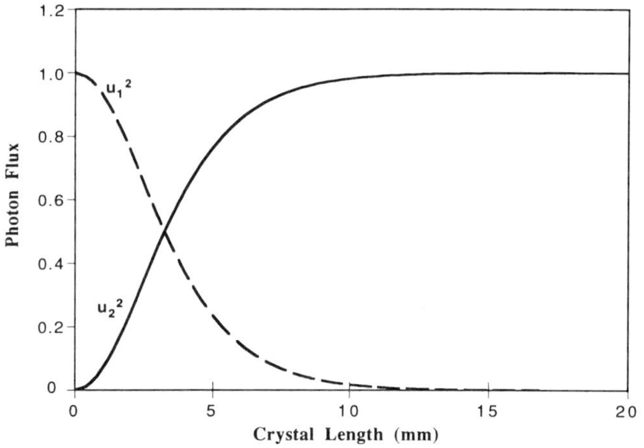

Figure 12.3. Photon flux, in units of $2\pi c\epsilon_0/h$, of the two waves in second harmonic generation ($2\omega_1 = \omega_2$) in $KTiOPO_4$ crystal with initial condition $u_1^2(0) = 1$, $u_2^2(0) = 0$ ($\lambda_1 = 1.064\ \mu m$, $\lambda_2 = 0.532\ \mu m$).

In the case when $E_2(0) = 0$, we have $c' = 0$ and the solutions become

$$|E_1|^2 = |E_1(0)|^2 \operatorname{sech}^2(|E_1(0)|\kappa z)$$
$$|E_2|^2 = |E_1(0)|^2 \tanh^2(|E_1(0)|\kappa z) \tag{12.2-35}$$

Figure 12.3 plots the power (or intensity) of the fundamental and the second harmonic as functions of z. We note that the rise of the second harmonic power depends on the power of the fundamental.

12.3 $\chi^{(2)}$-HOLOGRAPHY

Consider a nonlinear medium whose nonlinear susceptibility $\chi^{(2)}$ is related to E^3 by the following relationship

$$\chi^{(2)} = \alpha\langle E^3\rangle \tag{12.3-1}$$

where α is a constant, and $\langle E^3\rangle$ is a time-averaged E^3. If, now, three beams of light with frequencies ω_1, ω_2 and $\omega_3 = \omega_1 + \omega_2$ are incident into the medium (see Figure 12.4). The electric field can be written

$$E = \mathrm{Re}\left\{\sum_{j=1}^{3} E_j \exp[i(\omega_j t - \mathbf{k}_j \cdot \mathbf{r})]\right\} \tag{12.3-2}$$

where E_j is the amplitude of the jth field, ω_j is the frequency and $\mathbf{k}_j$ is the wave vector.

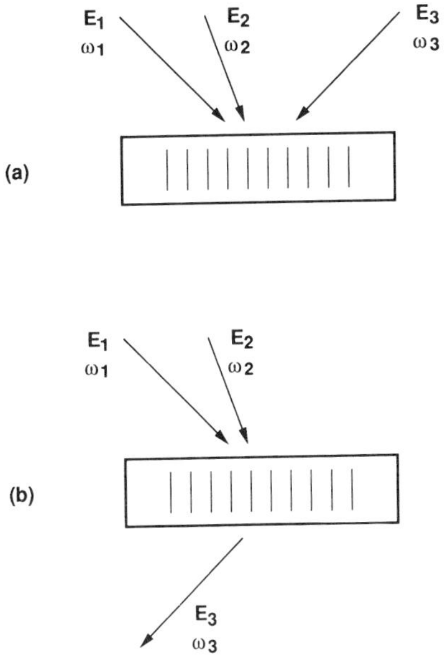

Figure 12.4. $\chi^{(2)}$-Holography. (a) Recording of $\chi^{(2)}$-hologram using three-beam interference in a $x^{(2)}$-holographic medium such as GeO-doped silica. (b) Readout of $\chi^{(2)}$-hologram. Beam E_3 is reconstructed when the $\chi^{(2)}$-hologram is illuminated with beams E_1 and E_2.

As a result of the nonlinearity, the $\chi^{(2)}$ of the medium can be written

$$\chi^{(2)} = \tfrac{1}{4}\alpha\{E_1 E_2 E_3^* \exp[i(\mathbf{k}_3 - \mathbf{k}_1 - \mathbf{k}_2)\cdot \mathbf{r}] + \text{c.c.}\} \tag{12.3-3}$$

where c.c. denotes complex conjugation. We note that wave front information about the three fields is now recorded in terms of volume $\chi^{(2)}$-gratings (or holograms). This wave front information can be reconstructed by the readout process.

To reconstruct E_3, the $\chi^{(2)}$ hologram is illuminated with E_1 and E_2 simultaneously. As a result of the illumination, a nonlinear polarization is produced

$$P^{\text{NL}} = \chi^{(2)} E_R^2 \tag{12.3-4}$$

where the electric field of the read out beams E_R is given by

$$E_R = \text{Re}\left\{\sum_{j=1}^{2} E_j \exp[i(\omega_j t - \mathbf{k}_j \cdot \mathbf{r})]\right\} \tag{12.3-5}$$

Using Eq. (12.3-3) for $\chi^{(2)}$, we obtain

$$P^{\text{NL}} = \tfrac{1}{8}\alpha\{|E_1 E_2|^2 E_3 \exp[i(\omega_3 t - \mathbf{k}_3 \cdot \mathbf{r})] + \text{c.c.}\} \tag{12.3-6}$$

By examining the above expression, we note that P^{NL} satisfies the wave

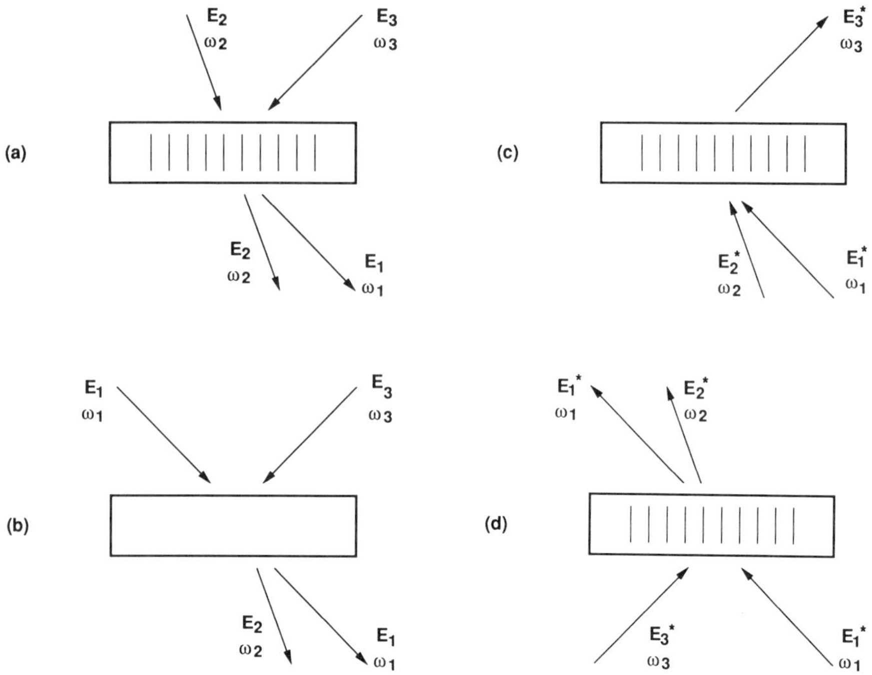

Figure 12.5. Other schemes of readout of $\chi^{(2)}$-hologram. (a) Beam E_1 is reconstructed when the $\chi^{(2)}$-hologram is illuminated with beams E_2 and E_3. (b) Beam E_2 is reconstructed when the $\chi^{(2)}$-hologram is illuminated with beams E_1 and E_3. (c) Phase conjugate beam E_3^* is reconstructed when the $\chi^{(2)}$-hologram is illuminated with beams E_1^* and E_2^*. (d) Phase conjugate beam E_2^* is reconstructed when the $\chi^{(2)}$-hologram is illuminated with beams E_1^* and E_3^*.

equation for $E_3 \exp[i(\omega_3 t - \mathbf{k}_3 \cdot \mathbf{r})]$ provided $|E_1 E_2| = \text{constant}$. Thus, such a nonlinear polarization will radiate efficiently and generate $E_3 \exp[i(\omega_3 t - \mathbf{k}_3 \cdot \mathbf{r})]$.

Similarly, E_1 (or E_2) can be reconstructed by illuminating the $\chi^{(2)}$-hologram with E_2 and E_2 (or E_1 and E_3). Phase conjugated wave fronts can also be constructed by reading the hologram with counterpropagating beams. Some of the readout schemes are shown in Figure 12.4(b) and Figure 12.5.

The formation of $\chi^{(2)}$-holograms in Ge-doped silica fibers is very slow. If the speed can be improved by using different materials, then real-time $\chi^{(2)}$-holography is possible. This can lead to applications such as wide-field optical parametric amplification.

PROBLEMS

12.1 Cubic Mixing. Consider a third-order photo-excitation in which the produced current density is proportional to the third power of the electric field,

$$J = \alpha \langle E^3 \rangle$$

where α is a constant and $\langle E^3 \rangle$ represents the time-averaged E^3 with E as the electric field. Show that J is periodic with a period given by Eq. (12.1-1), provided E consists of a fundamental part at ω and a second harmonic at 2ω.

12.2 **Electro-Optical Nonlinearity.** Starting from

$$\mathbf{P} = \epsilon_0 \chi^{(1)} \mathbf{E} + \epsilon_0 \chi^{(2)} \mathbf{E}^2 + \epsilon_0 \chi^{(3)} \mathbf{E}^3 + \cdots$$

Show that a dc field can cause a change in $\chi^{(2)}$ in media possessing an inversion symmetry.

12.3 **Conservation of Energy.** Show that Eq. (12.2-11) is valid even if $\Delta k \neq 0$.

12.4 Derive Eqs. (12.2-13), (12.2-14), and (12.2-16).

12.5 **Elliptic Functions.** The elliptic function

$$y = \operatorname{sn} x$$

is defined as

$$(y')^2 = (1 - y^2)(1 - k^2 y^2)$$

where the prime (′) is a differentiation with respect to x, $0 \leq k^2 < 1$, with the boundary conditions

$$y = 0 \quad \text{and} \quad y' > 0 \quad \text{at } x = 0$$

The function $\operatorname{sn} x$ is limited between $(-1, 1)$ and is very similar to $\sin x$. It reduces to $\sin x$ at $k = 0$.

a. Show that

$$x = \operatorname{sn}^{-1}(y, k) = \int_0^y \frac{dz}{[(1 - z^2)(1 - k^2 z^2)]^{1/2}}$$

b. Show that $\operatorname{sn} x$ is a periodic function with

$$\operatorname{sn}(x + 4K) = \operatorname{sn}(2K - x) = \operatorname{sn} x$$

where

$$K = \int_0^1 \frac{dy}{[(1 - y^2)(1 - k^2 y^2)]^{1/2}}$$

c. Show that

$$y = \mathrm{sn}(x, k) = \begin{cases} \sin x\,, & \text{for } k = 0 \\ \tanh x\,, & \text{for } k = 1 \end{cases}$$

12.6 Optical Parametric Amplification. In the difference frequency generation, let ω_1 be the signal frequency and ω_3 be the pump frequency. ω_3 must be greater than ω_1.

a. Show that the maximum signal gain is

$$g \equiv \frac{I_{1L}}{I_{10}} = \frac{I_{10} + I_{30}}{I_{10}}$$

where I_{10} is the input signal intensity, I_{30} is pump intensity, and I_{1L} is the output signal intensity.

b. In the case of sum-frequency generation with ω_2 as the pump, the signal gain is limited by

$$g \equiv \frac{I_{3L}}{I_{10}} = \frac{\omega_3}{\omega_1}$$

where I_{3L} is the output at ω_3.

12.7 Conservation of Energy in SHG. Show that Eq. (12.2-32) is valid even if $\Delta k \neq 0$.

12.8 Third-Order Optical Rectification. Show that in the presence of both the fundamental and the second harmonic wave, a dc nonlinear polarization can be produced.

Let the electric field be written

$$E = E_1 \exp[i(\omega_1 t - k_1 z)] + E_2 \exp[i(\omega_2 t - k_2 z)]$$

with $\omega_2 = 2\omega_1$.

a. Show that

$$p_{\mathrm{NL}}^{\mathrm{dc}} = \tfrac{3}{4}\epsilon_0 \chi^{(3)}\{E_1^2 E_2^* \exp[i(k_2 - 2k_1)z] + \text{c.c.}\}$$

b. Show that the dc polarization is periodic with a period given by Eq. (12.1-1).

REFERENCES

1. K. O. Hill, Y. Fujii, D. C. Johnson and B. S. Kawaski, "Photosensitivity in Optical Fiber Waveguides: Application to Reflection Filter Fabrication," Appl. Phys. Lett. **32**, 647 (1978).

2. U. Österberg and W. Margulis, "Dye Laser Pumped by Nd:YAG Laser Pulses Frequency Doubling in a Glass Optical Fiber," Opt. Lett. **11**, 516 (1986); "Experimental Studies on Efficient Frequency Doubling in Glass Optical Fibers," Opt. Lett. **12**, 57 (1987).
3. See, for example, A. Yariv, and P. Yeh, Optical Waves in Crystals (Wiley, New York, 1984).
4. R. H. Stolen and W. K. Tom, "Self-Organized Phase-Matched Harmonic Generation in Optical Fibers," Opt. Lett. **12**, 585 (1987).
5. See, for example, P. Yeh, A. Yariv, and C.-S. Hong, "Electromagnetic Propagation in Periodic Stratified Media: I. General Theory; II. Birefringence, Phase Matching and X-Ray Lasers," J. Opt. Soc. Am. **67**, 423–448 (1977).
6. In type I second harmonic generation (SHG-I), the two fundamental photons are in the same mode (i.e., either ordinary mode or extraordinary mode). The fundamental photons are polarized orthogonal to the second harmonic photons. In type II second harmonic generation (SHG-II), the two fundamental photons are in different modes, one is an ordinary wave and the other is an extraordinary wave. This can be achieved by polarizing the fundamental photons at 45 degrees with respect to the polarization state of the second harmonic photons.

Appendix A

Convergence Proof of Perceptron

Consider M sample vectors which are linearly separable. For the sake of simplicity, we assume a threshold value of $w_0 = 0$ and an arbitrary weight vector $\mathbf{w}_1 = 0$. We define a set of new vectors $\mathbf{y}^1, \mathbf{y}^2, \ldots, \mathbf{y}^M$ such that

$$\mathbf{y}^i = \begin{cases} \mathbf{x}^i, & \mathbf{x}^i \in \text{Class 1} \\ -\mathbf{x}^i, & \mathbf{x}^i \in \text{Class 2} \end{cases} \tag{A-1}$$

Assume there exists a solution weight vector $\mathbf{w}$ such that for any vector $\mathbf{y}^i$

$$\mathbf{w} \cdot \mathbf{y}^i > 0 \tag{A-2}$$

The perceptron algorithm (Eq. 10.4-4) now becomes

$$\mathbf{w}_{k+1} = \mathbf{w}_k + \mathbf{y}^k \tag{A-3}$$

where $\mathbf{y}^k$ is misclassified by $\mathbf{w}_k$. If we start with $\mathbf{w}_1 = 0$, we obtain

$$\mathbf{w}_{k+1} = \sum_{i=1}^{k} \mathbf{y}^k \tag{A-4}$$

The inner product between the solution vector $\mathbf{w}$ and $\mathbf{w}_{k+1}$ is

$$\mathbf{w} \cdot \mathbf{w}_{k+1} = \sum_{i=1}^{k} \mathbf{w} \cdot \mathbf{y}^i \tag{A-5}$$

Let β be the minimum value of all $\mathbf{w} \cdot \mathbf{y}^i$ for $i = 1, 2, \ldots, M$. According to Eq. (A-5),

$$\mathbf{w} \cdot \mathbf{w}_{k+1} > k\beta \tag{A-6}$$

Therefore

$$|\mathbf{w}_{k+1}|^2 > \frac{k^2\beta^2}{|\mathbf{w}|^2} \tag{A-7}$$

On the other hand, using Eq. (A-3), we obtain

$$|\mathbf{w}_{k+1}|^2 = |\mathbf{w}_k|^2 + |\mathbf{y}^k|^2 + 2\mathbf{w}_k \cdot \mathbf{y}^k \tag{A-8}$$

Since $\mathbf{y}^k$ is misclassified by $\mathbf{w}_k$, $\mathbf{w}_k \cdot \mathbf{y}^k < 0$ and

$$|\mathbf{w}_{k+1}|^2 < |\mathbf{w}_k|^2 + |\mathbf{y}^k|^2 \tag{A-9}$$

Using the iteration relationship given by Eq. (A-9), we obtain

$$|\mathbf{w}_{k+1}|^2 < \sum_{i=1}^{k} |\mathbf{y}^i|^2 \tag{A-10}$$

Let α be the maximum value of all $|\mathbf{y}^i|^2$ for $i = 1, 2, \ldots, M$. According to Eq. (A-10),

$$|\mathbf{w}_{k+1}|^2 < k\alpha \tag{A-11}$$

Using Eqs. (A-7) and (A-11), we obtain

$$\frac{k^2\beta^2}{|\mathbf{w}|^2} < k\alpha \tag{A-12}$$

or equivalently

$$k < \frac{\alpha|\mathbf{w}|^2}{\beta^2} \tag{A-13}$$

This result shows that the perceptron algorithm will converge to a solution $\mathbf{w}$ within the maximum number of iterations given by

$$k_{\max} = \frac{\alpha|\mathbf{w}|^2}{\beta^2} \tag{A-14}$$

Appendix B

Storage Capacity of Associative Memory

Suppose that we have a set of M independently chosen N-dimensional vectors, $\{\mathbf{x}^m\}$, each with independent binary (± 1) components. In addition, $P(x_i = 1) = P(x_i = -1) = 1/2$, where P is probability distribution function. Consider the linear discriminant function

$$g(\mathbf{x}^m) = \mathbf{w} \cdot \mathbf{x}^m \tag{B-1}$$

where

$$\mathbf{w} = \sum_{m=1}^{M} \alpha^m \mathbf{x}^m \tag{B-2}$$

with

$$\alpha^m = \begin{cases} 1\,, & \text{if } \mathbf{x}^m \in \text{Class 1} \\ -1\,, & \text{if } \mathbf{x}^m \in \text{Class 2} \end{cases} \tag{B-3}$$

and a probability distribution of

$$P(\alpha^m = 1) = P(\alpha^m = -1) = 1/2 \tag{B-4}$$

In this case, the expectation value is given by

$$\langle g(\mathbf{x}^m) \rangle = \alpha^m N \tag{B-5}$$

and the variance is given by

$$\langle [g(\mathbf{x}^m) - \langle g(\mathbf{x}^m) \rangle]^2 \rangle = N(M-1) \tag{B-6}$$

The random variable $g = g(\mathbf{x}^m) = \Sigma_{i=1}^{M} \alpha^i \mathbf{x}^i \cdot \mathbf{x}^m$ satisfies Gaussian distribution, according to the Central Limit Theorem. Therefore

$$p(g \mid \alpha^m) = \frac{1}{[2\pi N(M-1)]^{1/2}} \exp\left[-\frac{(g - \alpha^m N)^2}{2N(M-1)}\right] \tag{B-7}$$

The probability of error can be written as

$$P_e = P(\alpha^m = 1)\int_{-\infty}^{0} p(g \mid \alpha^m = 1)\, dg + P(\alpha^m = -1)\int_{0}^{\infty} p(g \mid \alpha^m = -1)\, dg \tag{B-8}$$

Assuming that $N \gg 2(M-1) \gg 1$, this integral can be approximately calculated as

$$P_e \approx \left(\frac{M-1}{2\pi N}\right)^{1/2} \exp\left[-\frac{N}{2(M-1)}\right] \tag{B-9}$$

We further assume that outputs due to different input vectors are independent. The probability of correctly classifying all M vectors is

$$P_c = (1 - P_e)^M \approx 1 - MP_e \tag{B-10}$$

where we have used $P_e \ll 1$ which is required by the associative memory. The maximum number of vectors that can be correctly classified is thus given by

$$M = \frac{N}{3 \ln M - \ln N - 2 \ln \epsilon} \tag{B-11}$$

where $\epsilon = MP_e$ is the probability of error for classifying M sample vectors. If $\ln \epsilon \ll \ln N$, Eq. (B-11) can be solved by iterations. First, we use $M = N$ as an approximate solution. Substituting $M = N$ into the right-hand side of Eq. (B-11), we obtain

$$M \approx \frac{N}{2 \ln N} \tag{B-12}$$

Further iterations will not change the order of magnitude of M as $N \to \infty$. When the output layer contains N_1 neurons, Eq. (B-10) becomes

$$P_c = (1 - P_e)^{MN_1} \approx 1 - MN_1 P_e \tag{B-13}$$

Similar calculations will result in

$$M \approx \frac{N}{2(\ln N + \ln N_1)} \tag{B-14}$$

Equation (B-14) gives the order of magnitude of the storage capacity of associative memories.

Appendix C

Useful Mathematical Formulae

1. $\nabla \cdot (f\mathbf{A}) = \mathbf{A} \cdot \nabla f + f \nabla \cdot \mathbf{A}$
2. $\nabla \times (f\mathbf{A}) = f \nabla \times \mathbf{A} - \mathbf{A} \times \nabla f$
3. $\nabla \cdot (\mathbf{A} \times \mathbf{B}) = \mathbf{B} \cdot (\nabla \times \mathbf{A}) - \mathbf{A} \cdot (\nabla \times \mathbf{B})$
4. $\nabla \times (\mathbf{A} \times \mathbf{B}) = \mathbf{A}(\nabla \cdot \mathbf{B}) - \mathbf{B}(\nabla \cdot \mathbf{A}) + (\mathbf{B} \cdot \nabla)\mathbf{A} - (\mathbf{A} \cdot \nabla)\mathbf{B}$
5. $\nabla(\mathbf{A} \cdot \mathbf{B}) = \mathbf{B} \times (\nabla \times \mathbf{A}) + \mathbf{A} \times (\nabla \times \mathbf{B}) + (\mathbf{A} \cdot \nabla)\mathbf{B} + (\mathbf{B} \cdot \nabla)\mathbf{A}$
6. $\mathbf{C} \cdot \nabla(\mathbf{A} \cdot \mathbf{B}) = \mathbf{A} \cdot (\mathbf{C} \cdot \nabla)\mathbf{B} + \mathbf{B} \cdot (\mathbf{C} \cdot \nabla)\mathbf{A}$
7. $\mathbf{C} \cdot \nabla(\mathbf{A} \times \mathbf{B}) = \mathbf{A} \times (\mathbf{C} \cdot \nabla)\mathbf{B} - \mathbf{B} \times (\mathbf{C} \cdot \nabla)\mathbf{A}$
8. $(\nabla \cdot \mathbf{A})\mathbf{B} = (\mathbf{A} \cdot \nabla)\mathbf{B} + \mathbf{B}\nabla \cdot \mathbf{A}$
9. $(\mathbf{A} \times \mathbf{B}) \cdot \nabla \times \mathbf{C} = \mathbf{B} \cdot (\mathbf{A} \cdot \nabla)\mathbf{C} - \mathbf{A} \cdot (\mathbf{B} \cdot \nabla)\mathbf{C}$
10. $(\mathbf{A} \times \nabla) \times \mathbf{B} = (\mathbf{A} \cdot \nabla)\mathbf{B} + \mathbf{A} \times (\nabla \times \mathbf{B}) - \mathbf{A}(\nabla \cdot \mathbf{B})$

Appendix D

Useful Integrals

1. $$\int \frac{dx}{\cosh x} = \tan^{-1} \sinh x = 2 \tan^{-1} e^x$$
$$= \sin^{-1} \tanh x$$

2. $$\int_{-\infty}^{\infty} \exp(-\alpha x^2 - \beta x)\, dx = \left(\frac{\pi}{4\alpha}\right)^{1/2} \exp\left(\frac{\beta^2}{4\alpha}\right)$$

3. $$\int_{-\infty}^{\infty} \exp(ixy) \exp\left(-\frac{x^2}{2}\right) H_n(x)\, dx = (2\pi)^{1/2} \exp\left(\frac{y^2}{2}\right) H_n(y) i^n$$

Author Index

Subject Index